THE TENSELESS THEORY OF TIME

SYNTHESE LIBRARY

STUDIES IN EPISTEMOLOGY,

LOGIC, METHODOLOGY, AND PHILOSOPHY OF SCIENCE

VOLUME 294

THE TENSELESS THEORY OF TIME

A Critical Examination

by

WILLIAM LANE CRAIG
Talbot School of Theology,
Marietta, GA, U.S.A.

KLUWER ACADEMIC PUBLISHERS
DORDRECHT / BOSTON / LONDON

Library of Congress Cataloging-in-Publication Data

Craig, William Lane.
The tensless theory of time : a critical examination / by William Lane Craig.
p. cm. -- (Synthese library ; v. 294)
Includes bibliographical references and index.
ISBN 0-7923-6635-2 (alk. paper)
1. Time--Philosophy. I. Title. II. Series.

BD638 .C74 2000
115--dc21

00-064722

ISBN 0-7923-6635-2

Published by Kluwer Academic Publishers,
P.O. Box 17, 3300 AA Dordrecht, The Netherlands.

Sold and distributed in North, Central and South America
by Kluwer Academic Publishers,
101 Philip Drive, Norwell, MA 02061, U.S.A.

In all other countries, sold and distributed
by Kluwer Academic Publishers,
P.O. Box 322, 3300 AH Dordrecht, The Netherlands.

Printed on acid-free paper

Printed in the Netherlands.

To

ALVIN PLANTINGA

who by his work and his life
has pointed the way

TABLE OF CONTENTS

PREFACE

The present book and its companion volume *The Tensed Theory of Time: a Critical Examination* are an attempt to adjudicate what one recent discussant has called "the most fundamental question in the philosophy of time," namely, "whether a static or a dynamic conception of the world is correct."[1]

I had originally intended to treat this question in the space of a single volume; but the study swelled into two. I found that an adequate appraisal of these two competing theories of time requires a wide-ranging discussion of issues in metaphysics, philosophy of language, phenomenology, philosophy of science, philosophy of space and time, and even philosophy of religion, and that this simply could not be done in one volume. If these volumes succeed in making a contribution to the debate, it will be precisely because of the synoptic nature of the discussion therein. Too often the question of the nature of time has been prematurely answered by some philosopher or physicist simply because he is largely ignorant of relevant discussions outside his chosen field of expertise. In these two complementary but independent volumes I have attempted to appraise what I take to be the most important arguments drawn from a variety of fields for and against each theory of time.

The two rival theories of time which are the subject of our examination have been known under a variety of names: the A- versus the B-Theory, the tensed versus tenseless theory, the dynamic versus static theory. None of these labels is wholly adequate. The terminology of A- and B-Theory has the advantage of being the traditional designations inspired by J. M. E. McTaggart; but these names are descriptively opaque. D. H. Mellor changed the vocabulary of the debate by speaking instead of tensed and tenseless theories, but he has now reverted to speaking of the A- and B-Theories because his labels aroused confusion in the minds of many concerning tense as an ontological category and tense as a linguistic phenomenon. Michael Tooley prefers to speak of dynamic versus static theories, but this terminology, too, can be misleading, since the vast majority of A-theorists do not think of time as literally moving. In these volumes, I use such labels interchangeably but have for the most part stuck with the traditional A and B terminology.

I have spoken of "the" tensed or tenseless theory of time, but this expression is purely stylistic. As we shall see, there is actually a family of A-Theories of time, and B-theorists, too, differ among themselves on certain key issues. I shall argue that many of these versions of the A- or B-Theory are, in fact, inconsistent and that a unique A- or B-theoretical paradigm exists; but I should not want to be thought to prejudice the issue in advance by my choice of words.

I am intellectually indebted in this study to too many persons to recall by name; but I should like to acknowledge my special gratitude to Quentin Smith, from whom

---

[1] Michael Tooley, *Time, Tense, and Causation* (Oxford: Clarendon Press, 1997), p. 13.

I heave learned a great deal about language and time, and to the late Simon J. Prokhovnik, the eminent Australian physicist, who helped me to see the wisdom of Lorentz. In this volume I have reproduced a considerable number of figures from textbooks and discussions of relativity theory. For their permission to reproduce such figures, I gratefully acknowledge *The Journal of Philosophy*, W. H. Freeman and Company, Akademiai Kiado, Oxford University Press, and MIT Press.

I am also deeply grateful to my wife Jan for her faithful labor in production of the typescript and to my research assistant Ryan Takenaga. I should also like to thank Edward White and the Day Foundation for their generous grant which helped to fund the production of the camera-ready copy and to Mark Jensen and Jennifer Jensen for meticulously bringing this book into its final form.

Atlanta, Georgia William Lane Craig

PART I

ARGUMENTS FOR A B-THEORY OF TIME

SECTION 1: THE SPECIAL THEORY OF RELATIVITY

CHAPTER 1

SR AND THE B-THEORY

INTRODUCTION

In Part I of our inquiry we wish to consider the principal arguments typically adduced on behalf of a tenseless or B-Theory of time. The reader may be surprised to find no mention of McTaggart's Paradox. This is because McTaggart's famous argument is, as I understand it, primarily an argument *against* the reality of tense and, hence, an A-Theory of time. Rather than being a positive demonstration of the B-Theory, it attempts to show the logical incoherence of the tensed theory and so is best considered as an objection to the A-Theory of time. Accordingly, I have discussed it as such in my companion volume *The Tensed Theory of Time: a Critical Examination.* In this volume we wish to consider arguments which seek to demonstrate more positively the truth of a B-Theory of time.

SR AND SPACETIME REALISM

Without a doubt the consideration most often adduced in support of a B-Theory of time is the claim that the Special Theory of Relativity (hereafter SR) demands a view of time according to which all events in spacetime are equally real, thereby excluding the objective reality of tense and temporal becoming. In a now famous paper presented to the American Physical Society on January 27, 1966, Hilary Putnam argued that SR implies that events future to us are real, that is to say, actual and existent.[1] Since this is a sufficient condition for a B-theoretic view of time, the B-Theory is implied by SR.

Putnam's argument is based on four assumptions:

I. I-now am real.

II. At least one other observer is real, and it is possible for this other observer to be in motion relative to me.

[1] Hilary Putnam, "Time and Physical Geometry," *Journal of Philosophy* 64 (1967): 240-247.

III. If it is the case that all and only the things that stand in a certain relation *R* to me-now are real, and you-now are also real, then it is also the case that all and only the things that stand in the relation *R* to you-now are real.

IV. The Special Theory of Relativity.

The most important of these assumptions, advises Putnam, is (III), which he dubs "the principle that There Are No Privileged Observers."[2] This nomenclature is somewhat misleading, since a similar principle is operative in SR itself to rule out a preferred reference frame which would function as a sort of aether frame. But Putnam's (III) is meant to be a One Common Reality principle, a statement of the belief that reality is not pluralistic and person-relative but is objective and person-independent. If you and I are both real, then any and only things real for me are also real for you. We do not each have our own reality. Putnam explicates this intuition in terms of a common relation in which you and I stand to things. He qualifies (III) by explaining that, first, the things referred to include all things past, present, and future, so that the verbs in (III) are tenseless; second, *R* is a physical relation independent of the choice of a coordinate system and is definable in terms of the fundamental notions of physics; and third, it must not be accidental that all and only the things standing in *R* to me-now are real. The first qualification makes it evident that Putnam's (III) is restricted to only and all *temporal* things and so does not apply to any timeless entities which one might wish to include in his ontology, such as God or abstract objects. If one conceives God or propositions, say, to be temporal, however, then it becomes a moot question whether there is any relation which is definable in terms of the fundamental notions of physics and not accidental which applies to such things, in which case (III) is vacuous. Perhaps, however, we could amend (III) so as to restrict its scope to only and all temporal, *physical* things.

Now Putnam takes *R* to be the relation of simultaneity. He does so because he believes that the "main principle"[3] underlying the convictions of "the man on the street" concerning the nature of time is that

1. All (and only) things that exist now are real.

So let *R* be the relation of *simultaneity-in-the-observer's-coordinate-system.* Putnam then assumes that a physical situation like the one illustrated in Figure 1.1 is actual:

[2] Ibid., p. 241.
[3] Ibid., p. 240.

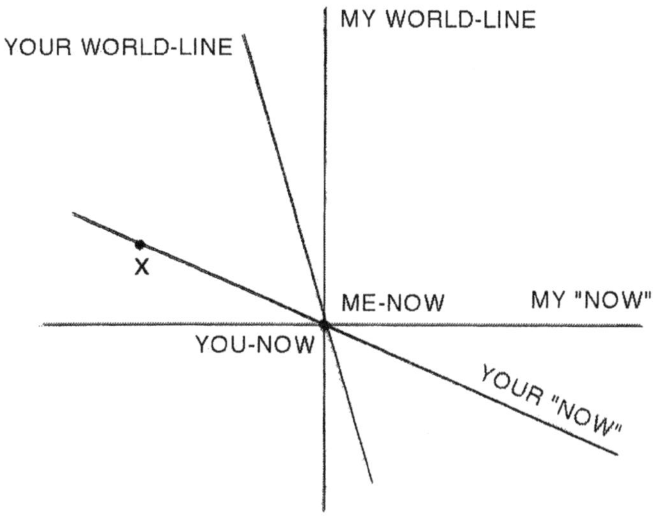

Figure 1.1. You-now and I-now are now at the same place, moving at a very large relative velocity. Event X is present for you-now but future for me-now.

If I assume that all and only those (temporal, physical) things which stand in this relation *R* to me-now are real, then, since you-now are also real, it follows that "everything that stands in the relation *R* to you—everything that is simultaneous to you-now in *your* coordinate system—is also real."[4] But it is a well-known consequence of SR that there are events which are future according to my coordinate system and which are present according to your coordinate system. Therefore, (III) requires me to say that such future things are real.

Unfortunately, as Putnam immediately realizes, his argument thus far is self-defeating. For the future things alleged to be real do *not* stand in *R* to me-now, so that one's initial assumption that *only* things standing in *R* to me-now are real was already inconsistent with (III). Hence, the simultaneity relation did not fulfill the conditions set down for *R* and should have been disqualified. Putnam reflects that (III) in effect requires that *R* be transitive, whereas the relation involved in "*x* is simultaneous with *y* in the coordinate system of *x*" is not transitive.

Putnam's attempt to repair his argument now takes a very obscure turn; indeed, his concluding summary seems to describe an argument which is inconsistent with the argument that appears at this point in his article. He proposes to replace (1) with

2. All things that exist now are real,

[4] Ibid., p. 242.

which is an attenuation of (1). The motivation seems to be to craft a relation which is transitive, as required of *R* by (III). Putnam says, "I take (2) to mean (at least when *I* assert it) that all things that exist *now* according to *my* coordinate system are real...."[5] He then supposes *X* to be an event which is simultaneous with you-now in your coordinate system but which is future according to my coordinate system. "Then, since the event *X* bears the relation *R* to you-now, and you-now bear the relation *R* to me-now, the event *X* bears the relation *R* to me now."[6] Putnam's inference seems manifestly incorrect. In order to avoid the difficulty encountered above, one must do more than replace (1) with (2); one must also remove the words "and only" from the antecedent and consequent of (III), so that we have

> III′. If it is the case that all the things that stand in a certain relation R to me-now are real, and you-now are also real, then it is also the case that all the things that stand in the relation R to you-now are real.

By so amending (III), we no longer require that the future event *X* stand in the relation of simultaneity with me in my reference frame. (III′) does not entail the transitivity of *R*, as did (III). But Putnam not only affirms transitivity in the above citation; he also makes it clear that he retains the "and only" phrase in (III): "we chose *R* to be such that all and only those events which bear *R* to me-now are real."[7] So long as Putnam retains (III), his argument remains self-defeating, for events in my future will not stand to me in the same relation in which they stand to you. They will be simultaneous with you-now in your reference frame but not simultaneous with me-now in my frame.

In his concluding summary Putnam seems to put a different face on his argument:

> If I accept (2), then I must say that all those things are real (but not *only* those things, as we saw) which bear to me the following relation: the relation *y* bears to *x* just in case *y* is simultaneous with *x* in the coordinate system of *x*.... But principle III then requires that I also count every thing and event which bears the *transitive closure of R* to me (*i.e.*, which bears *R* to me, or which bears *R* to something that bears *R* to me, or which bears *R* to something that bears *R* to something that bears *R* to me, or...) as real. But every thing and event in spacetime bears the transitive closure of the above relation *R* to me ...[8]

Here Putnam says that *all* the things bearing *R* to me are real and explicitly denies that *only* the things bearing *R* to me are real. That leaves it open that events not bearing *R* to me are real, which amounts to a rejection of (III) in favor of (III′). But what, then, of his affirmation concerning *R's* transitive closure? Notice that he does not say that an event which bears *R* to something that bears *R* to me itself also bears

---

5 Ibid., p. 243.
6 Ibid.
7 Ibid.
8 Ibid., p. 246.

R to me. Rather he says that an event bearing *R* to something which bears *R* to me is real. That is an affirmation of (III′). It is to affirm that if *X* is simultaneous with you in your reference frame and you are simultaneous with me in my reference frame, then *X* is real. Thus, even if *X* is a future event relative to me, it is nonetheless real.

A sympathetic reading of Putnam will take him to be implicitly endorsing (III′) and proposing for *R* the relation *simultaneous in a reference frame with.* His premiss (2) is a mere tautology which plays no role in the argument. A generalization of the argument shows that all events, whether future or past, are as real to an observer as events which are in his present.

Putnam concludes that "the notion of being 'real' turns out to be coextensive with the *tenseless* notion of existence"[9]—a vindication of the B-Theory of time. Apparently losing self-restraint in the heady atmosphere of a scientific convention, an ebullient Putnam declares,

> I conclude that the problem of the reality and the determinateness of future events is now solved. Moreover, it is solved by physics and not by philosophy. We have learned that we live in a four-dimensional and not a three-dimensional world, and that space and time...are just two aspects of a single four-dimensional continuum. ...Indeed, I do not believe that there are any longer any philosophical problems about Time; there is only the physical problem of determining the exact physical geometry of the four-dimensional continuum that we inhabit.[10]

Thus, the four-dimensional spacetime implied by the Special Theory has removed any basis for distinguishing objectively between past, present, and future, for these are merely observer-dependent notions and all events in the spacetime manifold are equally existent and actual.

AN A-THEORETICAL RESPONSE

Putnam's argument would not phase the hybrid A-B theorist, who attempts to wed objective tense with a B-theoretical ontology. For he affirms the equal reality of all spacetime events and conceives of presentness as a property shifting along the world line of any observer in a future-directed fashion. I have elsewhere argued that he thereby entangles himself in the snare of McTaggart's Paradox;[11] but at least he need not worry about the objection based on SR. Since he does not equate *being real* with *being present,* he may accept Putnam's conclusion that *X* is real for me-now without inferring that *X* is present for me-now. He will probably equate the class of events which are present for me-now with the simultaneity class of events for me-now. Since simultaneity is relative to reference frames, so is presentness; but that fact does not imply that non-present events (anymore than non-simultaneous events) are unreal.

---

9 Ibid., p. 247.

10 Ibid.

11 See my *The Tensed Theory of Time: a Critical Examination*, Synthèse Library (Dordrecht: Kluwer Academic Publishers, forthcoming), chap. 6.

Rather Putnam's argument strikes against the pure A-theorist or presentist, who denies the equal reality of all spacetime events. The A-theorist wants to affirm the common-sense intuition of the man in the street that only events which are present are real; but how can he coherently do so in the face of SR? The most obvious move for him to make would be to relativize presentness and, hence, reality to reference frames. Indeed, I suspect that this is the solution preferred by most A-theorists.[12] If one holds that presentness is not absolute, but frame-relative, then which events in a given frame are most plausibly to be taken as present? The answer seems obvious: only and all those events which are simultaneous with any given event when that event becomes present. Since simultaneity is frame-relative in SR, so is presentness, and, given the tight connection in the A-theorist's thinking between presentness and reality, so is the reality of all temporal events. Only and all things which are simultaneous with me-now are present for me-now and therefore real for me-now.

The A-theorist will regard (III′) as too weak to embody the intuitions behind the common-sense view of time, which are captured in (1). It is critical to the presentist metaphysic that not merely all, but *only* events which stand to me in the relation *present to* are real. So if *R* is a simultaneity relation, it is only and all events which are simultaneous with me-now which are real. Does this fact imply that only and all things that stand in *R* to you-now are real, as (III) states? How are we to avoid the self-defeating situation engendered by the transitivity of *R* implicit in (III)? The A-theorist could plausibly argue that (III) is true but that Putnam's self-defeating situation does not arise because Putnam misunderstood the simultaneity relation in SR. In SR simultaneity is not a dyadic, but a triadic relation: *x* is simultaneous with *y* relative to a reference frame *F*, but *x* is not simultaneous with *y* relative to a different frame *F′*. These are two *different* relations. Thus, principle (III), if true, concerns only the things which are simultaneous to me-now and you-now relative to *F*, and these are the *same* things. Therefore, by SR's own lights, Putnam's attempt to make events absolutely real which are simultaneous with you-now relative to your reference frame *F′* is a misapplication of (III). Oddly enough, Putnam himself says that *R* must be a relation which is independent of the choice of a coordinate system; but in SR simultaneity is not so independent. If we stipulate that *R* is just the bare, dyadic simultaneity relation, so that only a single relation is in view in (III), then (III) is self-defeating, as we have seen. If we revert then to (III′) to avoid self-defeat and insist on a single, dyadic simultaneity relation (to avoid having two triadic relations), then we shall have no choice, given SR, but to relativize reality: things standing in *R* to me-now are real relative to me and things standing in *R* to

---

12 See, for example, Milic Capek, "Do the New Concepts of Space and Time Require a New Metaphysics?" in *The World View of Contemporary Physics*, ed. with an Introduction by Richard F. Kitchener (Albany: State University of New York Press, 1988), pp. 90-104. See also the related attempt to relativize becoming to spactime points advocated by Howard Stein, "On Einstein-Minkowski Spacetime," *Journal of Philosophy* 65 (1968): 5-23; Rob Clifton and Mark Hogarth, "The Definability of Objective Becoming in Minkowski Spacefime," *Synthèse* 103 (1995): 355-387.

you-now are real relative to you. What will not follow from such a revised principle is that things real relative to you are real relative to me.

The A-theorist, then, may happily accept Putnam's linkage of simultaneity, reality, and "now-ness" or presentness. Because only events which are present are real, and only events which are simultaneous with an event which is present on the world line of some entity are present, and simultaneity is relative to reference frames, presentness and reality are relative to reference frames as well. Presentness cannot, within the context of SR, be possessed by an event absolutely because then that event would be present relative to *every* reference frame, whether or not that event is defined as simultaneous relative to a given frame according to SR's clock synchronization procedure. Given a plurality of absolutely present events, a privileged present or absolute now would exist in contradiction to SR. Given his presentist commitments, the A-theorist's relativizing reality to reference frames is quite natural.

The B-theorist might object at this point, as D. H. Mellor has done, that the simultaneity relation in SR is conventional, so that events determined to be simultaneous relative to a reference frame at any given point on the world line of some observer cannot be regarded as objectively present or real.[13] The objector might seek to force the A-theorist to choose between three equally unacceptable alternatives: either (i) to regard the entire "elsewhere" region as present and real to some observer, or (ii) to shrink the present down to a single spacetime point at which the hypothetical observer is located or (iii) to choose arbitrarily one frame of reference to define simultaneity and, hence, present reality at any time. According to alternative (i), all events having a space-like separation from me-now, that is, all events lying outside the forward and backward light cones of me-now, are present and real for me-now. The trouble with this construal of simultaneity on the A-Theory, Mellor rejoins, is that events co-existent with me-now are not co-existent with each other. For example, if e_1 and e_2 are events outside the light cone structure of me-now and e_2 lies in the absolute future of e_1, then both e_1 and e_2 exist with me-now, since they lie in my "elsewhere" region. Nonetheless, e_1 and e_2 are not co-existent, since e_2 lies in e_1's absolute future. Thus, two events which are equally real for me nonetheless do not co-exist. In one and the same reference frame, e_1 and e_2 both exist (being space-like separated from me-now) and do not both exist (one being absolutely future to the other). Hence, one cannot escape this contradiction by relativizing reality to reference frames. Similarly, Robert Weingard argues that if all events outside my light cone structure are equally real, then this result can be easily generalized to yield Putnam's conclusion that all spacetime events are real.[14] For e_1's "elsewhere" region will include events in my absolute future and past, which must therefore be real, too. Again it does no good to relativize reality to reference

[13] D. H. Mellor, "Special Relativity and Present Truth," *Analysis* 34 (1973-74): 75-76.

[14] Robert Weingard, "Relativity and the Reality of Past and Future Events," *British Journal for the Philosophy of Science* 23 (1972): 119-121; similarly, Vesselin Petkov, "Simultaneity, Conventionality, and Existence," *British Journal for the Philosophy of Science* 40 (1989): 69-76.

frames because in this case all the events are *ex hypothesi* being regarded from the standpoint of the same frame. Moreover, Mellor points out, on this alternative tensed sentences can turn out to be both true and false. Suppose that e_1 is rain in London and e_2 is clear weather. Then the tensed sentence "It is raining in London" is both true and false for me-now. Therefore, alternative (i) is untenable.

According to alternative (ii), we are to take as present only events that are present in every acceptable frame of reference. On this view, only events are present which every observer at some point in space agrees are present. Thus, I-now and you-now, though associated with different reference frames at the same location, will agree that our "now's" are coincident and, hence, equally present. But then, Mellor objects, the present becomes shrunk down to what is here at this point-instant. As Lawrence Sklar complains,

> We first reduce 'reality' to the lived experience of the observer; that is, we first fall...into solipsism. Then seeing that our own future experiences and past experiences are as remote *now* from us as the spatially distant, the non-immediately sensed, etc., we fall from solipsism into solipsism of the present moment. Reality has now been reduced to a point![15]

If this is not self-evidently absurd, Mellor also points out that if I-now am never located in London, then the bizarre conclusion follows that the tensed sentence "It is raining in London" is never true. Sklar notes the further strange consequence of this alternative that we must say that there will be events which are now such that they will be in my real past at some future time but which will never have a present reality to me at all! That is because they are now in my future light cone and will be in my past light cone but have no reality in between. Since neither past nor future events exist, how is it that such events become past without ever having been actualized in the present? Indeed, is it not self-contradictory to assert that the sentence "The Battle of Waterloo occurred" is true and yet that it was never true that "The Battle of Waterloo is occurring"? Thus, this second alternative is also untenable.

According to alternative (iii), we choose arbitrarily one frame of reference to define simultaneity and, hence, present existence and truth at any time. In other words, we deny that there are no privileged observers but single out one reference frame as privileged and, hence, definitive for the "now." But Mellor objects that such a move is *ad hoc*, picking out a present merely to preserve the A-theorist's account of existence and truth. Special Relativity allows no more than a conventional distinction between one such present and any other; but the distinction between what exists and what does not is surely more than conventional. Mellor concludes that for A-theorists, therefore, "it becomes conventional what exists at any time, and hence, for tense-logic, it becomes conventional what is presently true."[16]

---

15 Lawrence Sklar, "Time, Reality and Relativity," in *Reduction, Time and Reality*, ed. Richard Healey (Cambridge: Cambridge University Press, 1981), p. 140.

16 Mellor, "Special Relativity," p. 75.

The A-theorist may agree with Mellor that the three proffered alternatives to the standard simultaneity relation are unacceptable. But he may plausibly dispute Mellor's presupposition that the standard simultaneity relation is conventional. For, as Malament has shown, the standard simultaneity relation is *not* conventional *within the context of SR.*[17] The simultaneity relation as calculated by the clock synchronization procedure is an intrinsic feature of Minkowski spacetime. The question of conventionalism arises only within the context of discovery of whether we inhabit a spacetime which is Minkowskian. But Putnam's assumption IV has already decided that question; otherwise the alternative (iii) could not be so easily dismissed. Since the objection to the A-Theory from SR is based on the assumption that spacetime is (locally) Minkowskian, the B-theorist can hardly try to force the A-theorist to embrace alternatives incompatible with that theory. Accordingly, if the A-theorist wishes to relativize presentness and temporal becoming to reference frames, then, given that events simultaneous with a present event are also present, he will regard the world-wide "now" and the edge of becoming as coincident with the hyperplane of simultaneous events inherent in the structure of Minkowski spacetime, relative to a reference frame, on which a locally present event is located.

TWO RELATIVISTIC ONTOLOGIES

Our assessment of the above A-theoretical attempt to reconcile the objectivity of tense and temporal becoming with SR is apt to depend on which of two competing relativistic ontologies we adopt. The decisive question in that choice is whether or not we take a realist attitude toward Minkowski spacetime.[18] The above A-theoretical position implies an anti-realist or instrumentalist interpretation of Minkowski spacetime. There is no tenselessly subsisting manifold of events, for such a manifold would be incompatible with the objective reality of temporal becoming. On the A-Theory of time, things come into being and pass away, which the constituents and contents of spacetime do not do. Spacetime is therefore a

---

[17] David Malament, "Causal Theories of Time and the Conventionality of Simultaneity," *Noûs* 11 (1977): 293-300.

[18] It is important to note that accepting spacetime realism does not without further ado commit one to spacetime substantivalism. Spacetime realists disagree among themselves whether spacetime should be construed substantivally or relationally, substantivalists maintaining that spacetime is a substance existing independently of objects and events located in it, relationalists holding that spacetime is not a substance but depends for its existence on the existence of objects and events. Both are united in opposition to what we shall call the relativity interpretation in contending that all spacetime points are equally real. For discussion see Paul Horwich, "On the Existence of Time, Space, and Spacetime," *Noûs* 12 (1978): 397-419. Embracing spacetime realism does not even commit one automatically to so-called four-dimensionalism or perdurantism in contrast to three-dimensionalism or endurantism with respect to familiar physical objects, for Mellor, a spacetime realist, espouses three-dimensionalism and endurantism. He presumably takes spacetime realism to be the doctrine that all points in spacetime are equally real, which does not entail that objects are spatio-temporal wholes. I shall argue that Mellor's union of spacetime realism with endurantism is, however, ultimately untenable (chap. 8). A consistent spacetime realist will thus view objects as spatio-temporal entities which perdure.

theoretical construct only, a geometrical representation of a theory which is really about physical objects enduring through time. SR is, just as Einstein originally formulated it, a 3+1 dimensional theory, not a 4-dimensional theory. A Minkowski diagram is a helpful tool but neither depicts reality nor implies an ontology.

Graham Nerlich calls such an interpretation of SR the *relativity interpretation* and characterizes it in the following way:

> It has a familiar, classical ontology of continuant spatial objects persisting in time. It uses the idea of a frame of reference, a relative space and a relative time. The frame defines (relatively) the *same* place at different times and thus the concepts of rest, motion, speed and velocity. It also defines (relatively) the same time (simultaneity) at different places. Light is marked out as having an *invariant speed* for every frame of the interpretation. Whereas one can do spacetime physics without coordinates, it is meaningless to speak of a frame-free relativity physics. The concepts of the relativity interpretation make sense, in SR, only with respect to some frame.[19]

On this interpretation, Minkowski's advance over Einstein was mathematical, not metaphysical. A good representative of this original Einsteinian perspective is the French physicist Henri Arzeliès. In his *Relativistic Kinematics*, Arzeliès asserts, "The Minkowski continuum is an abstract space of four dimensions, the sole role of which is to interpret in geometrical language statements made in algebraic or tensor form.... The four-dimensional continuum should therefore be regarded as a useful tool, and not as a physical 'reality'."[20] While it is true that relativity theory banishes the notions of absolute spatial and temporal intervals from physics, nonetheless "It is perfectly clear that in relativity, the ordinary three-dimensional space (which is Euclidian in special relativity) and the time of pre-relativistic physics is employed."[21]

There is no gainsaying Arzeliès insofar as Einstein's original formulation of SR is concerned. But it is also indisputable that once having encountered Minkowski's geometrical formulation of the theory, Einstein became an outspoken realist concerning spacetime. Regarding reality as four-dimensional seemed more natural to him than the complicated maneuver of relativizing presentness to reference frames. He wrote,

> Inertial spaces, with their associated times, are only privileged four-dimensional co-ordinate systems, that are linked together by the linear Lorentz transformations. Since there exist in this four-dimensional structure no longer any sections which represent 'now' objectively, the concepts of happening and becoming are indeed not completely suspended, but yet complicated. It appears therefore more natural to think of physical reality as a four-dimensional existence, instead of, as hitherto, the evolution of a three-dimensional existence.[22]

---

19 Graham Nerlich, *What Spacetime Explains* (Cambridge: Cambridge University Press, 1994), p. 63.

20 Henri Arzeliès, *Relativistic Kinematics*, rev. ed. (Oxford: Pergamon Press, 1966), p. 258.

21 Ibid.

22 Albert Einstein, *Relativity: The Special and the General Theory*, 15th ed. (New York: Crown Trade Paperback, 1961), p. 150.

Again, in a work co-authored with Leopold Infeld, Einstein rejects his own original formulation of SR in 3+1 dimensions in favor of the spacetime ontology depicted in two dimensions on a Minkowski diagram: "We must not consider space and time separately in determining the time-space co-ordinates in another CS [coordinate system]. The splitting of the two-dimensional continuum into two one-dimensional ones seems, from the point of view of the relativity theory, to be an arbitrary procedure without objective meaning."[23] True, "Even in the relativity theory we can still use the dynamic picture if we prefer it. But we must remember that this division into time and space has no objective meaning since time is no longer 'absolute'."[24] Thus, relativity theory was "distinctly in favor of the static picture and found in this representation of motion as something existing in time-space a more convenient and more objective picture of reality."[25] As Ludwig Kostro has shown,[26] Einstein came to view the spacetime described in the General Theory of Relativity as a relativistic ether, uncomposed of either particles or moments, in which all things exist and light is propagated, but which is not itself a reference frame and so does not serve to establish absolute simultaneity and length as did the classical aether. It is a breath-taking metaphysical vision of reality, truly justifying the title "Philosopher-Scientist" for Albert Einstein.[27] The seriousness with which Einstein took this conception may be seen in the fact that when his life-long friend Michael Besso died, Einstein sought to comfort his bereaved family by reminding them that for physicists Besso had not ceased to exist, but exists tenselessly as a permanent feature of the spacetime reality.[28]

Nerlich calls this second approach to SR the *spacetime interpretation*, and he characterizes it as follows:

> It has an ontology of events (alternatively of four-dimensional objects) and speaks of a spacetime (but not of a space nor a time). Spacetime in SR has zero curvature and Minkowski metric in which physical quantities appear as 4-vectors or 4-tensors. A very wide range of coordinate systems may be used in the description of spacetime and its physics, but assigning the same time coordinate (or the same space coordinates) to different events is not interpreted as reflecting, even relatively, the concepts of

---

23 Albert Einstein and Leopold Infeld, *The Evolution of Physics* (New York: Simon & Shuster, 1938), p. 219.

24 Ibid., p. 220.

25 Ibid., p. 217.

26 Ludwig Kostro, "Einstein's New Conception of the Ether," paper presented at the International Conference of the British Society for the Philosophy of Science, "Physical Interpretations of Relativity Theory," Imperial College of Science and Technology, London, 16-19 September, 1988.

27 When his friend Michael Besso inquired if Einstein never concerned himself with the flow of time, Einstein wrote back, indicting Besso for not taking seriously four-dimensional reality. What Besso calls the "world" is a spacelike section for which relativity theory denies objective reality: the "now" is eliminated in the objective construction of the world (Albert Einstein and Michele Besso, *Correspondance 1903-1955*, trans. with Notes and an Introduction by Pierre Speziali [Paris: Hermann, 1979], pp. 276-277).

28 Of Besso's death Einstein wrote, "This signifies nothing. For us believing physicists the distinction between past, present, and future is only an illusion, even if a stubborn one" (Cited in Banesh Hoffmann with Helen Dukas, *Albert Einstein: Creator and Rebel* [London: Hart-Davis, MacGibbon, 1972], p. 258).

> simultaneity (or rest) since these are no part of the scheme of concepts of this interpretation. Further, the use of coordinate systems is not essential to spacetime physics, coordinate-free treatments of it being quite standard. Light is distinguished neither as invariant in speed nor as the fastest signal, since ideas like 'speed' and 'fastest' do not belong in this interpretation. Light is picked out in that light trajectories always lie in cones of null geodesics, one of which is defined at each spacetime point.[29]

This characterization makes evident how remarkable and unusual such an interpretation is, since all the familiar notions of SR as Einstein originally formulated it, such as reference frames, simultaneity, and the speed of light, find no place in it. Rather the central feature of the spacetime interpretation of SR is the lightcone structure, which is independent of reference frames or coordinate systems. In 1911 A. A. Robb was able to recover all the geometric structure of Minkowski spacetime on the basis of the single relation *after* among its points, conjoined with several conditions of that relation.[30] Taking Robb's relation to be extensionally equivalent to some sort of causal relation, recent theorists have defined causally the Lorentz group of transformation equations,[31] orthogonality to a time-line in Minkowski spacetime,[32] and the metrical congruence of intervals in that spacetime.[33] Spacetime realists debate intramurally whether causality is truly constitutive, rather than merely (at best) co-extensive, with Robb's fundamental relation,[34] but the point remains that the familiar physical entities of the relativity interpretation make no appearance in the spacetime interpretation.

When it comes to the General Theory of Relativity, these two interpretations present very different pictures. The relativity interpretation sees the key advance over SR as the introduction of "accelerated" frames. Gravitation is construed as a force acting on objects, including clocks, in three-dimensional space. But in the spacetime interpretation it is the introduction of the curvature of spacetime that marks the departure from SR. Gravitation is interpreted, not as a force acting between bodies in space, but as a warping of spacetime itself.[35]

These two interpretations of relativity theory thus present strikingly different metaphysical visions of reality; they are as radically divergent in their ontologies as

---

[29] Nerlich, *What Spacetime Explains*, p. 63. Cf., idem, "Time as Spacetime," in *Questions of Time and Tense*, ed. R. LePoidevin (Oxford: Oxford University Press, 1998), pp.119-134.

[30] A. A. Robb, *A Theory of Time and Space* (Cambridge: Heffer and Sons, 1913).

[31] E. C. Zeeman, "Causality Implies the Lorentz Group," *Journal of Mathematical Physics* 5 (1964): 490-493.

[32] David Malament, "Causal Theories of Time," pp. 293-300.

[33] John Winnie, "The Causal Theory of Spacetime," in *Foundations of Spacetime Theories*, ed. John S. Earman, *et al.* (Minneapolis: University of Minnesota Press, 1977), pp. 134-205.

[34] On this score Nerlich finds himself at odds with theorists such as Reichenbach, Grünbaum, van Fraassen, Salmon, and Winnie. On Nerlich's view the heart of SR is the second-order constraint on all physical laws of Lorentz invariance, which springs from the inherent symmetries of Minkowski spacetime. Temporal relations in that spacetime cannot be reduced to causal relations because the latter presuppose the former.

[35] For discussion see Martin Carrier, "Physical Force or Geometrical Curvature?" in *Philosophical Problems of the Internal and External Worlds*, ed. John Earman, Allen I. Janis, Gerald J. Massey, and Nicholas Rescher (Pittsburgh: University of Pittsburgh Press, 1993), pp. 3-21.

is relativity theory itself in comparison with the Newtonian physics of absolute time and space.

Now the A-theoretical attempt to relativize tense and temporal becoming to reference frames presupposes, as I say, the relativity interpretation of SR with its instrumentalist understanding of Minkowski spacetime. But while such a relativization seems quite natural within the context of an A-Theory of time, given the close connection between presentness and existence in that theory, nevertheless it is arguable that the resultant interpretation of SR is implausible and deficient.[36] That is to say, Einstein's SR ought to be construed along the lines of the spacetime interpretation. Two reasons may be given in support of this conclusion.

First, the pluralistic fragmentation of reality into distinct spaces and times associated with reference frames is an ontology which is fantastic. This was, I think, the point which Putnam was struggling to express with what I called his One Common Reality principle. He may have bungled the formulation of this intuition, but the root idea that there is a single world, an objective reality independent of observers, that if we both exist then what co-exists with me co-exists with you, is a powerful intuition. It is fantastic to think that you and I, occupying the same location in space and time, but in relative motion, should in virtue of that motion literally dwell in two different worlds, which intersect only at a point. Yet SR requires that even if we are merely passing each other in automobiles, our hyperplanes of simultaneity do not coincide, and at sufficient distances empirically distinguishable events and things are occurring and exist for me which are future and therefore unreal for you. Other events which are in my future and therefore unreal are already actual for you. But if I decelerate and we come to relative rest, then we share the same reality; events which were once present and real in relation to me are now non-existent and future. They remain real relative to my former reference frame, but I have switched frames, and in my new frame they do not yet exist. According to SR, after all, at any point there are an infinite number of reference frames with which hypothetical observers can be imagined to exist, and there is a different reality, a different space and time, relative to each one of these reference frames. One can change frames and, hence, realities just by changing one's relative motion.

By contrast, on the spacetime interpretation, which does not link simultaneity to reality, events do not pop in and out of existence as I switch reference frames. All

---

[36] A point forcefully made by some advocates of the spacetime interpretation. See in particular Yuri Balashov, "Relativistic Objects," *Noûs* (forthcoming), idem, "Enduring and Perduring Objects in Minkowski Space-Time," *Philosophical Studies* (forthcoming). Balashov argues that the relativity interpretation permits no coherent, plausible co-existence relation to hold between spatially separated objects. Interestingly, Milic Capek, a defender of the relativity interpretation has embraced just this conclusion: "It is entirely inaccurate to present Neptune and the Earth as two objects *co-existing* in instantaneous space;...*the only real continuity* which joins them is not a geometrical and instantaneous distance but the concrete and temporal continuity of light signals and gravitational attractions" (Milic Capek, "Bergson's Theory of Matter and Modern Physics," in *Bergson and the Evolution of Physics*, ed. P. A. Y. Gunter [Knoxville: University of Tennessee Press, 1969], p. 312).

that changes is which class of events is orthogonal to my worldline in spacetime at a designated point and, hence, which events I reckon to be simultaneous with my present. All the events subsist tenselessly, and different hyperplanes in spacetime serve merely to mark out which events count as simultaneous relative to my inertial frame. There is a shared, objective reality which exists independently of observers or reference frames, and we all inhabit the same spacetime world; we just reckon different events in that one world to stand in the relation of simultaneity with one another.

On a spacetime ontology, there is thus a unified, independent reality which is merely *measured* differently by observers using different coordinate systems.[37] But on the relativity interpretation reality literally falls apart, and there is no one way the world is.

What all this amounts to saying is that under the relativity interpretation of SR the ontology implied by Einstein's denial of the existence of Newton's absolute space or the aether frame is unbelievable. On the spacetime interpretation the change from Newtonian spacetime to Minkowski spacetime is not really so radical ontologically, chiefly involving the denial of any preferred foliation of spacetime into hyperplanes of simultaneity.[38] There is an attendant change in the metric of spacetime, but the reality of spacetime points or events remains unaffected. But in the absence of any all-embracing spacetime, the "dynamization of space," to borrow Capek's phrase,[39] which is involved in the change from Newtonian to relativistic physics is so radical as to be fantastic. Space is replaced by an infinity of disjoint spaces, each associated with a reference frame and each constituting a different reality. There is no encompassing space in which all these separate spaces exist and of which they are but three-dimensional slices. Time is similarly replaced by an infinity of separate times, each associated with a reference frame and each containing different sets of events at every respective moment. Certain events will actually be instantiated in reality in different sequences depending on the relevant reference frames, and this is not a matter of mere bookkeeping: reality is literally actualized in a reverse order in some frames in comparison to others. It is little wonder that Swinburne can complain that a universe to which the concept of absolute simultaneity has no application is nearly inconceivable;[40] coming from an anti-realist interpretation of spacetime he is surely correct. As Christensen observes, "Simultaneity is not seen in ordinary thought as being relative for the simple reason that existence is not. To the question, 'What events in different regions of the universe are simultaneous with one that is now happening here?' the ordinary view

---

37 See Edwin F. Taylor and John Archibald Wheeler, *Spacetime Physics* (San Francisco: W. H. Freeman, 1966), pp. 1-2.

38 See Michael Friedman, *Foundations of Space-Time Theories* (Princeton: Princeton University Press, 1983), p 125.

39 Milic Capek, *The Philosophical Impact of Contemporary Physics* (Princeton: D. Van Nostrand, 1961), p. 170.

40 Richard Swinburne, *Space and Time*, 2d ed. (London: Macmillan, 1981), p. 201.

answers, 'All and only those that exist!'"[41] But on the relativity interpretation there is no single answer as to what exists now because existence, like simultaneity, is relative. At the distance of the planet Neptune from Earth, for example, any event occurring there within a span of about eight hours could be regarded by us as happening now. On the relativity interpretation, this is not merely a matter of persons situated at the Earth's location using different coordinate systems to calculate what currently exists or is happening on Neptune; rather for different Earthlings in relative motion different states of affairs are actual on Neptune; there are now literally different realities on Neptune for these various Earth observers. Nor does this relativity of reality concern only greatly distant events. At the distance of the Earth's diameter, anything happening within about 1/10 of a second could be happening now for us; there is no unique instant corresponding to our instant here. On the relativity interpretation someone on the other side of the Earth, the Chinese premier Zhu Rongji, for example, may literally be dead for some of us and alive for others who are in relative motion at our shared location. I suspect that it is due to the counter-intuitive ontology of the relativity interpretation—coupled, perhaps, with a lack of serious appreciation of the spacetime interpretation—, that most of us are, as Capek said, unconsciously Newtonians, even when we profess to be relativists.[42] Indeed, I venture to speculate that much of the initial skepticism toward SR—Lorentz and Poincaré were never convinced of it—lay in the implicit assumption that SR presupposed the pluralistic ontology of the relativity interpretation, Minkowski's spacetime being construed instrumentally as a mere formalism. An anti-realist construal of spacetime, conjoined with a justifiable incredulity concerning the ontology of the relativity interpretation, left one with no choice but to reject Einsteinian relativity. Thus, for example, although he argued (notoriously) against the consistency of SR, Herbert Dingle's conversion from being an expositor to being a determined opponent of relativity theory sprang fundamentally from his anti-realism concerning spacetime and his realization that the relativity interpretation required more than the "pure relativity" he formerly expounded. He found it incredible that absolute effects should arise from merely relative motion, and his abortive attempts to prove an inconsistency within SR were based on an inability to assimilate the strange, fragmented world assumed by the relativity interpretation. As for Minkowski spacetime, he dismissed this as "essentially metaphorical,"[43] "a purely gratuitous interpretation of an arbitrarily adopted mathematical formula."[44] "To symbolize that association [of time and

---

41 F. M. Christensen, *Space-like Time* (Toronto: University of Toronto Press, 1993), p. 270. Cf. the scepticism voiced by A-theorist A. N. Prior concerning the relativity of existence to inertial frames in his "Some Free Thinking about Time," in *Logic and Reality*, ed. B. J. Copeland (Oxford: Clarendon Press, 1996), p. 50.

42 Capek, *Philosophical Impact of Contemporary Physics*, pp. 190-191.

43 Herbert Dingle, *The Special Theory of Relativity* (London: Methuen, 1940), p. 86.

44 Herbert Dingle, *Science at the Crossroads* (London: Martin Brian & O'Keefe, 1972), p. 172; cf. idem, "Time in Philosophy and Physics," *Philosophy* 54 (1979): 103-104, where he inveighs against Minkowski's conclusion as a "famous piece of mysticism" which has wrought inestimable harm in

space] by the image of co-ordinates in an imaginary continuum and then to endow the symbol with reality and call what is symbolized a shadow, is a proceeding which has only to be understood to be at once condemned."[45]

The difficulty posed by such an anti-realist reading of spacetime is that without real spacetime SR's denial of absolute space and time (in the sense of a privileged reference frame and attendant proper time) yields an ontology which seems truly outrageous.

Second, the relativity interpretation is explanatorily deficient. Nerlich complains,

> Without the realism of a physical Euclidean space, the ontic foundation of the relativity of uniform motion was anything but transparent. Nothing sustained it.... Special relativity before Minkowski deepened this obscurity since, in an ontology of enduring three-dimensional objects and time-extended processes there is this anomaly: three-dimensional objects have none of their three-dimensional properties (shape, mass or duration—though charge is an exception) well-defined intrinsically, despite the fact that each takes its existence, so to speak, as a shaped or massive or enduring thing. The fundamental material entities were continuants, but had no intrinsic continuant (metrical) properties. Each property had to be related to one of an infinite set of privileged frames of reference, yet the basis of this privilege remained mysterious.[46]

On the relativity interpretation physical objects have properties of shape, mass, and duration only extrinsically, relative to inertial frames, yet why this is so is not explained.

Moreover, it is unclear why three-dimensional objects enduring through time suffer relativistic effects such as length contraction and time dilation in virtue of their being in relative motion. It is important to realize that on the 3+1 dimensional ontology of the relativity interpretation these relativistic effects are *just as much real, physical effects* as on aether compensatory theories, such as Lorentz's. On the relativity interpretation length contraction and clock retardation cannot be dismissed as merely apparent phenomena on the analogy of the mutual observation of diminishing size when two observers retreat from each other. Admittedly, since length contraction and time dilation are reciprocal and the result of merely relative motion, it does seem incredible that they could be anything more than mere appearances, just as the so-called "pure relativists" insisted. But Einstein realized

---

general thought, in philosophy, and in the understanding of physics.

[45] Dingle, *Special Theory of Relativity*, p. 86.

[46] Nerlich, *What Spacetime Explains*, p. 5. Cf. Christensen, *Space-like Time*, p. 260:

> "Without the Minkowskian 'absolutes', we may be forced to hold that there are no intrinsic features in the world at all. For once again, the only non-invariant feature in original SR is electric charge, and its very uniqueness suggests that even it is really observer-dependent, being merely the same relative to all observers.... An entity must, under SR, have an *identity* that is not observer-relative, since we have to say that different observers get the same or different results for numerically the same particulars. But all of the *qualitative features* of an entity would be extrinsic, in the sense involved here. The Minkowski version of Relativity, with its world that is ultimately observer-independent, might seem highly preferable to any who find this prospect troubling."

right from the start that these effects described in his theory were real, not apparent, and could be shown to be real by various *Gedankenexperimente*. When V. Varicak asserted in 1911 that length contraction in Einstein's theory is, in contrast to Lorentz's theory, "only an apparent, subjective appearance," "psychological, not physical,"[47] Einstein responded with a *Gedankenexperiment* designed to show that the contraction does not have its roots in the arbitrary establishment of clock synchronization and length measurement procedures.[48] Consider the following situation (Figure 1.2):

Figure 1.2. Twin rod experiment. The length A*B*, marked by the coincidence of A′,A″ and B′,B″ is shorter than either A′B′ or A″B″.

Measuring rods A′B′ and A″B″ are two rods having the same length at rest and moving uniformly along the same axis of a stationary system in opposite directions. Now as they pass each other, let A* be marked in the stationary system as the point at which A′ and A″ coincide and B* be marked as the point in the stationary system at which B′ and B″ coincide. In contradiction to Newtonian physics, relativity theory predicts that the length A*B* will be shorter than either A′B′ or A″B″, which can be verified by placing either of the rods at rest along the length A*B*. Hence, the rods in motion really do contract, just as the Lorentzian theory also predicts, and that contraction has nothing to do with synchronization procedures.

In a discussion of Einstein's twin-rod experiment, John Winnie points out the subtlety of this *Gedankenexperiment.* To require the rods to move at equal speeds would have presupposed a definition of simultaneity, but Einstein does not require this. Nor does Einstein say that the length A*B* will be the length of one of the rods multiplied by a factor of $(1 - v^2/c^2)^{1/2}$, but only that A*B* < A′B′ (or A″B″). Regardless of the speed of the rods, classical kinematics and SR predict conflicting physical results. "According to classical kinematics the marked off length A*B* will always be equal to A′B′ (or A″B″), regardless of the speeds of the moving rods, while according to the special theory, A*B* will be less than A′B′ (or A″B″), again regardless of the (nonzero) speeds of the moving rods."[49]

---

47 V. Varicak, "Zum Ehrenfestschen Paradoxon," *Physikalische Zeitschrift* 12 (1911): 169.

48 Albert Einstein, "Zum Ehrenfestschen Paradoxon," *Physikalische Zeitschrift* 12 (1911): 509-510.

49 John A. Winnie, "The Twin-Rod Thought Experiment," *American Journal of Physics* 40 (1972): 1091-1094.

Even more engaging thought experiments have been devised to illustrate the point. Consider, for example, the case of a rod which is too long to fall through a hole when both are at rest.[50] Now let the rod be placed in motion with the velocity v along the x axis of an inertial frame which is considered to be at rest, and a plate with a hole in it centered on the y axis be moved upward along that axis with the velocity u. The center of the rod arrives at the origin of the inertial system at the same time in that system as the rising hole reaches the plane y=0 (Figure 1.3).

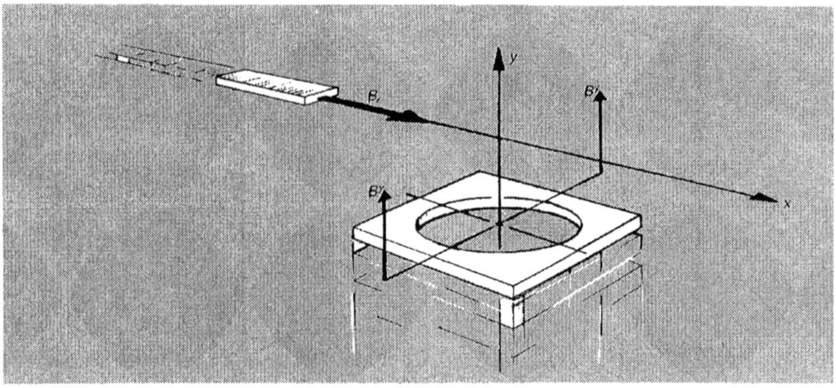

Figure 1.3. The camel and the needle's eye. Relativistic length contraction avoids a collision between the measuring rod and the plate when they both arrive at the origin of the co-ordinate system simultaneously.

Classically, there should be a collision at that point between the rod and the plate. But when relativistic phenomena are taken into account, the result is strangely different. To an observer at rest in the x, y, z inertial frame, the length of the moving rod is Lorentz-contracted, while the diameter of the hole is not, and so the rod neatly passes through the hole in the resting plate and continues on its path. Similarly, an observer associated with the plate will observe the descending rod to be contracted so that it will pass through the hole in the plate. From the reference frame of the rod, however, an observer calculates that it is the diameter of the hole that is contracted rather than himself, so that the hole which was already too small for the rod to fit through when both were at rest has now become even smaller, thus apparently ensuring a collision. But in fact no collision occurs because the observer associated with the rod will determine that the rising plate is not in fact parallel to

---

50 Adapted from R. Shaw, "The Length Contraction Paradox," *American Journal of Physics* 30 (1962): 72, this example is explained in Dieter Lorenz, "Über die Realität der FitzGerald-Lorentz Kontraction," *Zeitschrift für allgemeine Wissenschaftstheorie* 13/2 (1982): 308-312.

the rod but is approaching it at an angle, so that despite the contraction of the hole, the rod can still pass through it (Figure 1.4).

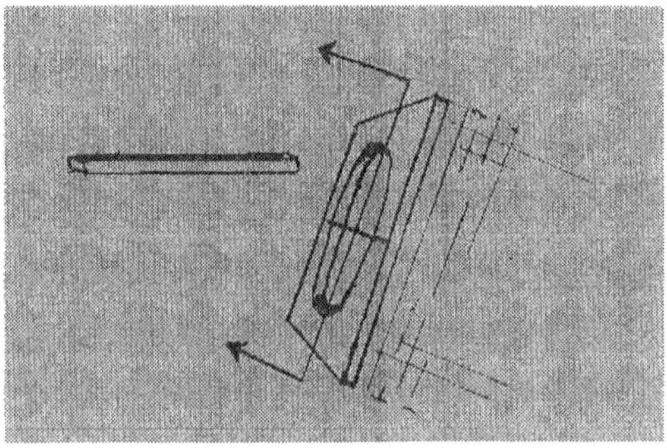

Figure 1.4. The scenario of Figure 1.3 as calculated from the inertial frame of the rod. Despite the contraction of the hole's diameter to less than the length of the rod, the collision is averted because the plate approaches the rod at an angle.

In this way, relativity theory enables the proverbial camel to pass through the eye of a needle!

Though examples of this sort are widely known, it seems that not all physicists have assimilated their significance. J. S. Bell recounts an interesting anecdote in this connection. He invites us to imagine three spaceships, A, B, and C, drifting freely in a region of space remote from other matter, without rotation and without relative motion, with B and C equidistant from A. On reception of a signal from A, the motors of B and C are ignited and they gently accelerate. Let ships B and C be similar and have similar acceleration programs. Then, as reckoned by an observer in A, they will have at every moment the same velocity and so remain separated from one another by the same distance. Suppose that a fragile thread is initially tied between B and C, just long enough to span the distance between them. Then, as the rockets speed up, the thread will be Lorentz-contracted and become too short to span the distance and so must finally break. Having explained the situation, Bell recounts,

> This old problem came up for discussion once in the CERN canteen. A distinguished experimental physicist refused to accept that the thread would break, and regarded my assertion, that indeed it would, as a personal misinterpretation of special relativity. We decided to appeal to the CERN Theory Division for arbitration, and made a (not very systematic) canvas of opinion in it. There emerged a clear consensus that the thread would not break!
>
> Of course many people who give this wrong answer at first get the right answer on further reflection. They usually feel obliged to work out how things look to observers B or C. They find that B, for example, sees C drifting further and further behind, so that a given piece of thread can no longer span the distance. It is only after working this out, and perhaps only with a residual feeling of unease, that such people finally accept a conclusion which is perfectly trivial in terms of A's account of things, including the Fitzgerald contraction. It is my impression that those with a more classical education,

knowing something of the reasoning of Larmor, Lorentz, and Poincaré, as well as that of Einstein, have stronger and sounder instincts.51

It is interesting that this very *Gedankenexperiment* is employed by the neo-Lorentzian physicist L. Jánossy as illustrative of the real Lorentz contraction in connected systems.[52] He points out that if the thread is strong enough, then the entire connected system will suffer a Lorentz-contraction, whereas if the thread is too fragile, then it will break, leaving the entire system uncontracted. He also furnishes an illustration involving two wheels on an axle (Figure 1.5).

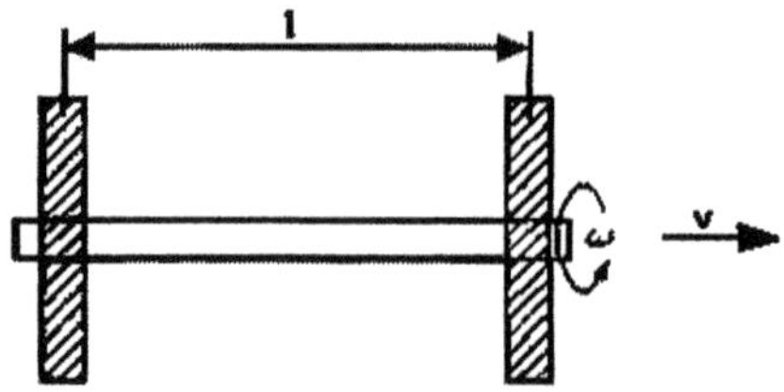

Figure 1.5. The Lorentz deformation in connected and non-connected systems.

The wheels are mounted on the axle at a distance l from each other and made to rotate with an angular velocity ω. If the system is accelerated in the direction of the axis so as to move finally with a velocity v in this direction, then the wheels slow down and their angular velocity becomes

$$\omega^* = \omega \sqrt{1 - v^2/c^2}$$

Provided the wheels are rotating freely on the axle, they react similarly to the accelerating forces and no shift arises between the phases of the rotating wheels. The angular velocities which the wheels take up at time t are thus always

$$\omega^*_1(t) = \omega^*_2(t) = \omega \sqrt{1 - v(t)^2/c^2}$$

But now suppose the wheels to be fixed on the common axle. The rotating axle when accelerated shows a tendency to twist and thus causes a shift in the phases of the rotating wheels. If the forces exceed the cohesive forces of the system, then the wheels may break off the axle and rotate freely; otherwise the axle breaks in two

[51] See J. S. Bell, "How to Teach Special Relativity," in *Speakable and Unspeakable in Quantum Mechanics* (Cambridge: Cambridge University Press, 1987), p. 68.

[52] L. Jánossy, *Theory of Relativity Based on Physical Reality* (Budapest: Akadémiae Kiadó, 1971), pp. 128-131.

between the wheels or the axle deforms permanently under the stress. In any of these cases, the reality of the Lorentz-contraction will be evident.

Examples such as these could be multiplied to prove that, perhaps contrary to expectation, the relativity interpretation of relativity theory involves real, physical length contraction and clock retardation, just as surely as does the Lorentzian theory. Podlaha concludes,

> In the relativity theory, the length contraction and time dilatation in all frames is often viewed as a consequence of a 'perspective of observation,' similarly as a rod seems to change its length as observed under different angles....however, it is seen that the results of relativistic experiments have their origin in the length contraction and time dilatation effects which are so real as a change of the length of a rod caused by the change of temperature.[53]

Podlaha's comments are quite correct, at least as far as the relativity interpretation of SR is concerned.

The reality of these relativistic effects under the relativity interpretation reinforces our first point about the fantastic ontology associated with this interpretation, for it requires us to say, for example, that a plate really does shrink up physically in one frame but experiences no contraction at all in another and that the reality of such a physical effect depends merely on which reference frame one chooses to occupy. But let that pass; the point I wish to make here is that the relativity interpretation is explanatorily deficient in that it provides, indeed, permits, no causal account of these real, physical distortions of three-dimensional continuants. In contrast to the Lorentzian theory, which sought for a causal explanation for these relativistic effects, in the primitive Einsteinian relativity interpretation they are just deduced from the theory's postulates. Lorentz himself remarked on this difference between the theories:

> His [Einstein's] results concerning electromagnetic and optical phenomena...agree in the main with those which we have obtained in the preceding pages. The chief difference being that Einstein simply postulates what we have deduced, with some difficulty and not altogether satisfactorily, from the fundamental equations of the electrodynamic field. By doing so, he may certainly take credit for making us see in the negative results of experiments like those of Michelson,...not a fortuitous compensation of opposing effects, but the manifestation of a general and fundamental principle. Yet, I think something may also be claimed in favor of the form in which I have presented the theory. I cannot but regard the ether which can be the seat of an electromagnetic field with energy and its vibrations, as endowed with a certain degree of substantiality, however different it may be from ordinary matter.[54]

According to Arthur Miller, "The principle of relativity of Bucherer, Lorentz, and Poincaré resulted from careful study of a large number of experiments, and it was on the basis of a theory in which empirical data could be explained to have been *caused*

---

[53] M. F. Podlaha, "Length Contraction and Time Dilatation in the Special Theory of Relativity—Real or Apparent Phenomena?" *Indian Journal of Theoretical Physics* 25 (1975): 74-75.

[54] H. A. Lorentz, *The Theory of Electrons* [1909] (rep. ed.: New York: Dover, 1952), pp. 229-230.

by electrons interacting with an ether. Einstein's principle of relativity excluded the ether of electromagnetic theory and did not explain anything."[55]

Arzeliès emphasizes that on the relativity interpretation "it is not possible to regard the phenomenon [of length contraction] as a deformation due to forces;" rather it is entirely a result of "the way in which the dimensions of a body in motion are measured."[56] He points out that in some cases the body undergoing contraction may not be subject to any forces and, moreover, even when it is, the pressure due to the force F and the resulting elastic deformation of the rod are the *same* in all Galilean systems. Similarly, with respect to clock retardation, "We no longer think of it as an elastic deformation due to certain forces. The body is subjected to no stress, no force, and thus undergoes no structural modification; only the way in which measurements are made introduces changes of shape as we transfer from one system of measurement to the other."[57] This might lead one to think these relativistic effects are merely apparent; but such an inference is erroneous. Arzeliès remarks,

> Some authors have stated that the Lorentz contraction only seems to occur, and is not real. This idea is false. So far as relativistic theory is concerned, this contraction is just as real as any other phenomenon. Admittedly...it is not absolute, but depends upon the system employed for the measurements; it seems that we might call it an apparent contraction which varies with the system. This is merely playing with the words, however. We must not confuse the reality of a phenomenon with the independence of this phenomenon of a change of system (absolute nature).[58]

In fact, we have seen in our above *Gedankenexperimente* that absolute structural changes do result from these relativistic effects, and these changes cannot be explained as the result of the accelerations involved (think of the Twin Paradox in which the periods of acceleration can be made increasingly negligible by extending the time of the journey or even eliminated by using three brothers instead of twins). These relativistic phenomena are, on the relativity interpretation, real deformations of three-dimensional objects enduring through time which have no causal explanation but are merely correlated with different reference frames. They come simply as deductions from the two postulates of the theory.

By contrast, on the spacetime interpretation, three-dimensional objects do not suffer length contraction or time dilation for the simple reason that three-dimensional objects do not exist. Reality is four-dimensional, and the account of relativistic phenomena is, on this interpretation, much more akin to the change of perspective imagined by the pure relativists. Given a spacetime ontology, the different length and time measurements given for specific objects and events by various observers is, as Taylor and Wheeler put it, just a matter of looking at them

---

55 Arthur I. Miller, "On Some Other Approaches to Electrodynamics in 1905," in *Some Strangeness in the Proportion*, ed. Harry Woolf (Reading, Mass.: Addison-Wesley, 1980), p. 85.

56 Arzeliès, *Relativistic Kinematics*, p. 121.

57 Ibid., p. 128.

58 Ibid., p. 120.

"from several angles."[59] A four-dimensional object viewed from a certain angle will appear fore-shortened in comparison with the object as viewed from a different angle. Less metaphorically, because coordinate time and length intervals, as opposed to proper time and length intervals, are not invariant under the Lorentz transformation equations, they vary from coordinate system to coordinate system. A Lorentz transformation may be regarded merely as a rotation of the coordinate axes in four-dimensional spacetime, with the result that different coordinate lengths and temporal intervals will be assigned to the same four-dimensional object. Although this rotation of the axes is obscured in a two-dimensional Minkowski diagram, it may be more perspicuously exhibited by means of a Loedel diagram.[60] Length contraction is not the mysterious (reciprocal!) shrinking of three-dimensional objects enduring through time, but merely the result of applying different coordinate systems to the unchanging, four-dimensional, spatio-temporal object. Similarly, time dilation does not involve a literal slowing down of clocks as they endure through time but rather results from the application of different co-ordinate systems to the changeless four-dimensional object and calculating the difference between the temporal coordinates of two events. Moreover, in the pseudo-Euclidian Minkowski spacetime a curved worldline between two points is the shortest distance, and, hence, the clock of an observer tracing out such a path through spacetime will record less time than a clock following a straight path. Thus in the Twin Paradox, although the path of the travelling twin appears in the diagram to be longer than that of his stationary counterpart, that impression is due to the two-dimensionality of the diagram. In four-dimensional spacetime, the path of the travelling sibling is actually shorter than that of the earth bound twin.[61] Therefore, it is not surprising that his clock records less time and that he is younger than his brother at their reunion. Since length measurements are ultimately dependent upon time measurements (length's being a matter not simply of computing x_2-x_1, but $x_2(t_2)$-$x_1(t_1)$ on the assumption that t_2=t_1), it is not surprising that a relatively moving rod, as viewed by a particular observer, should be calculated to suffer contraction.

Advocates of the relativity interpretation (and neo-Lorentzians) are apt to regard the spacetime account as a pseudo-explanation masquerading as a genuinely explanatory account. But that is to fail to take seriously the four-dimensional ontology of the spacetime realist. Given that such an ontology of physical objects is correct, as opposed to the ontology of continuants presupposed by proponents of the

---

[59] Taylor and Wheeler, *Spacetime Physics*, p. 4. Cf. Balashov's remark: "The explanation is that one is dealing with a 4D object presenting its various 3D parts in different *spatio-temporal* perspectives associated with the state of motion of different inertial reference frames" (Balashov, "Relativistic Objects").

[60] For an illustration see Roger B. Angel, *Relativity: The Theory and Its Philosophy*, Foundations and Philosophy of Science and Technology (Oxford: Pergamon Press, 1980), p. 89.

[61] See Taylor and Wheeler, *Spacetime Physics*, p. 34; cf. Herman Bondi, *Relativity and Common Sense* (New York: Dover, 1964), p. 67; L. Marder, *Time and the Space-Traveller* (London: George Allen & Unwin, 1971), p. 78, who paradoxically also endorses Rindler's interpretation that the traveller's path is Lorentz-contracted.

relativity interpretation and neo-Lorentzians, one can appreciate why Nerlich, an enthusiastic spacetime realist, finds "the elegant spacetime interpretation of the Fitzgerald contraction and the 'slowing of moving clocks' extraordinarily insightful and intuitive."[62] By contrast the relativity interpretation is markedly explanatorily deficient.

But are we justified in assuming that the relativity interpretation in fact precludes causal explanations of the sort envisioned by neo-Lorentzians? Dennis Dieks, while admitting that it is not natural to seek for causal explanations of relativistic phenomena in Einstein's SR, nevertheless protests that there is no bar to such an investigation:

> Some authors have suggested that it is illegitimate to ask...for an explanation of the length contraction. There seems to be no justification for such a point of view. The questions as to the factors determining the contraction can meaningfully be asked within the framework of the special theory of relativity; and the theory is perfectly able to provide an answer. In fact, it would be a serious shortcoming of the theory if no such answer were forthcoming. We would then be in the situation that the theory would predict a contraction but would not be able to account for it on the basis of physical laws; in that case a relativistic molecular theory of matter would be impossible.[63]

Dieks maintains that the Lorentz contraction in the Einsteinian interpretation can be understood as the consequence of physical processes, the result of changes in the intermolecular forces which occur when a body is set in motion, in exactly the same way as it is in the neo-Lorentzian interpretation. Similarly many authors, in explaining time dilation in Einstein's SR, appeal to the illustration of the light clock to show how clocks in motion do actually run slower than relatively stationary clocks. Hence, the Einsteinian version is not deficient in causal explanatory power.

Kroes, however, disputes Dieks's claim in this regard. He charges that Dieks has overlooked the fact that in Einstein's theory these relativistic effects are *reciprocal*. Lorentz's dynamical interpretations of the shortening of rods and of time dilation only make sense because they are not reciprocal.

> The shortening is an objective feature of a rod which is due to a change in the forces acting between the constituents of the rod and this change in the forces is caused by the motion relative to the ether. But since this motion with respect to the ether is absolute, the shortening of the rod is not reciprocal. And because of this, the Lorentz contraction and its dynamical interpretation cannot be transplanted just like that from Lorentz's to Einstein's theory.
>
> ...the reciprocity of length contraction and time dilation rules out dynamical interpretations of the Lorentzian type.[64]

---

[62] Graham Nerlich, personal communication, June 20, 1991.

[63] Dennis Dieks, "The 'Reality' of the Lorentz Contraction," *Zeitschrift für allgemeine Wissenschaftstheorie* 15/2 (1984): 340.

[64] Peter Kroes, "The Physical Status of Time Dilation within the Special Theory of Relativity," *Physical Interpretations of Relativity Theory*, Proceedings of the International Conference of the British Society for the Philosophy of Science, Imperial College of Science and Technology, London, September 16-19, 1988.

With regard to time dilation, Kroes calls the well-known illustration of the light clock (Figure 1.6) a "pseudo-explanation."

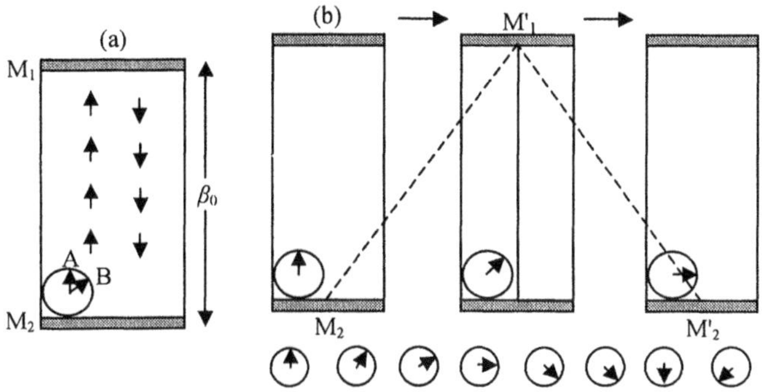

Figure 1.6. The light-pulse clock. (a) When at rest a light pulse is reflected back and forth from the mirrors M_1 and M_2; each return of the pulse to one end counts as a unit of time. The dial hand moves from A to B in that time. (b) When the clock is in motion the light travels the distance $M_2M'_1M'_2$ between ticks, so causing it to go slow. The dials underneath the moving clock are those stationary clocks distributed in space and used for comparison with the moving clock times.

For a vital element in the "explanation" of the time dilation of a light clock is the constancy of the velocity of light for all inertial observers, which is precisely the crucial point involved in the synchronization of distant clocks. "So it is more or less a coincidence that in the case of the light clock time dilation can be 'explained': the operation of the light clock and the synchronization of distant clocks are based upon the same kind of physical process (and its properties), namely the propagation of light in vacuum."[65] If Kroes is correct in this analysis, then the relativity interpretation of SR not only does not seek but actually excludes causal explanations for the relativistic effects in question.

But how does the reciprocity of length contraction and time dilation evacuate dynamical explanations of meaning? In the absence of metaphysical space, any inertial frame can be conventionally considered to be at rest. The dynamical explanations of contraction and dilation phenomena relative to this frame seem to make as much sense as they do in Lorentz's theory relative to the aether. They only lack the same absoluteness. For the moving frame could be considered to be at rest and the stationary frame as moving, and then the dynamical explanation would be reversed. To think that dynamical explanations cannot be reciprocal seems to revert surreptitiously to the conception of Newtonian metaphysical space with respect to

[65] Ibid.

which alone dynamical accounts are valid. But if no such fundamental frame exists, then why should, odd as it seems, reciprocal dynamical explanations be ruled out?

Kroes admits that it is possible to achieve reciprocity within Lorentz's theory by introducing Einstein's definition of simultaneity. This is because according to Einstein's method of clock synchronization, it is true for *all* clocks that $t_B - t_A = t'_A - t_B$, whereas for Lorentz this formula and, hence, the synchronization procedure based on it, holds good only for clocks at rest relative to the aether. By introducing Einstein's method into Lorentz's theory one arrives at judgements of simultaneity relations which the Lorentzian will regard as skewed, and this will in turn affect judgements concerning length measurements, since in order to measure the length of a moving rod one must determine the positions of the endpoints at the same time. A rod at rest in the aether would, if a Lorentzian synchronization procedure were used, appear to be expanded to a moving observer but would, given an Einsteinian procedure, appear contracted (just as the moving rod will appear contracted from the reference frame of the aether-based observer).[66] Hence, the Einsteinian synchronization procedure will result in reciprocal relativistic phenomena even though dynamical causes are at work. Of course, for the Lorentzian this reciprocity will be in appearance only, and so Kroes protests that Einstein's synchronization procedure would be unacceptable to a Lorentzian. No doubt; but that is because an absolute frame exists. But what would happen if the aether frame were regarded as no longer absolute? Here we enter into the realm of counterfactual discourse, where intuitions can become very unclear. For example, is it true that

3. If the aether frame were not absolute, then the reciprocal relativistic effects which would obtain between it and other frames would still have dynamical causes,

as Dieks's view allows, or is it true that

4. If the aether frame were not absolute, then the reciprocal relativistic effects which would obtain between it and other frames would no longer have dynamical causes,

as Kroes seems to hold? In favor of (3) it might be said that the dynamical causal factors present in the aether-based theory would not vanish just because the aether frame would no longer be absolute, for their existence does not depend on the absoluteness of that frame, but on properties of the rods and clocks themselves. On the other hand, in favor of (4) it might be said that the dynamical causes do depend in some way on the existence of an absolute frame. For if the aether frame were not absolute, then it would be in motion relative to other frames and relativistic effects would occur in it which would be as real as those obtaining in other frames. But

---

[66] See discussion in Karl R. Popper, "A Note on the Difference between the Lorentz-Fitzgerald Contraction and the Einstein Contraction," *British Journal for the Philosophy of Science* 16 (1966): 332-333; Jan Dorling, "Length Contraction and Clock Synchronization: The Empirical Equivalence of the Einsteinian and Lorentzian Theories," *British Journal for the Philosophy of Science* 19 (1968): 67-69.

then on the above reasoning the same dynamical causes would be present in it as in other frames. But where did they come from? The defender of (3) must believe that the dynamical causes sprang into being even though no intrinsic property of things in the frame was changed, but only its absoluteness. Since the relativistic effects in this frame are not due to intrinsic, dynamical causes, neither are similar effects with respect to all other frames.

What seems evident from this dispute is that both parties must agree that the existence of dynamical causes is counterfactually (N.B., not causally) dependent upon the absoluteness of a certain frame. For given the truth of the Lorentzian theory, the defender of (3) must hold that

> 5. If the aether frame were not absolute, then dynamical causes would be as present in it as they are in all other frames,

and the defender of (4) must accept that

> 6. If the aether frame were not absolute, then dynamical causes would be as absent from all other frames as they are from it.

But which is true? Under the condition of the relativization of the aether frame, would the dynamical causes be universally present or absent? Conversely, given the truth of the Einsteinian theory, the defender of (3) must maintain that

> 7. If an arbitrary frame were to be absolute, then dynamical causes would no longer be present in it,

and the defender of (4) must hold that

> 8. If an arbitrary frame were to be absolute, then dynamical causes would no longer be absent from all other frames.

But which of these is true? Would the absolutization of an arbitrary frame counterfactually imply that dynamical causes are no longer present in it or still absent from it?

Perhaps some headway could be made by recalling Einstein's illustration of the magnet and the conductor in his original relativity paper of 1905.[67] There an electric current is generated by the relative motion of the conductor and the magnet. The Lorentzian and Einsteinian are in complete agreement about this. But, according to the Lorentzian, if the magnet is at rest, then the current is induced by the magnet's force on the conductor's electric charges, whereas if the conductor is at rest the current is caused by an electric field associated with the magnet. The causal explanations for the observed effect are very different. The Lorentzian and Einsteinian would both affirm counterfactuals such as

---

[67] Albert Einstein, "On the Electrodynamics of Moving Bodies," trans. Arthur I. Miller, Appendix to Arthur I. Miller, *Albert Einstein's Special Theory of Relativity* (Reading, Mass.: Addison-Wesley, 1981), p. 392.

9. If the magnet were at rest in the aether, the current would be caused by its force on the conductor's charges.

and

10. If the conductor were at rest in the aether, the current would be caused by the magnet's electric field.

But suppose there were no absolute state of rest, as the Einsteinian claims. Then the question as to whether the field or the force causes the current becomes malformed. The cause of the current is simply the magnet and the conductor's relative motion. To ask for intrinsic causes is simply not to understand the situation.

Now something very much like this seems to be going on in the search for causal explanations of length contraction and time dilation in the context of the relativity interpretation. The Einsteinian cannot affirm that such phenomena are the products of intrinsic, dynamical causes because such a statement is malformed. These effects, like the electric current, are the result of relative motion. If there were absolute rest, then intrinsic causes would be identifiable. But in the absence of metaphysical space or a fundamental frame, the question disappears because it makes no sense. Therefore, the relativity interpretation does, indeed, preclude causal explanations of relativistic phenomena.[68]

Now some SR theorists will respond that within the context of SR no causal explanation of relativistic phenomena is necessary, and, therefore, failure to provide such explanations is no shortcoming. Grünbaum, for instance, argues that it is inept to demand of a theory explanations for phenomena which are simply considered natural in the context of that theory. Questions which may be well-posed in the context of one theory may be entirely out of place in the context of another theory. He gives the example of the demand for a cause of uniform motion in Aristotelian as opposed to Newtonian physics:

> According to Aristotle's physics, an external force is needed as the cause of a body's uniform motion in a straight line. In his physics the demand for such a perturbational (disturbing) external cause to explain uniform motion arises from the following assumption: when a body is not acted on by an external force, its *natural*, spontaneous unperturbed behavior is to be at rest. Yet, as we know, Galileo's analysis of the motions of spheres on inclined planes led him to conclude that the empirical evidence speaks against just this Aristotelian assumption. As Newton's First Law of Motion tells us, uniform motion does not require any external force as its cause; only accelerated motion does.... But, if so, then the Aristotelian demand for an *explanation* of uniform motion *by reference to an EXTERNAL, perturbing force* begs the explanatory question by means of *a false underlying assumption*, rather than asks a well-posed legitimate question as to the 'why' of uniform motion. By the same token, Galileo and Newton could only shrug their shoulders or throw up their hands in despair, if an Aristotelian

---

[68] See further Alfons Grieder, "Relativity, Causality and the 'Substratum'," *British Journal for the Philosophy of Science* 28 (1977): 35-48, who concludes: "...Einstein's theory gives rise to a causal anomaly: not only is it unable to give rise to a causal explanation, but it also blocks the way to a substratum theory of relativity and hence to what appears to be the most attractive causal approach to relativistic phenomena" (Ibid., p. 46).

> told them that he has a solution to the 'problem' of the external cause of uniform motion, whereas they do not.[69]

Applying this lesson to the case at hand, Grünbaum asserts that Lorentz made the mistake of thinking that Newtonian physics defines what is natural and so postulated *causes* to explain the result of the Michelson-Morley experiment, whereas Einstein saw that these results do not require causes after all because they are integral to the normal behavior of things.[70] Arthur Miller comments, "Einstein's two axioms of relativity theory do not explain the failure of the ether-drift experiments or, equivalently, why the measured velocity of light always turns out to be *c*, or why one cannot catch up with a light wave. Rather it is axiomatic that the space of every inertial reference system is homogeneous and isotropic for the propagation of light."[71] From this relativistic effects follow naturally.

Writing specifically on the reality of the Lorentz contraction, Dieks argues that while the Lorentzian and Einsteinian theories are empirically equivalent, there is a crucial philosophical difference between them in their ontology that renders questions about length contraction which are well-posed for the Lorentzian ill-posed for the Einsteinian:

> The philosophical difference between Lorentz's electron theory and the special theory of relativity is exactly the different ontology of the two theories. Whereas an adherent of the electron theory works with the classical spacetime relations, and therefore feels obliged to *explain* deviations from these relations which occur if moving rods are used, the relativist will, following Einstein's postulates, take it for granted that this world is relativistic, and not classical.[72]

According to Dieks the Principle of Relativity assumes a central explanatory role within Einstein's SR comparable to the role of the classical ontology in the electron theory. The question why moving rods do not behave classically therefore does not arise in a natural manner within Einstein's theory. Since there is no role in the theory for the classical ontology, there is no need to ask for explanations of deviations from it. The transformation properties are simply taken for granted and require no explanation.

In a fine discussion of the reality of time dilation in SR, Kroes similarly objects to the language of clock *retardation* (and length *contraction*). For although time dilation is real, it is not in any way the result of dynamical causes, of moving clocks'

---

69 Adolf Grünbaum, "The Pseudo-Problem of Creation in Physical Cosmology," *Philosophy of Science* 56 (1989): 386.

70 Adolf Grünbaum, "Science and Ideology," *Scientific Monthly* 79 (July 1954), pp. 16-17.

71 Miller, *Einstein's Special Theory*, p. 167. At the same time, Miller concedes that Einstein, like Langevin, was troubled by the question of the cause of relativistic effects and preferred a dynamical cause for time dilation, to be sought in GR, but that Builder demonstrated that general relativistic treatments of the clock paradox succeeded in only complicating the problem and often lacked physical significance (Ibid., pp. 272-273). See Albert Einstein, "Dialog über Einwände gegen die Relativitätstheorie," *Naturwissenschaften* (1918): 697-702; Geoffery Builder, "The Resolution of the Clock Paradox," *Australian Journal of Physics* 10 (1957): 246-262.

72 Dennis Dieks, "The 'Reality' of the Lorentz Contraction," p. 341.

running more slowly than stationary ones, but arises out of Einstein's synchronization procedure for distant clocks. In the Twin Paradox, the travelling clock (clock 2) does not run at a slower rate than the stationary clock (clock 1):

> The idea that the rate of clock 2...must somehow have undergone a slowing down during its motion in order to be able to explain the time difference at [the point of their reunion] C, can only be saved by making a tacit appeal to the notion of absolute time. This goes as follows. Suppose that a clock which behaves according to classical kinematics moves along with clock 2. This classical clock will show the same reading at C as clock 1. With regard to this classical clock, which measures absolute time, it does make sense to claim that the rate of clock 2 is slowed down.
>
> But as soon as the notion of absolute time is abandoned, the claim that the rate of clock 2 has slowed down, becomes meaningless. In my opinion, therefore, it is best to forget within the context of relativity theory about 'changing rates of moving clocks' altogether. Abandoning the notion of absolute time precisely implies that ideal, *i.e.* undisturbed, clocks can measure different amounts of time between events A and C. It also implies giving up the search for a cause for the time difference between the two clocks at C.... The difference in time between the two clocks at C is a 'brute' fact about the behavior of clocks, just as the triangle inequality is a brute fact about the behaviour of rods. In both cases, the search for an explanation of these facts is superfluous and pointless.[73]

The claim that the relativity interpretation of SR is explanatorily deficient because it fails to give causal explanations for length contraction and time dilation is therefore misconceived.

But this counter-rebuttal on behalf of the relativity interpretation has to be very carefully weighed. It is true that length contraction and time dilation follow automatically from the theory's denials of absolute time and space and its re-definition of key concepts. Given the theory's presuppositions and postulates, these phenomena are deducible. But the very fact that there is an interpretation of SR, namely, the spacetime interpretation, which does provide an explanation for these phenomena demonstrates that the quest for an explanation of them is neither inappropriate nor misguided. It is true that *within* the relativity interpretation, it is inappropriate (as we have seen) to look for causes of the relativistic phenomena. Similarly, *within* the Lorentzian interpretation, causal explanations should and can be sought. But presumably the issue here is not whether theoreticians are acting consistently *within* the framework of their respective theories or interpretations; rather the question is a meta- or extra-theoretical query posed by someone *without*, or outside, the theories. It is a question of theory-assessment, a question of which theory or interpretation is to be adopted.

Here the claim on behalf of the spacetime interpretation is that it possesses in some sense greater explanatory power than the relativity interpretation. Not that the relativity interpretation does not account adequately for all the phenomena, for it does. But a theory can account for all the phenomena without having much explanatory power. For example, the phenomena could be simply postulated in the axioms of the theory itself, in which case no phenomena would go unaccounted for,

---

73 Peter Kroes, "Physical Status of Time Dilation."

but the explanatory power of the theory *vis à vis* those phenomena would be nil.[74] Similarly, by gratuitously postulating as *natural* some phenomenon that would be regarded extra-theoretically as puzzling, one can avoid having to give an explanation of virtually anything. The question is why we should adopt the theory *with* its assumptions. One reason would be that the theory under consideration increases our explanatory power, it involves an increase of intelligibility over a rival theory which has to appeal to more brute facts or givens. If what is simply a brute fact in one theory can be given an explanation in another theory, then we have an increase in intelligibility that counts in favor of the second theory. Now something of this sort can be plausibly claimed on behalf of the spacetime interpretation versus the relativity interpretation. What is simply deduced from basic postulates and definitions in the one is given an explanation in the other, facts that are simply "brute" and "taken for granted" are accounted for in terms of spacetime structures in the other.

Of course, if there are good empirical grounds for accepting the postulates of a theory, then we may be justified or even forced to regard certain phenomena deduced therefrom as natural and not in need of any explanation. Grünbaum's example of Newtonian versus Aristotelian accounts of inertial motion would be a case in point. But it hardly needs to be said that there is nothing in Einstein's postulates which provides empirical grounds for preferring the relativity interpretation, which takes length contraction and time dilation as brute facts, over the spacetime interpretation, which explains such phenomena in terms of spacetime structure.[75]

Thus, it seems to me that the spacetime interpretation of SR has greater explanatory power than the original relativity interpretation. Phenomena which are simply brute facts according to the relativity interpretation can be explained within the spacetime interpretation, thus increasing our understanding of the physical world.

CONCLUSION

In summary, if what I have argued is correct, the relativity interpretation of SR, with its pluralistic ontology and contracting and retarded three-dimensional continuants, is fantastic and explanatorily impoverished. Of course, if I am wrong,

---

[74] As John Worrall points out,

> "...*any* theory can be made to have correct empirical consequences by 'writing those consequences into it'; the cases that have traditionally induced realist-inclinations in even the most hard-headed are cases of theories, designed with one set of data in mind, that have turned out to predict entirely unexpectedly some further general phenomena" (John Worrall, "How to Remain (Reasonably) Optimistic: Scientific Realism and the 'Luminiferous Ether'," in *PSA 1994*, ed. David Hull, Micky Forbes, and Richard M. Burian [East Lansing, Mich.: PSA, 1994], p. 335).

[75] This is not to say that the explanation must be causal; see Robert Di Salle, "Spacetime Theory as Physical Geometry," *Erkenntnis* 42 (1995): 317-337.

then SR poses no challenge to the A-theorist, and one can accept both SR and the objective reality of tense and temporal becoming. But if I am correct, then the move to relativize tense and temporal becoming and, hence, reality to reference frames is a desperate expedient. If, as a consequence, we adopt spacetime realism, then the B-Theory of time has been vindicated. It is therefore incumbent upon the A-theorist to probe more deeply into the philosophical foundations of SR and the classical concept of time to see if in fact Einstein has so subverted that conception as to make a presentist metaphysic untenable.

CHAPTER 2

TIME AND ITS MEASURES

Probably not too many physicists and philosophers of science would disagree with Wolfgang Rindler's judgement that with the development of SR Einstein took the step that would "destroy the classical concept of time."[1] But what is or was the classical concept of time, and how did Einstein's critique render it untenable? In

---

[1] Wolfgang Rindler, "Einstein's Priority in Recognizing Time Dilation Physically," *American Journal of Physics* 38 (1970): 1112. Cf. the verdicts of W. G. V. Rosser: "The theory of special relativity necessitated the abandonment of the concept of absolute time..." (W. G. V. Rosser, *An Introduction to the Theory of Relativity* [London: Butterworths, 1964], p. 397); Stephen Hawking: "...the theory of relativity put an end to the idea of absolute time! ...The theory of relativity does...force us to change fundamentally our ideas of space and time. We must accept that time is not completely separate from and independent of space, but is combined with it to form an object called spacetime" (Stephen W. Hawking, *A Brief History of Time: from the Big Bang to Black Holes*, with an Introduction by Carl Sagan [New York: Bantam Books, 1988], pp. 21, 23); E. F. Taylor and J. A. Wheeler, who note that the invariance of the spacetime interval "forces one to recognize that time cannot be separated from space" (Edwin F. Taylor and John Archibald Wheeler, *Spacetime Physics* [San Francisco: W. H. Freeman, 1966], p. 3); Fritjof Capra, who opines that relativity theory "shattered...the notion of absolute space and time," for space and time are not separate entities, but are both "intimately connected and form a four-dimensional continuum, 'spacetime'" (Fritjof Capra, *The Tao of Physics*, 2d ed. [London: Fontana, 1983], pp. 69-71); W. V. O. Quine, who concludes that relativity theory "leaves no reasonable alternative to treating time as space-like" (Willard Van Orman Quine, *Word and Object* [Cambridge, Mass.: MIT Press, 1960], p. 172); A. d'Abro: "...relativity compels us to abandon our traditional understanding of space and time....this discovery of the relativity of simultaneity marks a date of the same momentous importance as did the discovery of the Copernican system in astronomy" (A. d'Abro, *The Evolution of Scientific Thought*, 2d rev. ed [n.p.: Dover Publications, 1950], pp. ix, 171); Howard Stein: "...the results of physics leave the *ordinary* notion of 'absolute simultaneity' as empty of real content...as is, the earth being round, the old ordinary notion of the 'absolutely' vertical" (Howard Stein, Critical notice of *The Language of Time* by Richard M. Gale, *Journal of Philosophy* 66 [1969]: 355); Milic Capek: "The special theory of relativity...decisively proved the impossibility of an absolute frame of reference, that is, of an absolute space, which would...be the support of the objective simultaneity of events" (Milic Capek, "Bergson's Theory of Matter and Modern Physics," in *Bergson and the Evolution of* Physics, ed. P. A. Y. Gunter [Knoxville: University of Tennessee Press, 1969], p.308); John Sinks: "...the special theory forces one to recognize that what is in the future is not in the future *simpliciter*, but in the future relative to some inertial reference frame" (John D. Sinks, "On Some Accounts about the Future," *Journal of Critical Analysis* 2 [1971]: 15); Paul Fitzgerald, who believes that relativity theory "forces us to revise the old views of the 'future'": "Crudely speaking, events that are 'now' in your future may be 'now' in my past..." (Paul Fitzgerald, "The Truth about Tomorrow's Sea Fight," *Journal of Philosophy* 66 [1969]: 310-311); Hermann Bondi, who concludes that SR "forces revisions in the common sense concept of time"—we have to get used to the idea that time is a private matter; "That is to say, that MY time is what MY watch tells me" (Hermann Bondi, *Relativity and Common Sense* [New York: Dover, 1964], p. 65); J. G. Taylor: "We have obtained a time distortion formula which destroys Newton's idea of a universal time" (J. G. Taylor, *Special Relativity*, Oxford Physical Series [Oxford: Clarendon Press, 1975], p. 15); P. C. W. Davies, who concurs that Newton's universal and absolute time was wrong because relativity theory reveals that clock rates depend on the motion and gravitational situation of the observer (P. C. W. Davies, "Spacetime Singularities in Cosmology and Black Hole Evaporations," in *The Study of Time III*,

order to answer those questions, we need to recur to the fountainhead of the classical concept of time: Isaac Newton and his *Philosophiae naturalis principia mathematica.*

NEWTON'S DISTINCTION BETWEEN ABSOLUTE AND RELATIVE TIME

The *Scholium* to his Definitions in the *Principia* is the *locus classicus* of Newton's exposition of his concepts of time and space.[2] Newton observes that "the common people" conceive of such quantities as time, space, place, and motion only in terms of "the relation they bear to sensible objects," and thence arise certain prejudices. In order to overcome these, Newton draws a dichotomy with respect to these quantities between "absolute and relative, true and apparent, mathematical and common." With regard to time and space he asserts:

> I. Absolute, true, and mathematical time, of itself, and from its own nature, flows equably without relation to anything external, and by another name is called duration: relative, apparent, and common time, is some sensible and external (whether accurate or unequable) measure of duration by the means of motion, which is commonly used instead of true time; such as an hour, a day, a month, a year.
>
> II. Absolute space, in its own nature, without relation to anything external, remains always similar and immovable. Relative space is some movable dimension or measure of the absolute space; which our senses determine by its position to bodies; and which is commonly taken for immovable space; such is the dimension of a subterraneous, an aerial, or celestial space, determined by its position in respect of the earth. Absolute and relative space are the same in figure and magnitude; but they do not remain always numerically the same. For if the earth, for instance, moves, a space of our air, which relatively and in respect of the earth remains always the same, will at one time be one part of the absolute space into which the air passes; at another time it will be another part of the same, and so, absolutely understood, it will be continually changed.[3]

---

ed. J. T. Fraser, N. Lawrence, and D. Park [Berlin: Springer, 1978], pp. 74-75); Motz and Weaver: "Relativity theory has shown us that the 'objective reality' of Newtonian absolute space and Newtonian absolute time are really subjective entities that have different meanings for different observers....the constancy of the speed of light for all inertial observers...destroys the Newtonian concepts of absolute space and absolute time...." (Lloyd Motz and Jefferson Hane Weaver, *The Concepts of Science from Newton to Einstein* [New York: Plenum Press, 1988], pp. 15-16, 230; cf. pp. 242-243); Julian Barbour: "Newton's absolute time was completely overthrown by the revolution of the Special Theory of Relativity" (Julian B. Barbour, *Absolute or Relative Motion?*, vol. 1: *The Discovery of Dynamics* [Cambridge: University Press, 1989], p. 633).

[2] Isaac Newton, *Sir Isaac Newton's 'Mathematical Principles of Natural Philosophy' and his 'System of the World,'* trans. Andrew Motte, rev. with an Appendix by Florian Cajori, 2 vols. (Los Angeles: University of California Press, 1966), 1: 6.

[3] Ibid.

> I. Tempus absolutum, verum, & mathematicum, in se & natura sua, sine relatione ad externum quodvis, aequabiliter fluit, alioque nomine dicitur Duratio: Relativum, apparens, & vulgare est sensibilis & externa quaevis durationis per motum mensura (seu accurata seu inaequabilis) qua vulgus vice veri temporis utitur, ut hora, dies, mensis, annus.
>
> II. Spatium Absolutum, natura sua sine relatione ad externum quodvis, semper manet similare & immobile: Relativum est Spatii hujus mensura, seu dimensio quaelibet

Though much misunderstood and greatly vilified, Newton's distinction deserves our thoughtful consideration.[4] The most evident feature of this distinction is the independence of absolute time and space from the relative measures thereof. Absolute time or simple duration exists regardless of the sensible and external measurements which we try, more or less successfully, to make of it. In other words, clock time may or may not register the true time. J. R. Lucas calls this the "rational theory of clocks."[5] We cannot measure time directly because we cannot take an interval of, say, an hour and lay it out against another interval and compare the two in the way that we compare lengths. Rather we must measure time indirectly by having rules which enable us to pick out pairs of instants and to say that the interval between one pair is equal to, or greater than, or twice as great as, the interval between another pair. If two intervals are isochronous, or equal in duration, it is not because of our fiat, but because they really are. The witness to this fact is that our clocks are corrigible. Thus, "...time is not what the clocks say, but what they are trying to tell, are there to tell."[6] It may seem rather puzzling that Newton

---

> mobilis, quae a sensibus nostris per situm suum ad corpora definitur, & à vulgo pro spatio immobili usurpatur: uti dimensio spatii subterranei, aerii vel coelestis definita per situm suum ad terram. Idem sunt spatium absolutum & relativum, specie & magnitudine; sed non permanent idem semper numero. Nam si terra, verbi gratia, moveatur, spatium aeris nostri, quod relative & respectu terrae semper manet idem, nunc erit una pars spatii absoluti in quam aer transit, nunc alia pars ejus; & sic absolute mutabitur perpetuo.

(The critical edition of the *Principia* is Isaac Newton, *Philosophiae Naturalis Principia Mathematica*, 3d ed. [1726], ed. Alexandre Koyré and I. Bernard Cohen, 2 vols. [Cambridge, Mass.: Harvard University Press, 1972]; for the *Scholium* on time and space see vol. 1, p. 46.)

4 With respect to the opprobrium heaped upon Newton at this point, John Earman remarks,

> "What I find especially disturbing about such condemnations of Newton is not the injustice they do to Newton but rather the fact that they are possible only after an abdication of philosophical responsibility. In all the philosophical literature with which I am acquainted, there is precious little attempt to give reasonably clear and precise answers to the questions which are central to the cluster of philosophical issues which revolve around Newton's conception of space and time....
>
> ...it seems to me that Newton demonstrated a much deeper understanding of the nature of space and time than Berkeley, Leibniz, and Mach. And so far as I can see, neither modern philosophers of science...nor the people identified by modern philosophers as major philosophical figures of the 17th, 18th, and 19th centuries, have succeeded in raising any compelling philosophical objections to absolute space, absolute time, or absolute spacetime..." (John Earman, "Who's Afraid of Absolute Space?" *Australasian Journal of Philosophy* 48 [1970]: 288, 317).

5 John R. Lucas, *A Treatise on Time and Space* (London: Methuen, 1973), pp. 62-64, 69.

6 Ibid., p. 64. He concludes,

> "The fact that we have a rational theory of clocks vindicates Newton's doctrine of absolute time. If we really regarded time simply as the measure of process, we should have no warrant for regarding some processes as regular and others as irregular... Even our best clocks are subject to correction. So long as we are prepared to assess the time-keeping qualities of a clock, and are prepared in principle to replace it by a more regular one, if it could be obtained, we are committed to an idea of absolute time which is not simply what the clocks actually say" (Ibid., p. 91).

Cf. the judgement of Richard Swinburne, *Space and Time*, 2d ed. (London: Macmillan, 1981), p. 202, who agrees that Newton was correct: "there is a true time which might or might not be recorded by actual measuring instruments."

calls only absolute time mathematical, since clock time certainly seems mathematical in character. But the contrast which Newton has in mind here is, I think, the same as the difference between a geometrical circle and a physical circle: the latter is only a more or less accurate approximation to the former, true circle. Like mathematical objects, absolute time is not constituted by or dependent upon sensible approximations thereto.

In stating that absolute time "flows equably without relation to anything external," Newton implies that time itself is not metrically amorphous. As Kroes explains, Newtonian absolute time possesses its own intrinsic metric with which our clocks seek to stay in harmony:

> To Newton it was self-evident that there exists just one fundamental metric for time: the intrinsic metric of absolute time. Physical processes of whatever kind could provide a more or less accurate 'sensible measure' thereof. A 'true' sensible measure of absolute time could only be reached in the case of a perfectly isolated, completely undisturbed periodical system which would constitute an ideal clock. In a certain sense, all physical processes had to obey, according to Newton, the rhythm of absolute time; an ideal clock, of whatever nature (mechanical, gravitational, etc.) could provide an exact measure of the unique, fundamental metric of absolute time.[7]

The clock retardation that occurs in the context of SR and GR would not have disturbed Newton, since, as we shall see, he freely concedes that we may not have any accurate measure of time. Newton, of course, did not anticipate the relativistic phenomenon of clock retardation, but it would have troubled him little to learn that clocks in motion or in gravitational fields run slowly.

Kroes concurs that if we reject Newton's distinction between time and its measures, then the statement that two time intervals are isochronous can only be conventionally true.[8] Without absolute time, our temporal metrics as determined by, say, mechanical, gravitational, or electromagnetic clocks, may not stay in synchronism, and the unity of time becomes a mere assumption.[9]

---

[7] Peter Kroes, *Time: Its Structure and Role in Physical Theories*, Synthèse Library 179 (Dordrecht: D. Reidel, 1985), p. 49. Cf. John Earman, *World Enough and Spacetime* (Cambridge, Mass.: MIT Press, 1989), p. 8.

[8] Thus Sklar observes that Newton is on to something "of vital importance" in drawing his distinction (Lawrence Sklar, "Real Quantities and their Sensible Measures," in *Philosophical Perspectives on Newtonian Science*, ed. Phillip Bricker and R. I. G. Hughes [Cambridge, Mass.: MIT Press, 1990], p. 61). Sklar notes that there are natural measures of time which yield simple, elegant laws of nature and that a wide variety of clocks will not only agree with each other in their metric of time but will measure time in a way that approximates the natural measure. If, on the other hand, time itself is not distinguished from its measures, then *any* process has equal right to the status of the standard measure, regardless of how sporadic it might be relative to the concordant "natural" measures. See further J. R. Lucas and P. E. Hodgson, *Spactime and Electromagnetism* (Oxford: Clarendon Press, 1990), p. 239.

[9] David Park seems oblivious to the vicious circularity of his reasoning when he says,

> "*Time is what is measured by a clock.* What is a clock?
>
> *A clock is a device whose law of motion is known*.... How is it that we can define time in terms of clocks and clocks in terms of time without running into trouble? There is an assumption of regularity in the world that underlies the definition. We assume, in fact, that there is a universal time that governs all motion.... And the basis for the assumption is our knowledge of the world. It might have been otherwise—at least one can easily imagine it otherwise—but it is not" (David Park, *The Image of Eternity* [Amherst, Mass.: University of Massachusetts Press, 1980], p. 40).

Similarly, Newtonian space is absolute in the sense that it is distinct from the relatively moving spaces associated with inertial frames. Max Jammer has emphasized that the absolute statement of Newton's first law of motion,

> I. *Every body continues in its state of rest, or of uniform motion in a right line, unless it is compelled to change that state by forces impressed upon it*[10]

requires the assumption of absolute space as a prerequisite of its validity.[11] The classical principle of inertia becomes meaningless apart from a state of absolute rest relative to which bodies may be said to be in a state of motion. Since in Corollary V Newton affirms Galilean relativity, however, the first law of motion, while necessitating the existence of absolute space, provides no means by which that space can be experimentally distinguished from relative spaces. Since the laws of motion hold in all inertial frames, "it turns out that the condition that absolute space be that space in which Newton's laws hold fails to specify a unique space as absolute space. The condition picks out an infinite set of spaces, the inertial spaces, which are the relative spaces of a set of observers moving uniformly with respect to one another in inertial motion."[12] Jammer muses that "If Newton had been a confirmed positivist he would have acknowledged all uniformly moving inertial systems as equivalent to each other. As it was, only one absolute space existed for him."[13]

Newton proposed to pick out experimentally the privileged frame of absolute space, not by his laws of motion, but by means of rotation, which is experientially distinguished from merely relative motion with respect to its effects.[14] Modern commentators have observed that Newton's *Gedankenexperimente* of the rotating bucket and the revolving globes only succeed in demonstrating the existence of absolute motion, not absolute space. A spinning bicycle tire, for example, will be rotating in every inertial frame, whether, for example, the bicycle is taken to be moving and the earth at rest, or the earth moving and the bicycle at rest; but such absolute motion does not serve to pick out any reference frame as preferred.

Absolute space, in contrast to the plurality of relative spaces, is one and immovable. In stating that absolute and relative space are the same in figure and magnitude, differing only in that a volume of relative space moves through different (equal) volumes of absolute space, Newton shows that he has no suspicion of what came to be known as the FitzGerald-Lorentz contraction. Sklar observes the oddity that Newton does not mention measuring rods as a more or less accurate measure of

---

10 Newton, *Principles of Natural Philosophy*, 1: 13. "Corpus omne perseverare in statu suo quiesciendi vel movendi uniformiter in directum, nisi quatenus illud a viribus impressis cogitur statum suum mutare."

11 Max Jammer, *Concepts of Space* (Cambridge, Mass.: Harvard University Press, 1954), pp. 99-103. Cf. Einstein's verdict that this realization was "one of Newton's greatest achievements" (Albert Einstein, "Foreword" to *Concepts of Space*, by Max Jammer, p. xiv). See also Fritz Rohrlich, *From Paradox to Reality* (Cambridge: Cambridge University Press, 1987), chap. 5.

12 John D. Norton, "Philosophy of Space and Time," in *Introduction to the Philosophy of Science*, ed. Merrilee Salmon (Englewood Cliffs, N. J.: Prentice-Hall, 1992), p. 181.

13 Jammer, *Concepts of Space*, pp. 100-101.

14 Newton, *Principles of Natural Philosophy*, 1: 10-12.

the absolute metric of space.[15] But it is implied that absolute space, like absolute time, is endowed with an intrinsic metric which determines whether the distances we measure are equal. Because "the parts of space cannot be seen," Newton explains, we use in their stead "sensible measures," so that "from the positions and distances of things from any body considered as immovable, we define all places."[16] On the other hand, "in philosophical disquisitions we ought to abstract from our senses."[17] Again, if we reject Newton's distinction, then it becomes impossible to hold that two spatial distances are objectively congruent, and we must swallow metric conventionalism with respect to space.[18]

Newton's distinction between absolute and relative quantities thus far discussed concerns only the "absolute-measured" distinction, according to which true time and space are distinct from measured time and space. But, of course, Newton also conceived of time and space as absolute in a more profound sense, which is expressed in the "absolute-relational" distinction; namely, he held that time and space are absolute in the sense that they exist independently of any physical objects whatsoever. He claims, as we have seen, that absolute time "of itself, and from its own nature flows equably without relation to anything external" and that absolute space "in its own nature, without relation to anything external, remains always similar and immovable."

Usually, this is interpreted to mean that time and space would exist even if nothing else existed, that there exists a possible world which is completely empty except for the container of absolute space and the flow of absolute time. But here we must be very careful. Modern secular scholars have tended frequently to forget how ardent a theist Newton was and how central a role this theism played in his metaphysical outlook. Noting that Newton considered God to be temporal and therefore time to be everlasting, David Griffin observes that "Most commentators have ignored Newton's heterodox theology, and his talk of 'absolute time' has been generally misunderstood to mean that time is not in any sense a relation and hence can exist apart from actual events."[19] In fact, Newton made quite clear in the General Scholium to the *Principia*, which he added in 1713, that absolute time and space are constituted by the divine attributes of eternity and omnipresence:

---

15 Sklar, "Real Quantities," p. 61.

16 Newton, *Principles of Natural Philosophy*, 1: 8.

> Verum quoniam hae Spatii partes videri nequeunt, & ab invicem per sensus nostros distingui; earum vice adhibemus mensuras sensibilis. Ex positionibus enim & distantiis rerum a corpore aliquo, quod spectamus ut immobile, definimus loca universa.

17 Ibid. "in Philosophicis autem abstrahendum est a sensibus."

18 For a compelling demonstration of just how indigestible metric conventionalism is see Graham Nerlich, *The Shape of Space*, 2d ed. (Cambridge: Cambridge University Press, 1994).

19 David Ray Griffin, "Introduction: Time and the Fallacy of Misplaced Concreteness," in *Physics and the Ultimate Significance of Time*, ed. David R. Griffin (Albany, N. Y.: State University of New York Press, 1986), pp. 6-7. See also W. H. Newton-Smith, "Space, Time, and Spacetime: A Philosopher's View," in *The Nature of Time*, ed. Raymond Flood and Michael Lockwood (Oxford: Basil Blackwell, 1986), p. 27, who errs, however, in saying that space and time are aspects of God, rather than concomitants of God.

> He is eternal and infinite...; that is, his duration reaches from eternity to eternity; his presence from infinity to infinity.... He is not eternity and infinity, but eternal and infinite; he is not duration or space, but he endures and is present. He endures forever, and is everywhere present; and, by existing always and everywhere, he constitutes duration and space. Since every particle of space is *always*, and every indivisible moment of duration is *everywhere*, certainly the Maker and Lord of all things cannot be *never* and *nowhere*.[20]

Because God is eternal, there exists an everlasting duration, and because He is omnipresent, there exists an infinite space. Since God exists in all possible worlds and is essentially eternal and omnipresent, absolute time and space also exist in all possible worlds, whether in those worlds there exists any creation at all. Thus, time and space are absolute in the sense that they exist independently of any physical objects, but that is not to say that they exist wholly non-relationally.[21]

In Newton's thinking, metaphysical time, God's time, is A-theoretic time and the foundation of becoming. We have seen that metaphysical time "from its own nature flows equably without relation to anything external." To Newton therefore it makes perfectly good sense to imagine that all ideal clocks and physical processes in the universe should start to run twice as fast as they previously had. God could simply speed up everything so that events occur as in a film being run on fast-forward. His time, metaphysical time, would remain unaffected by this alteration in physical time. In the absence of metaphysical time, such a speed-up becomes literally meaningless. Since clocks would define what a temporal interval is, it would be nonsense to say that clocks could all run twice as fast as they had, for the minute hand's circling the face of a clock one time *is* the temporal interval we call an hour. The hand would not circle the clock's face twice in an hour; rather two hours would elapse. That is

---

20 Newton, *Principles of Natural Philosophy*, 1: 545.

> Aeternus est & Infinitus,...id est, durat ab aeterno in aeternum, & adest ab infinito in infinitum.... Non est aeternitas & infinitas, sed aeternus & infinitus; non est duratio & spatium, sed durat & adest. Durat semper, & adest ubique; & existendo semper & ubique, durationem & spatium constituit. Cum unaquaeque spatii particula sit *semper*, & unumquodque durationis indivisibile momentum *ubique*, certe rerum omnium Fabricator ac Dominus non erit *nunquam, nusquam.*

21 According to Earman, by "without relation to anything external," Newton meant without relation to material bodies (Earman, "Absolute Space," p. 289).

It is worth drawing attention to the fact that while the absolute-relational distinction implies the absolute-measured distinction, the reverse is not the case. That is to say, if time is absolute in the non-relational sense, then it is also absolute in the sense that its metric is the standard for isochronous intervals; but it is not obviously the case that if time transcends our attempts to measure it then time exists non-relationally, wholly independently of all events. One could quite consistently maintain, as Leibniz did, that time would not exist in the utter absence of events but is in some sense constituted by the fact of change (perhaps in God's actions or thoughts) and also maintain that there is a true time (God's time) which our physical clocks approximate. Similarly, one could maintain that space is absolute in the sense that it is approximated but not constituted by our measurements of it without holding that it is absolute in the sense of either "non-relational" or "non-dynamical." Newton never provides any argument, so far as I can determine, as to why God's existence must be spatial. *Contra* Newton, one could consistently hold that in the absence of physical objects space would not exist and also that the metaphysical space which does exist is non-Euclidean globally and locally, and yet maintain that it is absolute in the sense that it constitutes a privileged fundamental frame which provides the background for local reference frames (relative spaces), against which absolute motion or rest is determined.

not to say that two hours would elapse in an interval previously designated as one hour, for that is to presuppose the standard of metaphysical time to which physical time is relative. Rather in the absence of metaphysical time, it is just two hours elapsing in two hours' time, and any talk of a speed-up is meaningless. For Newton, however, it would seem obvious that it lies within God's power to make all physical processes occur more quickly, which implies the existence of metaphysical time.

NEWTON'S THEISM AND THE CLASSICAL CONCEPT OF TIME

Scholars writing during the positivist era tended to treat Newton's theological disquisitions dismissively as incidental intrusions into otherwise sober scientific thinking. E. W. Strong, for example, writing in 1952, contended that the absence of any references to God in the original editions of the *Principia* and *Opticks* showed that God was irrelevant to works written by a scientist for a scientific purpose:

> Newton nowhere asserts that space and time are postulated as absolute because they are the sensorium of God and hence cannot be other than absolute in natural philosophy without impropriety.... When it is said that he constitutes duration and space, it may, with equal logic be said that he constitutes bodies also in their composition and in their motions relative to one another.[22]

Similarly Jammer seems to see Newton's references to God as indications of the onset of senility: "So a comparison of the first and later editions of the *Principia* shows that the identification of absolute space with God, or with one of his attributes, came into the foreground of Newton's thought only toward the end of his life, that is, at the beginning of the eighteenth century."[23] Even a contemporary philosopher like Howard Stein insists that Newton's ideas are not based in theology; rather a certain theology was acceptable because its conceptions agreed with those required by mechanics.[24]

With the decline of positivism and the awakening of interest in the history of science, contemporary historians and philosophers of science have come to realize that Newton's statements concerning God need to be taken seriously, and this has led to a new appreciation of the role of Newton's theology in his scientific thinking. Observing that "Metaphysics is both a starting point and the final aim of Newton's physics" and that his declared commitments on the subject of God's place in his system must therefore be taken at face value, Zev Bechler reflects the consensus of contemporary Newtonian historiography when he writes,

> The effort to present Newton as a sober positivistic, no-nonsense scientist fit for incorporation into the venerable origins of the twentieth century tradition leads scholars

---

22 E. W. Strong, "Newton and God," *Journal of the History of Ideas* 13 (1952): 154.

23 Jammer, *Concepts of Space*, p. 108.

24 Howard Stein, "Newtonian Spacetime," *Texas Quarterly* 10 (1967): 198. Cf. the emotional reactions in L. A. Whitt, "Absolute Space: Did Newton Take Leave of His (Classical) Empirical Senses?" *Canadian Journal of Philosophy* 12 (1982): 709-710, to Newton's giving theological reasons for absolute space (which the author mislocates as being in the Scholium to the Definitions!). Whitt huffs: "Newton has given way (or perhaps vent) to his theological scruples and has, for the brief space of this scholium, gone off the (empirical) rails" (Ibid., p. 711).

> to attribute to him ideas about the autonomy of science which are strangely out of tune with his declared commitments.[25]

In editing the *Principia*, Bernard Cohen discovered that in Book III, Proposition 8, Corollary 5 of the first edition, Newton explicitly states that God placed the planets in their orbits at appropriate distances, but that this reference to the deity was dropped in subsequent editions.[26] "The result of this alteration has been that almost all commentators of Newton have erroneously assumed that Newton mentioned God in the *Principia* only in the later editions," specifically in the General Scholium, Cohen notes; "We may now reject wholly the view that Newton's introduction of God into the *Principia* was a result of senility, intellectual decline, or even a later development" after the first edition.[27] In fact, on 10 December 1692, Newton confided to Richard Bentley, "When I wrote my Treatise about our System, I had an Eye upon such Principles as might work with considering Men, for the Belief of a Deity, and nothing can rejoice me more than to find it useful for that purpose."[28] Similarly, in the Latin edition of the *Opticks* (1706), Newton declares space to be "the Sensorium of a Being incorporeal, living and intelligent, who sees the things themselves intimately, and thoroughly perceives them, and comprehends them wholly by their immediate presence to himself...."[29] But the most important factor in the reassessment of the role of Newton's theology in his science was undoubtedly the scholarly excavation of Newton's unpublished manuscripts, roughly a third of which are theological in character, which disclose that the themes of the General Scholium were lifelong concerns of Newton. The Yahuda manuscripts 15.3, f. 59 and 15.5, f. 98, housed at the Jewish National and University Library in Israel, are in part virtually identical to the text of the General Scholium.[30] Newton's "On the Gravity and Equilibrium of Fluids" (1666-70) is an especially rich resource for his ideas on space, time, and divinity, which are echoed in the General Scholium.[31] McGuire has also identified an untitled manuscript on time, place, and God in the

---

[25] Zev Bechler, "Introduction: Some Issues of Newtonian Historiography," in *Contemporary Newtonian Research*, ed. Zev Bechler, Studies in the History of Modern Science 9 (Dordrecht: D. Reidel, 1982), p. 13. Cf. Alan Gabbey, "Newton and Natural Philosophy," in *Companion to the History of Modern Science*, ed. R. C. Olby, G. N. Cantor, J. R. R. Christie, and M. J. S. Lodge (London: Routledge, 1990).

[26] Newton, *Principia*, E_1{405.35}, p. 383.

[27] I. Bernard Cohen, *Introduction to Newton's 'Principia'* (Cambridge: Cambridge University Press, 1971), p. 156. See also Michael J. Buckley, "God in the Project of Newtonian Mechanics," in *Newton and the New Direction in Science*, ed. G. V. Coyne, M. Heller, and J. Zycinski (Vatican City: Specola Vaticana, 1988), pp. 85-105.

[28] Isaac Newton to Richard Bentley, December 10, 1692, in *Isaac Newton's Papers and Letters on Natural Philosophy*, 2d ed., ed. with a General Introduction by I. Bernard Cohen (Cambridge, Mass.: Harvard University Press, 1978), p. 280.

[29] See Richard S. Westfall, *Never at Rest: A Biography of Isaac Newton* (Cambridge: Cambridge University Press, 1980), p. 647.

[30] See Westfall, *Never at Rest*, p. 749.

[31] Isaac Newton, "On the Gravity and Equilibrium of Fluids," [*De gravitatione et aequipondio fluidorum*] in *Unpublished Scientific Papers of Isaac Newton*, ed. A. Rupert Hall and Marie Boas Hall (Cambridge: Cambridge University Press, 1962), pp. 89-156.

Cambridge University Portsmouth Collection which constitutes an intermediate stage between these two.[32]

In *De gravitatione* Newton argues that space (and by implication time) is neither substance, nor accident, nor nothing at all. It cannot be nothing because it has properties, such as infinity and isotropy. It cannot be an accident because it can exist without bodies. Neither is it a substance:

> It is not substance; on the one hand, because it is not absolute in itself, but is as it were an emanent effect of God, or a disposition of all being; on the other hand, because it is not among the proper dispositions that denote substance, namely actions, such as thoughts in the mind or motions in the body.[33]

Contrary to the conventional understanding, Newton here declares explicitly that space is *not* in itself absolute (*non absoluta per se*) and therefore not a substance. Rather it is an emanent—or emanative—effect of God (*Dei effectus emanativus*). The notion of emanative causality had played a key role in the metaphysic of the Cambridge Platonist Henry More. In his Axiom XVI More explained, "By an *Emanative Cause* is understood such a Cause as merely by Being, no other activity or causality interposed, produces an Effect."[34] Correlatively, More explained in his Axiom XVII:

> *An Emanative Effect is coexistent with the very Substance of that which is said to be the Cause thereof.*
>
> This must needs be true, because that very substance which is said to be the Cause, is the adequate and immediate Cause, and wants nothing to be adjoyned to its bare essence for the production of the Effect; and therefore by the same reason the Effect is at any time, it must be at all times, or so long as that substance does exist.[35]

Three times in *De gravitatione* Newton calls space an emanative effect of God. It is uncreated and co-existent with God and yet ontologically dependent upon Him for its being. Newton calls space a disposition or affection of being (*entis affectio*). To be is to be spatially and, as Newton goes on to explain, temporally.

> Space is a disposition of being *qua* being. No being exists or can exist which is not related to space in some way. God is everywhere, created minds are somewhere, and body is in the space that it occupies: and whatever is neither everywhere nor anywhere does not exist. And hence it follows that space is an effect arising from the first existence of being [*entis primario existentis effectus emanativus*], because when any being is postulated, space is postulated. And the same may be asserted of duration: for certainly both are dispositions of being or attributes according to which we denominate

---

32 Manuscript Add. 3965, section 13, folios 541^r-542^r; 545^r-546^r. See J. E. McGuire, "Newton on Place, Time, and God: An Unpublished Source," *British Journal for the History of Science* 11 (1978): 114-129.

33 Newton, *De gravitatione*, p. 132.

> Non est substantia tum quia non absolute per se, sed tanquam Dei effectus emanativus, et omnis entis affectio quaedam subsistit; tum quia non substat ejusmodi proprijs affectionibus quae substantiam denominant, hoc est actionibus, quales sunt cogitationes in mente et motus in corpore.

34 Henry More, "The Immortality of the Soul," in *Philosophical Writings of Henry More*, ed. with an Introduction and Notes by Flora Isabel MacKinnon, Wellesley Semi-Centennial Series (New York: Oxford University Press, 1925), p. 74.

35 Ibid.

> quantitatively the presence and duration of any existing individual thing. So the quantity of the existence of God was eternal, in relation to duration, and infinite in relation to the space in which he is present; and the quantity of the existence of a created being was as great, in relation to duration, as the duration since the beginning of its existence, and in relation to the size of its presence as great as the space belonging to it.[36]

Both space and time are thus inherent to being. God's infinite being thus has as its consequence infinite time and space, which represent the quantity of His duration and presence.

In the Neo-Platonic tradition represented by More the doctrine of emanation is associated with pantheism or panentheism. But, as Newton makes clear in the *General Scholium*, he does not conceive of space or time as in any way aspects of God Himself. Similarly, in the manuscript on time and place, Newton cautions,

> By reason of its eternity and infinity space will neither be God nor wise nor powerful nor alive, but will merely be increased in duration and magnitude; whereas God by reason of the eternity and infinity of his space (that is, by reason of his eternal omnipresence) will be rendered the most perfect being.[37]

In this same manuscript Newton foreshadows the famous passage in his *Opticks* concerning space and God's *sensorium*. He writes,

> The most perfect idea of God is that he be one substance, simple, indivisible, live and making live, necessarily existing everywhere and always, understanding everything to the utmost, freely willing good things, by his will effecting all possible things, and containing all other substances in Him as their underlying principle and place; a substance which by his own presence discerns and rules all things, just as the cognitive part of man perceives the forms of things brought into the brain, and thereby governs his own body; which always and everywhere can bring to act all possible things, which most freely brings about all things that are best and most accord with reason, and cannot be induced to act otherwise by error or blind fate.[38]

---

[36] Newton, *De gravitatione*, pp. 136-137.
Spatium est entis quaetenus ens affectio. Nullum ens existit vel potest existere quod non aliquo modo ad spatium refertur. Deus est ubique, mentes creatae sunt alicubi, et corpus in spatio quod implet, et quicquid nec ubique nec ullibi est id non est. Et hinc sequitur quod spatium sit entis primario existentis effectus emanativus, quia posito quolibet ente ponitur spatium. Deque Duratione similia possunt affirmari: scilicet ambae sunt entis affectiones sive attributa secundum quae quantitas existentiae cujuslibet individui quoad amplitudinem praesentiae et perseverationem in suo esse denominatur. Sic quantitas existentiae Dei secundum durationem aeterna fuit, et secundum spatium cui adest, infinita; et quantitas existentiae rei creatae secundum durationem tanta fuit quanta duratio ab inita existentia, et secundum amplitudinem praesentiae tanta ac spatium cui adest.

[37] See McGuire, "Newton on Place, Time, and God," p. 119.
Spatium ex aeternitate et infinitate <sua> nec Deus erit nec sapiens nec potens nec vivum sed duratione et magnitudine tantum augebitur, Deus autem ex aeternitate et infinitate spatii sui (id est ex aeterna sua omnipraesentia) reddetur ens perfectissimum.

[38] Ibid., p. 123.
Perfectissima Dei Idea est ut sit substantia una, simplex, indivisibilis, viva et vivifica, ubique semper necessario existens, summe intelligens omnia, libere volens bona, voluntate efficiens possibilia, et substantias omnes alias in se continens tanquam eorum principium substans & locus; substantia quae per praesentiam suam cernit et regit omnia

God is here declared to contain all other substances, not merely as their first principle, but also as the place where they exist. Since God fills all of space, physical objects literally exist in God. As Newton was to write in the General Scholium, all things are contained in God, yet He neither affects nor is affected by the motion of bodies in Him. Since all things are immediately present to Him in this way, everything is immediately discerned and ruled by Him. It is like, Newton says, the immediate presence of perceptible forms to a man's mind, whereby he may rule his physical body. In a manuscript dating from the early 1690's Newton similarly speaks of God as "decreeing and ruling all things by means of his substantial presence (as the thinking part of man perceives the appearances of things brought into the brain and thence its own body)."[39] When he wrote the Latin edition of the *Opticks,* Newton incautiously queried,

> Is not infinite Space the Sensorium of a Being [*Annon Spatium Universum, Sensorium est Entis*] incorporeal, living and intelligent, who sees the things themselves intimately, and thoroughly perceives them, and comprehends them wholly by their immediate presence to himself...?[40]

As if having a premonition of the controversy to follow, Newton tried to recall the entire edition and subsequently changed the passage to read as follows:

> Is not the Sensory of Animals that place to which the sensitive Substance is present, and into which the sensible Species of Things are carried through the Nerves and Brain, that there they may be perceived by their immediate presence to that Substance? And these things being rightly dispatch'd, does it not appear from Phaenomena that there is a Being incorporeal, living, intelligent, omnipresent, who in infinite Space, as it were in his Sensory, sees the things themselves intimately, and thoroughly perceives them, and comprehends them wholly by their immediate presence to himself: Of which things the Images only carried through the Organs of Sense into our little Sensoriums, are there seen and beheld by that which in us perceives and thinks.[41]

The addition of the phrase *tanquam Sensorio sua* did not deter Leibniz in his famous correspondence with Clarke from attacking Newton's views on this score. Leibniz apparently thought that for Newton space was a sort of sense organ for God: "Sir Isaac Newton says, that space is an organ, which God makes use of to perceive things by. But if God stands in need of any organ to perceive things by, it will follow, that they do not depend upon him, nor were produced by him."[42] This is obviously a misinterpretation on Leibniz's part. The sensory is not a physical organ,

---

> sicut hominis pars cogitans sentit species rerum in cerebrum delatas et illinc regit corpus proprium; quaeque possibilia omnia semper et ubique in actum deducere potest, liberime agit quae optima & rationi maxime consentanea sunt, et errore vel fato caeco adduci non potest ut aliter agat.

[39] David Gregory MS. 245, fol. 14a, Library of the Royal Society, London, cited in J. E. McGuire, "Force, Active Principles, and Newton's Invisible Realm," *Ambix* 15 (1968): 190.

[40] Cited in Westfall, *Never at Rest*, p. 647, cf. Alexandre Koyré and I. Bernard Cohen, "The Case of the Missing *Tanquam*," *Isis* 52 (1961): 555-566.

[41] Sir Isaac Newton, *Opticks, or a Treatise of the Reflections, Refractions, Inflections and Colours of Light*, based on 4th ed., 1730, with a Foreword by Albert Einstein, an Introduction by Sir Edmund Whittaker, and a Preface by I. Bernard Cohen (New York: Dover, 1952), Bk. III, Pt. 1, Q. 28, p. 370.

[42] G. W. Leibniz, "Mr. Leibniz's First Paper," [1715] in *The Leibniz-Clarke Correspondence*, ed. with an Introduction and Notes by H. G. Alexander (Manchester: Manchester University Press, 1956), p. 11.

but an ideal place where sense images appear to the mind. Hence, Clarke rightly replies,

> Sir Isaac Newton doth not say, that space is the organ which God makes use of to perceive things by; nor that he has need of any medium at all, whereby to perceive things: but on the contrary, that he, being omnipresent, perceives all things by his immediate presence to them, in all space wherever they are, without the intervention or assistance of any organ or medium whatsoever. In order to make this more intelligible, he illustrates it by a similitude: that as the mind of man, by its immediate presence to the pictures or images of things, form'd in the brain by the means of the organs of sensation, sees those pictures as if they were the things themselves; so God sees all things, by his immediate presence to them; he being actually present to the things themselves, to all things in the universe; as the mind of man is present to all the pictures of things formed in his brain. Sir Isaac Newton considers the brain and organs of sensation, as the means by which those pictures are formed: but not as the means by which the mind sees or perceives those pictures, when they are so formed. And in the universe, he doth not consider things as if they were pictures, formed by certain means, or organs; but as real things, form'd by God himself and seen by him in all places wherever they are, without the intervention of any medium at all. And this similitude is all that he means, when he supposes infinite space to be (as it were) the sensorium of the Omnipresent Being.[43]

This strikes me as a careful and accurate exegesis of Newton's statements on this head.[44] But as the illustrious correspondence proceeded, Clarke began to stray from Newton's position. For example, in his third reply Clarke asserts, "Space is not a being, an eternal and infinite being, but a property, or a consequence of the existence of a being infinite and eternal. Infinite space, is immensity; but immensity is not God: and therefore infinite space, is not God."[45] Here the words "a property" intrude inaccurately into this otherwise faithful Newtonian exposition. By his fourth reply Clarke's error has clearly broken out:

> Space void of body, is the property of an incorporeal substance.... Space is not a substance, but a property, and if it be the property of that which is necessary, it will consequently (as all other properties of that which is necessary must do) exist more necessarily (though it be not itself a substance,) than those substances themselves which are not necessary.[46]

Similarly, in his fifth reply Clarke infers that since space and duration are not substances, "It remains, therefore, by necessary consequence that space is a property, in like manner as duration is."[47] Clarke has here departed from Newton and is expounding his own position, enunciated in his Boyle Lectures, that eternity and immensity "must needs be Modes or Attributes of a necessary Being actually existing."[48] Newton was scandalized, and when the Des Maiseaux edition of the

---

43 Samuel Clarke, "Dr. Clarke's First Reply," in *Correspondence*, pp. 12-13.

44 See Buckley, "God in the Project of Newtonian Mechanics," pp. 85-105, esp. p. 100 for a similar account.

45 Samuel Clarke, "Dr. Clarke's Third Reply," in *Correspondence*, p. 31.

46 Samuel Clarke, "Dr. Clarke's Fourth Reply," in *Correspondence*, p. 47.

47 Samuel Clarke, "Dr. Clarke's Fifth Reply," in *Correspondence*, p. 121.

48 Samuel Clarke, *A Discourse Concerning the Being and Attributes of God, the Obligations of Natural Religion and the Truth and Certainty of the Christian Revelation* [1704-05], in *The Works of Samuel Clarke*, vol. 2: *Sermons on Several Subjects* (London: John & Paul Knapton, 1738), 2: 530.

Leibniz-Clarke correspondence appeared in 1720, it carried the following *avertissement au lecteur:*

> Since the terms *quality* or *property* have normally a sense different from that in which they must be taken here, M. Clarke has asked me to warn his readers that when he speaks of infinite space or immensity and infinite duration or eternity, and gives them, through an inevitable imperfection of language, the name of qualities or properties of a substance which is immense or eternal, he does not claim to take the term *quality* or *property* in the same sense as they are taken by those who discuss logic or metaphysics when they apply them to matter; but that by this name he means only that space and duration are modes of existence of the Substance which is really necessary, and substantially omnipresent and eternal.[49]

On the basis of multiple drafts of this passage preserved in the Portsmouth Collection 82-84, 96-97, Koyré and Cohen have shown that it was Newton himself who had written this correction.[50] In the drafts Newton had spoken of space and time once more as "consequents of the existence of a Being which is really, necessarily & substantially Omnipresent & Eternal."[51] Newton thus emphatically repudiated the view that space and time are attributes of God. Rather absolute space and time are concomitant consequences of God's existence.

Although on Newton's view space (and by implication time) is not a substance in the sense that it is an independently existing entity, nevertheless it is very much like a substance. McGuire explains,

> Newton refers to the nature of space as an existing thing, and has in mind a decided list of properties which he considers belong properly to its nature: it has inherent parts that constitute its structure; in all directions it possesses unlimited extension; it is actually infinite; independently of things external to it, its inherent parts maintain the same eternal and immutable order; all true 'positions, distances, and local motions' have reference to these parts alone; and lastly, space is passively receptive to being occupied, and can offer no resistance to motion. It is clear that Newton considers space to be an individual in the sense of being the proper subject of certain sorts of predications.... Newton believes space to be a positive individual embodying real properties which derive from its actual extension in three dimensions. As such, it is...an individual fully actual at all time, and endowed with a rich inherent structure.[52]

Space and time thus have a sort of quasi-substantival status for Newton.

What becomes clear from this historical excursus is that the classical, Newtonian concept of time is rooted in a theistic metaphysic. Far from being after-the-fact theological reflections on concepts arising solely from physical considerations, the views of the General Scholium on time and space are the fruit of a long study which

---

[49] Cited in H. G. Alexander, "Introduction," in *Correspondence*, p. xxix. For discussion, see John Carriero, "Newton on Space and Time: Comments on J. E. McGuire," in *Perspectives on Newtonian Science*, pp. 122-124.

[50] Alexandre Koyré and I. Bernard Cohen, "Newton and the Leibniz-Clarke Correspondence," *Archives internationales d'histoires des sciences* 15 (1962): 63-126.

[51] Draft A, cited in Koyré and Cohen, "Newton," Cf. the French text: "*l'Espace* & la *Durée* sont des *Modes d'existence* dans tous les Etres; & des *Modes infinis*, & des Conséquences, de *l'existence* de la Substance qui est réellement, nécessairement, & substantiellement *toute-presente*, & éternelle" (Ibid., p. 83).

[52] J. E. McGuire, "Space, Infinity and Indivisibility: Newton on the Creation of Matter," in *Contemporary Newtonian Research*, p. 163.

from start to finish was metaphysical and theological in character.[53] James Force rightly concludes,

> Newton would have been astonished to learn that some of his interpreters, following Hume's lead, have claimed that theology, metaphysics, and epistemology have no necessary, integrated, synthetic relationship in themselves, much less that he himself has been placed into this school. Newton's own thought is in fact a seamless unity composed of theology, metaphysics, and epistemology all mixed together because, at their base, is the Lord God of supreme dominion.[54]

Our exposition of Newton's views on God, space, and time makes it evident that when Newton speaks of divine eternity, he does not, like scholastic theologians in the Augustinian tradition, mean a state of timelessness, but rather infinite and everlasting temporal duration. In a preliminary draft of the General Scholium, Newton had explicitly rejected the conception of God's eternity as an eternal now: "His duration is not a *nunc stans* without duration, nor is his presence nowhere."[55] Writing to Des Maiseaux in 1717, he elaborated,

> The schoolmen made a *Nunc stans* to be eternity & by consequence an attribute of God & eternal duration hath a better title to that name, though it be but a mode of his existence. For a *nunc stans* is a moment which always is & yet never was nor will be, which is a contradiction in terms.[56]

---

[53] Westfall states,

> "Composed virtually at the end of his active life, the General Scholium contained a vigorous reassertion of those principles which Newton had adopted in his rebellion against the perceived dangers of Cartesian mechanical philosophy. The same principles had continued to govern his scientific career as he followed the consequences of his rebellion into a new natural philosophy and a new conception of science" (Westfall, *Never at Rest*, p. 749).

[54] James E. Force, "Newton's God of Dominion: The Unity of Newton's Theological, Scientific, and Political Thought," in *Essays on the Context, Nature, and Influence of Isaac Newton's Theology*, ed. James E. Force *et al.*, (Dordrecht: Kluwer Academic Publishers, 1990), p. 90. Cf. Bechler's comment, "The view of physics and metaphysics as one indivisible block was inevitable to him, for physics was the inquiry into the nature of bodies in space and time, but since space and time were the abode of God, there was no possible way to demonstrate separate compartments of inquiry within all this" (Bechler, "Introduction," p. 14); similarly McGuire: "...his ontology of space and time cannot be understood without fully appreciating how it relates to the nature of divine existence" (McGuire, "Existence, Actuality, and Necessity: Newton on Space and Time," *Annals of Science* 35 [1978]: 463); and Brooke, "Newton himself provides one of the most spectacular examples of the integration of scientific and religious interests in one and the same mind" (John Hedley Brooke, "Science and Religion," in *History of Modern Science*, p. 775). Elsewhere Brooke observes, "If theistic discourse constitutes part of natural philosophy, it becomes anachronistic to ask how Newton *reconciled* his science with his religion. The right question is more likely to be: *how did his distinctive view of God's dominion, in both nature and history, affect his interpretation of nature?*" (John Brooke, "The God of Isaac Newton," in *Let Newton Be!*, ed. John Fauvel, *et al.* [Oxford: Oxford University Press, 1988], pp. 171-172). So also Ivor Leclerc, "The Relation between Natural Science and Metaphysics," in *The World View of Contemporary Physics*, ed. with an Introduction by Richard F. Kitchener (Albany: SUNY Press, 1988), pp. 26-27.

[55] Cited in McGuire, "Predicates of Pure Existence: Newton on God's Space and Time," in *Philosophical Perspectives on Newtonian Science*, ed. Phillip Bricker and R. I. G. Hughes (Cambridge, Mass.: MIT Press, 1990) p. 93.

[56] Newton to Des Maiseaux, in *Unpublished Papers*, p. 357. See also his rejection of God's existing *totum simul* in "Place, Time, and God" (ms. add. 3965, sect. 13, f. 545^r-546^r, in McGuire, "Newton on Place, Time, and God," p. 121).

Far from being atemporal, God's now or present is thus the present of absolute time. Since God is not "a dwarf-god" located at a place in space,[57] but is omnipresent, every indivisible moment of duration is everywhere, as we saw in the General Scholium. There is thus a worldwide moment which is absolutely present. Dieks effectively captures the implications of Newton's theism when he writes,

> In classical mechanics absolute simultaneity can easily be interpreted in terms of the '*flowing now*' we know from introspection. Just like [*sic*] our personal '*now*' is connected with our progressing history, the universal 'now' determined by the absolute simultaneity relation may be regarded as related to one history, *e.g.* the history of the Universe as a whole. For Newton *God* endures, and thereby constitutes time. This naturally fits in with the conception that there must be a universal succession of moments, determining the existence of one entity...there is *one* history, with *one* '*now,*' dividing past and future.[58]

Newton's temporal theism thus provides the foundation for both absolute simultaneity and absolute becoming. These are features first and foremost of metaphysical time, God's time, and derivatively of measured or physical time.

NEWTONIAN TIME AND RELATIVISTIC TIME

Returning to Newton's Scholium to his Definitions, we find that Newton, after making further distinctions with regard to absolute and relative place and absolute and relative motion, proceeds to grant freely that due to Galilean relativity we cannot for the most part know whether something is at absolute rest or in absolute motion.

> But because the parts of space cannot be seen, or distinguished from one another by our senses, therefore in their stead we use sensible measures of them.... And so, instead of absolute places and motions, we use relative ones; and that without any inconvenience in common affairs; but in philosophical disquisitions, we ought to abstract from our senses.[59]

But these commonly used relative quantities should not be confused with their absolute counterparts:

> Wherefore relative quantities are not the quantities themselves, whose names they bear, but those sensible measures of them (either accurate or inaccurate), which are commonly used instead of the measured quantities themselves. And if the meaning of words is to be determined by their use, then by the names time, space, place, and motion, their [sensible] measures are properly to be understood; and the expression will be unusual, and purely mathematical, if the measured quantities themselves are meant. On this account those violate the accuracy of language, which ought to be kept precise,

---

57 Newton, "Place, Time, and God," in McGuire, "Newton on Place, Time, and God," p. 123.

58 Dennis Dieks, "Newton's Conception of Time in Modern Physics and Philosophy," in *Newton's Scientific and Philosophical Legacy*, ed. P. B. Scheurer and G. Debrock (Dordrecht: Kluwer Academic Publishers, 1988), pp. 156-157.

59 Newton, *Principles of Natural Philosophy*, 1: 8.

> Verum quoniam hae Spatii partes videri nequeunt, & ab invicem per sensus nostros distingui; earum vice abhibemus mensuras sensibilis.... Sic vice locorum & motuum absolutorum relativis utimur; nec incommode in rebus humanis: in philosophicis autem abstrahendum est a sensibus.

> who interpret these words for the measured quantities. Nor do those less defile the purity of mathematical and philosophical truths who confound real quantities with their relations and sensible measures.[60]

Newton believed that there were certain properties, causes, and effects which served to distinguish absolute motion and rest from relative motion and rest.[61] It is noteworthy, however, that some of these do not manifest themselves empirically. Others do, such as rotation, and Newton concluded his Scholium with the promise, "...how we are to obtain the true motions from their causes, effects, and apparent differences, and the converse, shall be explained more at large in the following treatise. For to this end it was that I composed it."[62]

How, then, we may ask, did Newton fall short in comparison with Einstein's analysis of time? A little reflection reveals that the shortcoming of Newton's analysis lay not in his distinguishing between metaphysical and physical time, as is so often alleged, but in his not realizing that the accuracy of physical time in its approximation to metaphysical time depends on the relative motion of one's clocks. Though Newton labeled his distinction between time and the sensible measures thereof "absolute" versus "relative," his physical time was in fact as absolute as metaphysical time in the sense that it, too, was well-defined independent of any reference frame apart from that of absolute space. The sense in which it was relative is that it was a more or less accurate approximation to metaphysical time. Because clocks may not be truly periodic, physical time may not register the true time:

> Absolute time, in astronomy, is distinguished from relative, by the equation or correction of the apparent time. For the natural days are truly unequal, though they are commonly considered as equal, and used for a measure of time; astronomers correct this inequality that they may measure the celestial motions by a more accurate time. It may be, that there is no such thing as an equable motion, whereby time may be accurately measured. All motions may be accelerated and retarded, but the flowing of absolute time is not liable to any change. The duration or perseverance of the existence of things remains the same, whether the motions are swift or slow or none at all: and therefore this duration ought to be distinguished from what are only sensible measures thereof; and from which we deduce it, by means of the astronomical equation.[63]

---

60 Ibid., 1: 11.

> Quantitates relativæ non sunt igitur eæ ipsæ quantitates, quarum nomina præ se serunt, sed sunt earum mensuræ illæ sensibiles (veræ an errantes) quibus vulgus loco quantitatum mensuratarum utitur. At si ex usu definiendæ sunt verborum significationes; per nomina illa Temporis, Spatii, Loci & Motûs propriè intelligendæ erunt hæ mensuræ sensibilis; & sermo erit insolens & purè mathematicus, si quantitates mensuratæ hic intelligantur. Proinde vim inferunt sacris literis, qui voces hasce de quantitatibus mensuratis ibi interpretantur. Neque minus contaminant Mathesin & Philosophiam, qui quantitates veras cum ipsarum relationibus & vulgaribus mensuris confudunt.

61 Ibid., 1: 8-11.

62 Ibid., 1: 12.

> Motus autem veros ex eorum causis, effectibus, & apparentibus differentiis colligere, & contra ex motibus seu veris seu apparentibus eorum causas & effectus, docebitur fusius in sequentibus. Hunc enim in finem tractatum sequentem composui.

63 Ibid., 1: 7-8.

> Tempus Absolutum a Relativo distinguitur in Astronomia per æquationem temporis

What Newton did not realize, nor could he have suspected, is that physical time is not only *relative,* but also *relativistic*, that the approximation of physical time to metaphysical time depends not merely upon the regularity of one's clock, but also upon its motion. Unless a clock were at absolute rest, it would not accurately register the passage of metaphysical time. *Moving clocks run slow.* This truth, unknown to Newton, only intimated by Larmor and Lorentz in the concept of "local time," was finally grasped by Einstein.

Where Newton fell short, then, was not in his analysis of absolute or metaphysical time—he had theological grounds for positing such a time—, but in his incomplete understanding of physical time. He assumed too readily that an ideal clock would give an accurate measure of metaphysical time independently of its motion. But the above quotation suggests that Newton, if confronted with relativistic evidence, would have welcomed this correction and seen therein no threat at all to his doctrine of metaphysical time. Lucas emphasizes, "The relativity that Newton rejected is not the relativity that Einstein propounded; and although the Special Theory of Relativity has shown Newton to be wrong in some respects,...it has not shown that time is relative in Newton's sense, and merely some numerical measure of process."[64] In short, relativity corrects Newton's concept of physical time, not his concept of metaphysical time.

Of course, it hardly needs to be said that there is a great deal of antipathy in modern physics and philosophy of space and time toward such metaphysical entities as Newtonian space and time, primarily because they are physically undetectable and serve no physical purpose. In their standard text *Gravitation*, Misner, Thorne, and Wheeler, for example, challenge Newton's metaphysic by asking, "But how does one give meaning to Newton's absolute space, find its cornerstones, mark out its straight lines?" Complaining that "his ideal geometry is beyond observation," they conclude, "Newton's absolute space is unobservable, non-existent."[65] Newton would have been singularly unimpressed with this verificationist challenge and the positivistic equation between unobservability and non-existence.[66] From his

---

> vulgi. Inæquales enim sunt dies naturales, qui vulgo tanquam æquales pro mensura temporis habentur. Hanc inæqualitatem corrigunt Astronomi, ut ex veriore tempore mensurent motus coelestes. Possibile est, ut nullus sit motus æquabilis, quo tempus accurate mensuretur. Accelerari & retardari possunt motus omnes, sed fluxus temporis absoluti mutari nequit. Eadem est duratio seu perseverantia existentiæ rerum, sive motus sint celeres, sive tardi, sive nulli: proinde hæc a mensuris suis sensibilibus merito distinguitur, & ex iisdem colligitur per æquationem astronomicam.

64 Lucas, *Treatise*, p. 90. Later he adds,

> "It is often said that relativity refuted Newton. But it is a misleading oversimplification. There is no straight opposition between relativity theories and Newtonian, absolute, theories.... Theologically (and not only theologically) speaking we may assign a preferred frame of reference,...which is at rest. There is no reason why we should not—God may have looked, and seen that it was so, and told Newton. Only there is no *physical* reason why we should" (Ibid., pp. 197-198).

65 C. Misner, K. S. Thorne, and J. A. Wheeler, *Gravitation* (San Francisco: W. H. Freeman, 1973), p. 19.

66 See Zev Bechler, *Newton's Physics and the Conceptual Structure of the Scientific Revolution*, Boston Studies in the Philosophy of Science 127 (Dordrecht: Kluwer Academic Publishers, 1991), p. 255, who

predecessor Isaac Barrow, he had been taught, "'*We do not perceive it, therefore it does not exist,*' is a fallacious inference...."[67] The grounds for metaphysical space and time were not physical, but philosophical, or more precisely, theological. Lucas emphasizes:

> The critics of Newtonian space, Leibniz (on occasion), Mach, and Einstein, urge an epistemological approach: 'How do we tell whether something is moving or at rest?' ...Newton did not feel the force of the epistemological criticism. He takes a 'God's-eye' view of the universe. God is present 'from infinity to infinity' and 'governs all things, and knows all things that are or can be done.' There are no epistemological problems for God. He is 'omnipresent, who...sees the things themselves intimately and thoroughly perceives them, and comprehends them wholly, by their immediate presence to Himself.' He knows, just knows, where everything is—more, Newton would say, He knows because He puts it there; God places each atom in its place by fiat of His will—and so knows where it is because He knows what He is doing, immediately and without any room for any epistemological problem to arise.[68]

Epistemological problems fail to worry Newton because, as Lucas nicely puts it, "He is thinking of an omniscient, omnipresent Deity whose characteristic relation with things and with space is expressed in the imperative mood."[69] Modern physical theories say nothing against the existence of such a God or the metaphysical space and time constituted, in Newton's thinking, by His eternity and omnipresence. What relativity theory did, in effect, was simply to cut God out of the picture and to substitute in His place a finite observer. "Thus," according to Holton, "the *RT* [Relativity Theory] merely shifted the focus of spacetime from the sensorium of Newton's God to the sensorium of Einstein's abstract *Gedanken*experimenter—as it were, the final secularization of physics."[70] But to a man like Newton, who wrote in his General Scholium, "and thus much concerning God; to discourse of whom from the appearances of things, does certainly belong to Natural Philosophy,"[71] such a secular outlook impedes rather than advances our understanding of the nature of reality. And even if we do not go so far as Newton in including discourse about God in scientific theorizing, still it is clear that if we are prepared to draw metaphysical inferences about the nature of space, time, and spacetime on the basis of physical science, then we must also be ready to entertain theistic metaphysical hypotheses such as Newton deemed relevant.

---

sees the distinction between testability on the one hand and truth and meaningfulness on the other to lie at "the basis of his ontology of space and time."

[67] Isaac Barrow, *The Geometrical Lectures of Isaac Barrow*, trans. with Notes by J. M. Child (Chicago: Open Court, 1916), p. 36.

[68] Lucas, *Treatise*, p. 143.

[69] Ibid.

[70] Gerald Holton, "On the Origin of the Special Theory of Relativity," in *Thematic Origins of Scientific Thought: Kepler to Einstein* by Gerald Holton (Cambridge, Mass.: Harvard University Press, 1973), p. 171. Cf. Capek's remark that it was the "alliance of theology and cosmology" from the Middle Ages through the nineteenth century that provided the basis for the concept of an absolute frame of reference—an alliance broken in the twentieth century (Milic Capek, "Introduction," in *The Concepts of Space and Time*, ed. M. Capek, Boston Studies in the Philosophy of Science 22 [Dordrecht: D. Reidel, 1976], p. xxii).

[71] Newton, *Principles of Natural Philosophy*, 2: 546. "Et haec de Deo; de quo utique ex Phaenomenis disserere, ad Philosophiam Naturalem pertinet."

CHAPTER 3

THE EPISTEMOLOGICAL FOUNDATIONS OF SR

By the end of the nineteenth century, physicists had already realized that there was something fundamentally wrong with Newton's analysis of physical time and space. The failure to detect the earth's motion through the aether, which constituted a relative or physical space at rest with respect to metaphysical space,[1] prompted a crisis in physics which compelled men like Lorentz, Larmor, and Poincaré to revise and then abandon the Galilean transformation equations in favor of the relativistic Lorentz transformations. In so doing, they had already sounded the death knell of Newtonian physics, for they had relativized the sensible measures of metaphysical time and space in a way undreamed of by Newton. But they did so without abandoning the notion that there really is a true time and a true space.

Einstein interrupted this research programme with a radically different approach. Foundational to Einstein's approach was his denial of absolute space and his consequent re-definition of time and simultaneity so as to deny their absolute status as well. What Einstein did, in effect, was to shave away Newton's metaphysical time and space, and along with them the aether, thus leaving behind only their sensible measures, so that physical time became the only time there is and physical space the only space there is. Since these are relativized to inertial frames, one ends up with the relativity of simultaneity and of length.

What justification did Einstein have for so radical a move? How did he know that metaphysical time and space do not exist? The answer, in a word, is positivism. Although one rarely finds this discussed in textbook expositions of the theory or even in discussions of the philosophical foundations of the theory, nevertheless historians of science have demonstrated convincingly that at the philosophical roots of Einstein's theory lies an epistemological positivism of Machian provenance which issues in a verificationist analysis of the concepts of time and space.[2]

---

1 Sir Oliver Lodge was the first to make this assumption explicit: "the whole of this subject [the aether problem] indicates that the aether is a physical standard of rest; and that motion relative to it, which is becoming cognizable by us, is in that sense an ascertained absolute motion" (O. J. Lodge, "Note on Mr. Sutherland's Objection to the Conclusiveness of the Michelson-Morley Aether Experiment," *Philosophical Magazine* 46 [1898]: 344). Even Lorentz did not make explicit the aether frame's privileged status until after Einstein's development of SR. For discussion see Tetu Hirosige, "The Ether Problem, the Mechanistic Worldview, and the Origins of the Theory of Relativity," *Historical Studies in the Physical Sciences* 7 (1976): 3-82; A. J. Knox, "Hendrik Antoon Lorentz, the Ether, and the General Theory of Relativity," *Archive for History of Exact Sciences* 38 (1988): 67-78.

2 I am especially indebted to the writings of Gerald Holton; see in particular Gerald J. Holton, "Mach, Einstein and the Search for Reality," in *Ernst Mach: Physicist and Philosopher*, ed. Robert S. Cohen and Raymond J. Seeger, Boston Studies in the Philosophy of Science 6 (Dordrecht: D. Reidel, 1970), pp. 165-199: idem, "Where Is Reality? The Answers of Einstein," in *Science and Synthesis*, ed. UNESCO

POSITIVIST INFLUENCES ON EINSTEIN

In 1905, when Einstein published his paper on the electrodynamics of moving bodies, and for several years thereafter, Einstein was a self-confessed epistemological pupil of the positivistic physicist Ernst Mach, and the epistemological analysis of space and time given in the opening section of that paper clearly displays this influence. According to Holton, "If one studies Einstein's earliest papers on special relativity from a philosophically critical point of view, one can discern the influence of many, partly contradictory, points of view.... But perhaps the most important of these influences was that of the empirico-critical positivism of the type of which the Austrian philosopher Ernst Mach was the foremost exponent."[3] Mach was a giant of nineteenth century German physics, of whom Einstein later remarked, "Even those who think of themselves as Mach's opponents hardly know how much of Mach's views they have, as it were, imbibed with their mother's milk."[4] Holton remarks, "The influence of Mach's point of view, particularly in German-speaking countries, was enormous—in physics, on physiology, on psychology, and on the fields of the history and the philosophy of science."[5]

Mach's philosophy of science was phenomenalist or sensationalist in character. In experience we are given various sensations, such as colors, sounds, pressures, and so forth, which Mach called "the elements", and the aim of scientific theorizing is to construct the simplest possible description of the connections among these sensations.[6] Mach had no use for theoretical entities or even for entities behind the sensations. Statements in theories were meaningful only if they were related directly to sensations. His attitude was militantly "anti-metaphysical," and in his *Die Mechanik in ihrer Entwicklung*, which Einstein studied carefully, Mach declared in his Preface to the work that "Its intention is...an enlightening one, or to put it more bluntly, an anti-metaphysical one."[7] In line with his phenomenalism,

---

(Berlin: Springer-Verlag, 1971), pp. 45-69; the essays collected together in idem, *Thematic Origins of Scientific Thought: Kepler to Einstein* (Cambridge, Mass.: Harvard University Press, 1973).

3 Holton, "Where Is Reality?" p. 45. It is interesting that Holton refers to Mach as a philosopher. Mach himself disclaimed that title, saying, "...I have already declared explicitly that *I am by no means a philosopher, but only a scientist*" (Ernst Mach, *Erkenntnis und Irrtum* 2d ed. [Leipzig: D. Reidel, 1906], pp. vii-viii).

4 Albert Einstein, "Ernst Mach," *Physikalische Zeitschrift* 17 (1916): 101, reprinted in Ernst Mach, *Die Mechanik in ihrer Entwicklung historisch-kritisch dargestellt*, ed. Renate Wahsner and Horst-Heino Borzeszkowski (Berlin, DDR: Akademie-Verlag, 1988), pp. 683-689. Einstein also refers to Mach as "the greatest influence on the epistemological orientation of the natural scientists of our time."

5 Holton, "Mach," p. 168.

6 See Philipp Frank, "The Importance of Ernst Mach's Philosophy of Science for our Times," [1917], rep. in *Ernst Mach: Physicist and Philosopher*, pp. 219-234.

7 Ernst Mach, *Mechanik in ihrer Entwicklung*, p. 13 (Ernst Mach, *The Science of Mechanics: A Critical and Historical Account of Its Development*, trans. Thomas J. McCormack [LaSalle, Ill.: Open Court, 1960], p. xxii). Here we may discern the even deeper influence of the father of positivism, Auguste Comte. Comte analyzed the evolution of thought along three stages: the theological, the metaphysical, and the positive. Positivism involves the determination to have nothing to do with theology and metaphysics. The classical concept of time as a metaphysical reality rooted in theology was

Mach held that "Space and time are well-ordered systems of series of sensations," comparable to sensations of color, sounds, and smells.[8] In his *Mechanics*, Mach assailed Newton's concepts of absolute space and time, which in his preface to the final seventh edition (1912) he denounced as "blunders" and "conceptual monstrosities."[9] According to Mach,

> No one is competent to predicate things about absolute space and absolute motion; they are pure things of thought, pure mental constructs, that cannot be produced in experience. All our principles of mechanics are...experimental knowledge concerning the relative positions and motions of bodies.... No one is warranted in extending these principles beyond the boundaries of experience. In fact, such an extension is meaningless, as no one possesses the requisite knowledge to make use of it.[10]

In Newtonian mechanics, absolute space served no purpose, since Newton's laws of motion held good in all inertial frames, not simply in frames at rest with respect to absolute space. The elimination of absolute space would not in any way reduce the empirical content of classical dynamics. It was therefore "metaphysical," and Mach proposed to do away with it along with any distinction between inertial and non-inertial frames.

Similarly, in his *Prinzipien der Wärmelehre*, Mach drew applications from his critique of the concept of temperature to Newton's concept of absolute time. Of temperature, Mach wrote,

> It is remarkable how long a period elapsed before it definitely dawned upon inquirers that the designation of thermal states by numbers rests on a convention. Thermal states exist in nature, but the conception of temperature exists only by virtue of our arbitrary *definition*, which might very well have taken another form. Yet until very recently inquirers in this field appear more or less unconsciously to have sought after a *natural* measure of temperature, a real temperature, a sort of Platonic Idea of temperature of which the temperatures read from the thermometric scales were only the imperfect and inexact expression.[11]

Mach thought that the Newtonian concepts of absolute time and space were as misconceived as the concept of an absolute measure of temperature:

> Newton's conceptions of 'absolute time,' 'absolute space,' *etc.*, which I have discussed in another place, originated in quite a similar manner. In our conceptions of time the sensation of duration plays the same part with regard to the various measures of time as the sensation of heat played in the instance just adduced. The situation is similar with respect to our conceptions of space.[12]

---

anathema. Mach refers favorably to Comte's three stages in his *Erkenntnis und Irrtum*, p. 99. For discussion see J. Bradley, *Mach's Philosophy of Science* (London: Athlone Press, 1971), pp. 204-207.

8 Mach, *Mechanik in ihrer Entwicklung*, p. 522 (*Science of Mechanics*, p. 611).

9 Ibid. pp. 23-24.

10 Ibid., p. 252 (*Science of Mechanics*, p. 280). In his obituary notice of Mach, it is to precisely these sections of the *Mechanics* (II. 6, 7) that Einstein draws attention (Einstein, "Mach," pp. 685-688).

11 Ernst Mach, *Prinzipien der Wärmelehre* (Leipzig: J. A. Barth, 1896), p. 154 (Ernst Mach, *Principles of the Theory of Heat*, trans. T. J. McCormack, rev. and compl. P. E. B. Jourdain and A. E. Heath, with an Introduction by M. J. Klein, ed. Brian McGuiness, Vienna Circle Collection 17 [Dordrecht: D. Reidel, 1986], pp. 53-54).

12 Ibid. Cf. Mach, *Mechanik in ihrer Entwicklung*, p. 249 (*Science of Mechanics*, p. 276).

The implication is that just as there is no absolute temperature behind our various conventional measures of our sensation of heat, so no absolute time exists behind the various conventional measures of our sensation of duration. Schaffner concludes that Mach's methodology in the *Wärmelehre* "seems to have influenced Einstein in his analysis of the concept of time."[13]

Mach's influence upon Einstein was reinforced by positivist physicists whom Einstein read while a student at the Eidgenössische Technische Hochschule in Zürich. Holton describes the phenomenalistic and revisionist programme they pursued:

> Their form of positivism—as against the logical positivism developed later in Carnap and Ayer's work—provided an epistemology for the new phenomenologically based science of correlated observations, linking energetics and sensationism.... "Hypothetical" quantities such as atomic entities were to be omitted; instead, these authors claimed they were satisfied, as Herz wrote around 1904, with "measuring such quantities as are presented in observation, such as energy, mass, pressure, volume, temperature, heat, electrical potential, etc., without reducing them to imaginary mechanisms or kinetic quantities." They condemned such conceptions as the ether, with properties not accessible to direct observation, and they issued a call "to consider anew the ultimate principles of all physical reasoning, notably the scope and validity of the Newtonian Laws of motion and of the conceptions of force and action, of absolute and relative motion."[14]

Chief among the influences on the young Einstein was August Föppl, whose *Introduction to Maxwell's Theory of Electricity* (1894) employed a Machian analysis of absolute space in dealing with experiments with moving conductors.[15] In the fifth chapter, entitled "The Electrodynamics of Moving Conductors," Föppl discusses the case of the magnet and the conductor with which Einstein opens his 1905 paper. Föppl replaced Newton's absolute space with the aether because the former was "not at all subject to experience."[16] But he struggled with the fact that insofar as the moving conductor and magnet were concerned, it is only their relative motion which matters. If both are absolutely moving in tandem, no electric current is produced, since they are at rest relative to each other. According to Holton, Föppl, with his "antimetaphysical and self-conscious empiricism," helped Einstein to realize that "the fundamental problem to be cracked is how to achieve a new point of view on the conceptions of time and space."[17] Another of these positivist physicists was W. Ostwald, with whom Einstein corresponded and whose findings were employed by Einstein in a paper of 1901. Ostwald went further than Föppl in rejecting the aether altogether, declaring, "There is no need to inquire for a carrier of

---

[13] Kenneth F. Schaffner, "Einstein versus Lorentz: Research Programmes and the Logic of Comparative Theory Evaluation," *British Journal for the Philosophy of Science* 25 (1974): 59.

[14] Gerald Holton, "Mach, Einstein and the Search for Reality," in *Thematic Origins*, p. 220.

[15] See Holton, "Influences," pp. 198-212; Arthur I. Miller, *Albert Einstein's Special Theory of Relativity* (Reading, Mass.: Addison-Wesley,1981), pp. 150-154.

[16] August Föppl, *Einführung in die Maxwell'sche Theorie der Elektrizität* (Leipzig: Teubner, 1894), p. 307.

[17] Holton, "Influences," pp. 207, 212.

[energy] when we find it anywhere. This enables us to look upon radiant energy as independently existing in space"—a position which Einstein was to adopt.[18]

Einstein himself freely expressed his indebtedness to Mach with respect to the epistemological foundations of Special Relativity. It was his student friend and colleague at the Bern patent office, Michael Besso—an enthusiastic Machist and the only person acknowledged in Einstein's 1905 paper[19]—who first introduced Einstein to Mach. In a letter of April 8, 1952, Einstein wrote, "My attention was drawn to Ernst Mach's *History of Mechanics* by my friend Besso while a student, around the year 1897. The book exerted a deep and persisting impression upon me..., owing to its physical orientation toward fundamental concepts and fundamental laws."[20] On December 8, 1947, Besso recalled in a letter to Einstein, "Is it not true that this introduction [to Mach] fell into a phase of development of the young physicist [*i.e.*, Einstein] when the Machist style of thinking pointed decisively at observables—perhaps even, indirectly, to clocks and meter sticks?"[21] In a reply to Besso the following month, Einstein conceded, "As for the influence of Mach on my thinking, it has certainly been very great. I remember very well how, during my early years as a student, you directed my attention to his treatise on mechanics and to his theory of heat, and how these two works made a deep impression on me."[22]

Einstein's deep admiration for Mach is evident in their surviving correspondence between 1909 and 1913. In the second edition of his *Conservation of Energy* (1909), Mach declared, "I subscribe, then, to the principle of relativity, which is also firmly upheld in my *Mechanics* and *Theory of Heat*."[23] In August of that year Einstein wrote to Mach upon receipt of the book, "You have had such a strong influence upon the epistemological conceptions of the younger generation of physicists that even your opponents today, such as Planck, undoubtedly would have been called Mach-followers by physicists of the kind that was typical a few decades ago."[24] Eight days later, Einstein wrote again in reply to a lost letter from Mach, saying, "I am very glad that you are pleased with the relativity theory" and signing off "Your admiring pupil."[25] As Einstein progressed toward the General Theory of

---

18 Wilhelm Ostwald, *Chemische Energie: Lehrbuch der allgemeinen Chemie*, 2d ed. (Leipzig: Verlag von Wilhelm Engelman, 1893), 2: 1016.

19 He wrote, "In conclusion I wish to say that in working at the problem here dealt with, I have had the loyal assistance of my friend and colleague M. Besso, and that I am indebted to him for several valuable suggestions" (Albert Einstein, "On the Electrodynamics of Moving bodies," trans. Arthur I. Miller, Appendix to Miller, *Einstein's Special Theory*, p. 415).

20 Albert Einstein to Carl Seelig, April 8, 1952, unpublished letter in the Einstein Archives at Princeton cited by Holton, "Where Is Reality?" pp. 47-48.

21 Michael Besso to Albert Einstein, December 8, 1947, in *Correspondance 1903-1955*, trans. with Notes and an Introduction by Pierre Speziali (Paris: Hermann, 1979), p. 228.

22 Albert Einstein to Michael Besso, January 6, 1948, in *Correspondance*, p. 231.

23 Ernst Mach, *Die Geschichte und die Wurzel des Satzes von der Erhaltung der Arbeit* (Prague: J. G. Calvé, 1872), in Ernst Mach, *Abhandlungen*, ed. J. Thiele (Amsterdam: E. J. Bonset, 1969), p. 60 (Ernst Mach, *History and Root of the Principle of the Conservation of Energy*, trans. Philip E. B. Jourdain [Chicago: Open Court, 1911], p. 95).

24 Albert Einstein to Ernst Mach, August 9, 1909, in Mach, *Mechanik in ihrer Entwicklung*, p. 679 (Holton trans.).

25 Albert Einstein to Ernst Mach, August 17, 1909, in Mach, *Mechanik in ihrer Entwicklung*, p. 680 (Holton trans.).

Relativity he still considered himself to be employing Mach's epistemology.[26] In fact, in the spring of 1911 Mach had helped to formulate a manifesto calling for the founding of a society for positivistic philosophy, and among the signatories of this document, such as David Hilbert, Sigmund Freud, and J. Petzoldt, was Albert Einstein.[27]

Eventually Mach came to repudiate relativity theory,[28] and Einstein threw off Mach's phenomenalism for a critical realism; but the founder of relativity theory continued to acknowledge Mach's influence on him during his early years. Writing off Mach's rejection of relativity theory to the intransigence of old age, Einstein insisted that "the whole direction of thought of this theory conforms with Mach's..."[29] While Mach's positivistic philosophy proved itself unfruitful in positive theory construction, still it served its purpose in eliminating unwanted metaphysical entities such as absolute time and space: "It cannot give birth to anything living, it can only exterminate harmful vermin."[30] Though converted into a "believing rationalist" through his work on gravitation, Einstein conceded that in his earlier work on the Special Theory he was "coming from sceptical empiricism of somewhat the kind of Mach's...."[31] In his later "Autobiographical Notes," Einstein connects his denial of absolute simultaneity with the critical reasoning of Ernst Mach and mentions the influence of another sceptical empiricist in this connection:

> Today everyone knows, of course, that all attempts to clarify this paradox satisfactorily were condemned to failure as long as the axiom of the absolute character of time, viz., of simultaneity, unrecognizedly was anchored in the unconscious. Clearly to recognize this axiom and its arbitrary character really implies already the solution of the problem. The type of critical reasoning which was required for the discovery of this central point was decisively furthered, in my case, especially by the reading of David Hume's and Ernst Mach's philosophical writings.[32]

---

26 See letters reprinted in Mach, *Mechanik in ihrer Entwicklung*, pp. 680-681 and trans. in Holton, "Where Is Reality?" pp. 51-52.

27 See Friedrich Herneck, "Nochmals über Einstein und Mach," *Physikalische Blätter* 17 (1961): 276.

28 Or so it has traditionally been believed. According to Brown, Gerion Wolters has shown Mach's rejection of relativity to have been a forgery by his son Ludwig (James Robert Brown, "Einstein's Brand of Verificationism," *International Studies in the Philosophy of Science* 2 [1987]: 36).

29 Albert Einstein to Armin Weiner, September 18, 1930, unpublished letter from the Archives of the Burndy Library in Norwalk, Connecticut, cited by Holton, "Where Is Reality?" p. 55. In his obituary notice "Ernst Mach," Einstein goes so far as to say, "It is not improbable that Mach would have discovered the theory of relativity, if, at the time when his mind was still young and susceptible, the problem of the constancy of the speed of light had been discussed among physicists" (Einstein, "Mach," p. 688 [Holton trans.]).

30 Albert Einstein to Michael Besso, May 13, 1917, in *Correspondance*, p. 68.

31 Albert Einstein to C. Lanczos, January 24, 1938, unpublished letter in the Einstein Archives at Princeton cited in Holton, "Where Is Reality?" p. 64.

32 Albert Einstein, "Autobiographical Notes," in *Albert Einstein: Philosopher-Scientist*, ed. P. A. Schilpp, Library of Living Philosophers 7 (LaSalle, Ill.: Open Court, 1949), p. 53. The paradox to which he refers is the question which occurred to him at the age of sixteen: if one pursues a light beam at the velocity *c*, one should observe it as an electromagnetic field at rest; but there seems to be no such thing. On the force of this reasoning, see Adolf Grünbaum, *Philosophical Problems of Space and Time*, 2d ed. Boston Studies in the Philosophy of Science 12 (Dordrecht: D. Reidel, 1973), pp. 73-75.

A similar mention is found in Einstein's memorial notice for Mach in 1916: referring to Mach's critique of Newton's absolute space and time, Einstein asserts, "No one can take it away from the epistemologists that here they paved the way for this development [of relativity theory]; for my own part at least I know that I have been greatly aided, directly and indirectly, through especially Hume and Mach."[33] In his letter to Besso, already quoted, which was written about the time of "Autobiographical Notes," he asserts that Hume's influence on his thinking was even *greater* than Mach's. Referring to the *Mechanics* and *Theory of Heat*, Einstein reflected, "The extent to which they influenced my own work is, to say the truth, not clear to me. As far as I am conscious of it, the immediate influence of Hume on me was greater.... But, as I said, I am not able to analyze that which lies anchored in unconscious thought."[34]

It is not too difficult to imagine what aspect of Hume's thought proved influential on Einstein at this point. In Hume's view, ideas which are not derived from sense impressions are empty of content and meaningless.

> When we entertain, therefore, any suspicion that a philosophical term is employed without any meaning or idea (as is but too frequent), we need but enquire, *from what impression is that supposed idea derived?* And if it be impossible to assign any, this will serve to confirm our suspicion.[35]

Propositions expressing analytic relations of ideas "are discoverable by the mere operation of thought, without dependence on what is anywhere existent in the universe."[36] Such *a priori* reasoning is valid only in the realm of mathematics; "...all attempts to extend this more perfect species of knowledge beyond these bounds are mere sophistry and illusion."[37] Propositions expressing matters of fact cannot be known or demonstrated *a priori* but can only be known on the basis of sense experience. "The existence, therefore, of any being can only be proved by arguments from its cause or its effect; and these arguments are founded entirely on experience. If we reason *a priori,* anything may appear able to produce anything."[38] According to Hume, unless we are able to furnish arguments based on sense experience for a purported matter of fact, that supposed fact is revealed to be vacuous.

> When we run over libraries, persuaded of these principles, what havoc must we make? If we take in our hand any volume; of divinity or school metaphysics, for instance; let us ask, *Does it contain any abstract reasoning concerning quantity or number?* No. *Does it contain any experimental reasoning concerning matter of fact and existence?* No. Commit it then to the flames: for it can contain nothing but sophistry and illusion.[39]

---

33 Einstein, "Mach," p. 686.

34 Einstein to Besso, January 8, 1948, in *Correspondance*, p. 231.

35 David Hume, *An Enquiry concerning Human Understanding,* in *Enquiries concerning Human Understanding and concerning the Principles of Morals*, ed. L. A. Selby-Bigge, 3d ed., rev. with Notes by P. H. Nidditch (Oxford: Clarendon Press, 1975), II.17 (p. 22).

36 Ibid., IV.i.20 (p. 25).

37 Ibid., XII.iii.131 (p. 163).

38 Ibid., XII.iii.132 (p. 164).

39 Ibid. (p. 165).

Although Hume's anti-metaphysical remarks were directed principally at scholastic philosophizing, Einstein would have seen in his words an equal indictment of Newton's metaphysical concepts of absolute time and space.[40]

A positivistic philosophy of science was also mediated to Einstein through his study of the writings of Poincaré. Einstein recalled, "In Bern I had regular philosophical reading and discussion evenings, together with K. Habicht and Solovine, during which we were mainly concerned with Hume.... The reading of Hume, along with Poincaré and Mach, had some influence on my development."[41] Einstein's biographer Abraham Pais comments, "I stress that Einstein and his friends did much more than just browse through Poincaré's writings. Solovine has left us a detailed list of books which the Akademie members read together. Of these he singles out one and only one, *La science et l'hypothèse*, for the following comment: '[this] book profoundly impressed us and kept us breathless for weeks on end!'"[42] In *Science and Hypothesis* Poincaré repudiated an absolute metric of time and objective simultaneity relations on the basis of a verificationist theory of meaning:

> There is no absolute time; to say two durations are equal is an assertion which has by itself no meaning and which can acquire one only by convention.... Not only have we no direct intuition of the equality of two durations, but we have not even direct intuition of the simultaneity of two events occurring in different places; this I have explained in an article entitled 'La mesure du temps.'[43]

Poincaré's positivism issues not only in his metric conventionalism (which is compatible with there being a privileged time) but also points to the conventionality of simultaneity. In his "Measure of Time"—Pais says it is "virtually certain" that Einstein also knew it[44]—Poincaré deals with two questions: (1) can we transform psychological time, which is qualitative, into quantitative time, and (2) can we reduce to one and the same measure facts which transpire in different worlds? The first is the problem of time's metric; the second is the question of distant simultaneity. The approach to both questions is thoroughly positivistic: with respect

---

40 Cf. Hume's phenomenalistic analysis of space and time in his *A Treatise of Human Nature*, ed. L. A. Selby-Bigge, rev. with Notes by P. H. Nidditch (Oxford: Clarendon Press, 1978), I.ii.3 (pp. 33-39), though Hume's analysis of time in terms of the succession of mental impressions does not succumb to Einstein's physicalism. See also Hirosige, "Origins of the Theory of Relativity," p. 58. Later in life Einstein blamed Hume for having fostered "a fateful 'fear of metaphysics'" which became "a malady of contemporary empiricistic philosophizing" (Albert Einstein, "Remarks on Bertrand Russell's Theory of Knowledge," in *The Philosophy of Bertrand Russell*, 3d ed., ed. P. A. Schilpp, Library of Living Philosophers [New York: Tudor Publishing, 1951], p. 289); but even so he urges that a system of concepts will not degenerate into metaphysics so long as enough propositions are connected with sensory experiences!

41 Albert Einstein to Michael Besso, March 6, 1952, in *Correspondance*, p. 464. The translation is Pais's.

42 Abraham Pais, *'Subtle is the Lord...': The Science and Life of Albert Einstein* (Oxford: Oxford University Press, 1982), pp. 133-134. The citation is from Albert Einstein, *Lettres à Maurice Solovine* (Paris: Gauthiers-Villars, 1956), p. viii.

43 Henri Poincaré, *Science and Hypothesis*, in *The Foundations of Science* (Science Press, 1913; rep. ed.: Washington, D. C.: University Press of America, 1982), p. 92.

44 Pais, *'Subtle is the Lord,'* p. 133.

to (1), since we have no direct intuition of the equality of temporal intervals, the affirmation that they are equal has no meaning at all, and so we must conventionally define a measure of time such that the equations of mechanics may be as simple as possible; with respect to (2) affirmations of the simultaneity of distant events "have by themselves no meaning" and we should ask ourselves how we came up with the idea of "putting into the same frame" events so far removed from one another.[45] It is not hard to discern here possible inspiration for Einstein's operational definition of time and the frame-dependence of simultaneity, but far more important, because unquestioningly presupposed, is Poincaré's positivistic epistemological approach to issues of space and time, a presupposition which underlay Einstein's own SR.

EINSTEIN'S POSITIVISM

When we turn to Einstein's 1905 paper on relativity, we discover that the introductory sections of the paper offer a verificationist analysis of space and time based on Machian positivism. Einstein's verificationism comes through most clearly in his operationalist re-definition of key concepts. Holton comments:

> ...the Machist component...shows up prominently in two related respects: first, by Einstein's insistence from the beginning of his relativity paper that the fundamental problems of physics cannot be understood until an epistemological analysis is carried out, particularly so with respect to the meaning of the conceptions of space and time; and second, by Einstein's identification of reality with what is given by sensations, the 'events,' rather than putting reality on a plane beyond or behind sense experience.
>
> From the outset, the instrumentalist, and hence sensationist, views of measurement and of the concepts of space and time are strikingly evident.
>
> ...We can say that just as the time of an event assumes meaning only when it connects with our consciousness through sense experience (that is, when it is subjected to measurement-in-principle by means of a clock present at the same place), so also is the place, or space co-ordinate, of an event meaningful only if it enters our sensory experience while being subjected to measurement-in-principle (that is, by means of meter sticks present on that occasion at the same time).[46]

Holton's insistence on the influence of Mach's sensationism is exaggerated; but his pinpointing Einstein's verificationist analysis of time and space as exhibiting the influence of positivism is justified.[47] Time and space are defined in terms of physical operations. Newton's distinction between time and the sensible measures thereof is quietly abolished. Einstein does not even attempt to justify his operationalism; it is just presupposed. He asserts,

---

[45] Henri Poincaré, "The Measure of Time," in *The Value of Science*, 1905, trans. G. B. Halstead, in *Foundations of Science*, p. 228.

[46] Holton, "Mach," pp. 169-170.

[47] See T. A. Ryckman, "'P(oint)-C(oincidence) Thinking': The Ironical Attachment of Logical Empiricism to General Relativity (and some lingering Consequences)," *Studies in History and Philosophy of Science* 23 (1992): 473, who observes that Holton's portrayal of Einstein as a convert from sensationism to arch-realism is no longer considered to be adequate, for Einstein was never a sensationist, and he remained a Machean or Humean up to the end in the sense that he never abandoned verificationism.

> Now we must bear carefully in mind that a mathematical description of this kind has no physical meaning unless we are quite clear as to what we will understand by 'time'. We have to take into account that all our judgments in which time plays a role are always judgments of *simultaneous events*. If, for instance, I say, 'The train arrives here at 7 o'clock,' I mean something like this: 'The pointing of the small hand of my watch to 7 and the arrival of the train are simultaneous events.'[48]

It is taken for granted that *all* our judgements in which time plays a role must have a physical meaning. When it comes to judgements concerning the simultaneity of distant events, the concern is to find a "practical arrangement" to compare clock times.[49] In order to "define" a common time for spatially separated clocks, we assume that the time light takes to travel from A to B equals the time it takes to travel from B to A—a definition which *presupposes* that absolute space does not exist. Thus, time is reduced to physical time (clock readings) and space to physical space (readings of measuring rods) and both of these are relativized to local frames. Simultaneity is defined in terms of clock synchronization via light signals. All of this is done by mere stipulation. With admirable bluntness, Heinz Pagels comments,

> These definitions, with their appeal to measurement, cut through all the excess philosophical baggage that the ideas of space and time had carried for centuries. The positivist insists that we talk only about what we can know through direct operations like a measurement. Physical reality is defined by actual empirical operations, not by fantasies in our head.[50]

Through Einstein's operational definitions of time and space, Mach's positivism triumphs in the Special Theory of Relativity. Reality is reduced to what our

---

[48] Einstein, "Electrodynamics of Moving bodies," p. 393.

[49] Ibid. Frank comments, "It is easy to see which lines of Mach's thought have been particularly helpful to Einstein. The definition of simultaneity in the special theory of relativity is based on Mach's requirement that every statement in physics has to state relations between observable quantities" (Philipp Frank, "Einstein, Mach, and Logical Positivism," in *Einstein: Philosopher-Scientist*, p. 272).

[50] Heinz Pagels, *The Cosmic Code* (London: Michael Joseph, 1982), p. 59. Less polemically, Katsumori makes the point:

> "In Einstein's view the concepts of space and time can by no means be defined prior to possible experience. Rather the meaning of any physical concept is inseparable from the procedure of measurement of the corresponding phenomena; the concepts of space and time, in particular, are endowed with physical meaning only through the prescription of measurement using rigid measuring-rods (*starre Maßstäbe*) and clocks (*Uhren*)" (Makoto Katsumori, "The Theories of Relativity and Einstein's Philosophical Turn," *Studies in History and Philosophy of Science* 23 [1992]: 560).

Tonnelat comments,

> "The notion of distant simultaneity—of simultaneity which is fundamental, universal, and independent of the movement of bodies—has always appeared, until now, to be a concept endowed with an evident meaning.
>
> For the physicist this notion must be capable of being experimentally defined in such a way as to permit the application of physical criteria. Any other idea of absolute simultaneity, however intuitive it may be, has no chance of being an object of experience if it does not at some moment enter into realizable operations.
>
> In other words, the simultaneity of two distant events and, in general, the time which an observer must ascribe to a distant phenomenon must be defined by *operations* which could be effectively accomplished, even if it is a matter of a process which can only be approximately realized" (Marie-Antoinette Tonnelat, *Histoire du principe de relativité*, Nouvelle Bibliothèque Scientifique [Paris: Flammarion, 1971], p. 142).

measurements read; Newton's metaphysical time and space, which transcend operational definitions, are implied to be mere figments of our imagination. "Thus," concludes Gutting, "Einstein's seemingly innocuous requirement that simultaneity be operationally defined led to the rejection of a basic concept of classical physics, the concept of an absolute time valid in all coordinate systems."[51]

In Einstein's other early papers on relativity, his verificationist theory of meaning comes even more explicitly to the fore. Concepts which cannot be given empirical content and assertions which cannot be empirically verified in principle are discarded as meaningless. In his article in the *Jahrbuch der Radioaktivität und Elektronik* of 1907, after giving his operational definitions for time and simultaneity, he states, "According to the definition of time given in section 1, a specific time has sense only in relation to a reference frame having a particular state of motion."[52] To refer to the time of an event without reference to its inertial frame thus has no sense (*Sinn*). In his piece in the *Physikalische Zeitschrift* of 1909, he muses on the apparent contradiction between the Relativity Principle and Lorentz's conviction that light travels at constant velocity in the aether and comments,

> The theorem of addition of velocities rests, however, on the arbitrary presupposition that time specifications, as well as statements about the shape of moving bodies, have a meaning independent of the state of motion of the relevant coordinate system. But one is convinced that the introduction of clocks which are at rest relative to the relevant coordinate system is required for a definition of time and the shape of moving bodies.[53]

Here Einstein asserts that statements about the time of an event have no meaning (*Bedeutung*) unless one refers to clocks at rest in the relevant inertial system. In his summary paper, "Die Relativitäts-Theorie," published in 1911, Einstein expresses himself at greater length concerning the meaning of statements about time and space. After laying out the "dreadful dilemma" confronting late nineteenth century physicists concerning the apparent incompatibility of the Relativity Principle and the Principle of the Constancy of the Velocity of Light, Einstein writes,

> Now it turns out that Nature is completely innocent with regard to this dilemma, but that this dilemma follows from the fact that in our thinking...we have silently and arbitrarily made presuppositions which must be dropped in order to arrive at a consistent and simple conception of things.
>
> ...The first and most important of these arbitrary presuppositions concerns the concept of time, and I shall try to describe wherein this arbitrariness consists. In order to do this well, I shall first deal with space, in order to put time into parallel with it. When we want to express the position of a point in space, that is, the position of a point relative to a coordinate system k, we give its right angle coordinates x, y, z. The meaning of these coordinates is the following: one constructs according to familiar procedures perpendiculars on the coordinate planes and sees how many times a given

---

[51] Gary Gutting, "Einstein's Discovery of Special Relativity," *Philosophy of Science* 59 (1972): 60. On Gutting's view, Einstein's two fundamental principles were the principle of relativity and the principle of operational definition. "Therefore, our assessment of the nature of Einstein's discovery will, to an important extent, depend on our view of the status of these principles" (Ibid., p. 63).

[52] Albert Einstein, "Über das Relativitätsprinzip und die aus demselben gezogenen Folgerungen," *Jahrbuch der Radioaktivität und Elektronik* 4 (1907): 417.

[53] Albert Einstein, "Über die Entwicklung unserer Anschauungen über das Wesen und die Konstitution der Strahlung," *Physikalische Zeitschrift* 10 (1909): 819.

> unit measure can be laid against these perpendiculars. The results of this counting are the coordinates. A spatial specification in coordinates is thus the result of certain manipulations. The coordinates which I assign have accordingly a very precise physical meaning; one can verify whether a certain given point really has the assigned coordinates or not.
>
> How do things stand in this connection with time? There we shall see that we are not in such a good position. Until now one was always content to say: time is the independent variable of the event. With such a definition the measurement of the time value of an actually present event can never be founded. We must therefore try to so define time that time measurements are possible on the basis of this definition.[54]

Einstein then proceeds to describe the familiar light signal synchronization procedure. In so doing, he asserts that since we are unable to measure the one-way velocity of light without making "arbitrary stipulations," we are therefore justified in making such stipulations. "We now stipulate that the velocity of propagation of light in a vacuum on the way from point *A* to point *B* is the same as the velocity of propagation of a light beam from *B* to *A*."[55] Having described the synchronization procedure, he concludes, "When we have fulfilled this prescription, we have thus attained a time determination from the standpoint of the measuring physicist. The time of an event is, namely, equal to the reading of these clocks regulated by the above prescription which are at the place of the event."[56] Einstein thinks that all this sounds so self-evident, but what is noteworthy, he states, is that "in order to arrive at time specifications of a very precise sense" we use a prescription that relates to clocks which are relative to a certain coordinate system *k*. We have not gained simply a time, but a time relative to a coordinate system. "It is not said that time has an absolute…meaning. That is an arbitrary element which was contained in our kinematics."[57]

We now come, Einstein proceeds, to the second arbitrary element in kinematics. "We speak of the shape of a body, for example, the length of a rod, and believe that we know exactly what its length is, even when it is in motion with respect to a reference system from which we are describing appearances."[58] But a little reflection shows that the length of a rod is not so simple a concept as we instinctively imagine. "We now ask: how long is this rod? This question can have only the meaning: what experiments must we carry out in order to discover how long the rod is?"[59] Einstein then proceeds to describe length measurement of a moving rod by means of synchronized clocks.

By abandoning the presuppositions of absolute time and space and substituting in their stead operational definitions, Einstein reduces time and space to our measurements of them. He concludes, "Since we have in a precise way physically defined coordinates and time, every relation between spatial and temporal entities

---

[54] Albert Einstein, "Die Relativitäts-Theorie," *Vierteljahrsschrift der naturforschenden Gesellschaft in Zürich* 56 (1911): 6-7.
[55] Ibid., p. 8.
[56] Ibid.
[57] Ibid., p. 9.
[58] Ibid.
[59] Ibid.

will have a very precise physical content."[60] Statements about spatial or temporal relations which are metaphysical in character, that is, are independent of clocks, rods, or reference frames, are nonsense.

It is frequently asserted that as Einstein labored on the General Theory, he came to see the bankruptcy of Mach's positivism.[61] But this claim needs to be carefully qualified. When one reads Einstein's 1916 paper laying the foundations of General Relativity, Mach's verificationism fairly shouts at one in section 2 of the article. Einstein presents here a "weighty argument from the theory of knowledge" in favor of the extension of the Postulate of Relativity to non-inertial frames of reference.[62] The purpose of this argument is to remedy "an inherent epistemological defect" in both classical mechanics and SR "which was...clearly pointed out by Ernst Mach."[63] The defect is that if we imagine two spatially isolated fluid bodies S_1 and S_2, so situated that observers at rest relative to each respectively will judge that the other is in rotation about their common axis, then there still emerges an absolute effect, namely, one of the bodies will be, say, spherical, while the other is ellipsoidal. "What is the reason for this difference in the two bodies? No answer can be admitted as epistemologically satisfactory, unless the reason given is an *observable fact of experience*."[64] In Newtonian mechanics, such rotational motion is said to reveal the existence of absolute space, but Einstein objects to such an account on the basis of the positivistic principle enunciated above. "But the privileged space R_1 of Galileo, thus introduced, is a merely *factitious* cause, and not a thing that can be observed."[65] This leads Einstein to embrace the rather outlandish view of Mach that "the mechanical behavior of S_1 and S_2 is partly conditioned, in quite essential respects, by distant masses which we have not included in the system under consideration."[66] According to Einstein, "These distant masses and their motions relative to S_1 and S_2 must then be regarded as the seat of the causes (which must be susceptible to observation) of the different behavior of our two bodies S_1 and S_2. They take over the role of the factitious cause R_1."[67] We cannot say that R_1 is privileged "without reviving the above-mentioned epistemological objection."[68] Einstein's defense of Mach's Principle (as this point of view is called) is clearly based on a verificationist epistemology. As Earman emphasizes, Mach's rejection of absolute space, like his much criticized rejection of atomism, stems from "his highly positivistic and operationalistic philosophy."[69] Fortunately, Mach's Principle

---

[60] Ibid., p. 11.
[61] See Holton, "Search for Reality," pp. 185-186.
[62] Albert Einstein, "The Foundations of General Relativity Theory," in *General Theory of Relativity*, ed. C. W. Kilmister, Selected Readings in Physics (Oxford: Pergamon Press, 1973), p. 144. The original paper appeared in *Annalen der Physik* 49 (1916): 769.
[63] Ibid., p. 143.
[64] Ibid.
[65] Ibid., pp. 143-144.
[66] Ibid., p. 144.
[67] Ibid.
[68] Ibid.
[69] John Earman, "Who's Afraid of Absolute Space?" *Australasian Journal of Philosophy* 48 (1970): 298.

does not belong essentially to the foundations of GR, and most textbook expositions of that theory pass over Einstein's argument in benign neglect, focusing instead on his subsidiary argument in section 2 on the equivalence of acceleration and gravitational fields.[70]

What Einstein's labor on General Relativity made clear to him was the inadequacy of Mach's phenomenalism. Scientific theorizing is not the mere linking of observation statements but involves a creative exercise of the mind, which is free to postulate theoretical entities not directly given in observation. Nevertheless, even after GR he continued to regard such theoretical terms as meaningless unless they could be somehow linked to observation statements. In his *Relativity: the Special and General Theory* (1920), for example, he asks his reader what sense (*Sinn*) there is in asserting that two lightning bolts simultaneously strike at two separate places. He explicitly rejects the answer that the meaning (*Bedeutung*) of the assertion is in and of itself clear, even if it might be difficult to show by means of observations that the two events occurred simultaneously. According to Einstein, the concept of simultaneity does not exist for the physicist until he can in a concrete situation discover whether the concept holds or not.

> We thus require a definition of simultaneity such that this definition supplies us with the means by which, in the present case, he can decide by experiment whether both lightning strokes occurred simultaneously. As long as this requirement is not satisfied, I allow myself to be deceived as a physicist (and of course the same applies if I am not a physicist) when I imagine that I am able to attach a meaning to the statement of simultaneity.[71]

For physicist and non-physicist alike the statement that two events occur simultaneously is meaningless unless an operational definition can be given for that concept.[72] Thus, he continued to cling to his rejection of metaphysical time and space:

> The only justification for our concepts and system of concepts is that they serve to represent the complex of our experiences; beyond this they have no legitimacy. I am convinced that the philosophers have had a harmful effect upon the progress of scientific thinking in removing certain fundamental concepts from the domain of empiricism, where they are under our control, to the intangible heights of the *a priori*. For even if it should appear that the universe of ideas cannot be deduced from experience by logical means, but is, in a sense, a creation of the human mind, without which no science is possible, nevertheless this universe of ideas is just as little independent of the nature of our experiences as clothes are of the form of the human body. This is particularly true of our concepts of time and space, which physicists have

---

[70] On GR's failure to render S_1 and S_2 indistinguishable see Michael Friedman, *Foundations of Spacetime Theories* (Princeton: Princeton University Press, 1983), pp. 211, 230-232.

[71] Albert Einstein, *Relativity, the Special and the General Theory*, trans. Robert W. Lauren (London: Methuen, 1920), p. 26.

[72] That Einstein was a verificationist outside the realm of physics is also evident from his remarks in an interview with Max Wertheimer that when someone uses the word "hunchback," "If this concept is to have any clear meaning, there must be some way of finding out whether or not a man has a hunched back. If I would conceive of no possibility of reaching such a decision, the word would have no real meaning for me" (Max Wertheimer, *Productive Thinking*, ed. Michael Wertheimer, enlar. ed. [London: Tavistock, 1961], p. 220).

> been obliged by the facts to bring down from the Olympus of the *a priori* in order to adjust them and put them in a serviceable condition.[73]

The Machist-Humean tones of this line of thinking are strikingly evident.

Even as late as his "Autobiographical Notes," it is clear that the incubus of Machism had not been exorcised from Einstein's soul. He had become a critical realist and emphasized the necessity of logical, mathematical, and other non-observation terms in scientific theory construction[74]; still the old positivism hung on. In stating his "epistemological credo," Einstein wrote,

> I see on the one side the totality of sense experiences, and, on the other, the totality of concepts and propositions which are laid down in books. The relations between the concepts and propositions among themselves and each other are of a logical nature, and the business of logical thinking is strictly limited to the achievement of the connection between concepts and propositions among each other according to firmly laid down rules, which are the concern of logic. The concepts and propositions get 'meaning,' viz., 'content,' only through their connection with sense-experiences. The connection of the latter with the former is purely intuitive, not itself of a logical nature. The degree of certainty with which this relation, viz., intuitive connection, can be undertaken, and nothing else, differentiates empty phantasy from scientific 'truth.'[75]

It seems evident that Einstein was still holding to a relaxed verificationist theory of meaning. It is extremely instructive to observe that when Einstein several pages later blasts Mach and Ostwald for their "positivistic philosophical attitude" which led them to deny the reality of atoms, he equates positivism with the view "that facts by themselves can and should yield scientific knowledge without free conceptual construction"[76]—a view which Einstein had long rejected. But as Frank points out in the same volume, Einstein's requirement that there must be some observational statements to give meaning to the body of concepts and propositions is positivistic.[77] In Victor Lenzen's essay, "Einstein's Theory of Knowledge," which Einstein hails as "convincing and correct" in everything it says,[78] Lenzen explains that "The concept of objective time is introduced through the intermediary of space. Just as

---

[73] Albert Einstein, *The Meaning of Relativity*, 6th ed. (1922; rep. ed.: London: Chapman and Hall, 1967), p. 2. Cf. his "Fundamental Ideas and Problems of the Theory of Relativity," [1923], in *Nobel Lectures, Physics: 1901-1921* (New York: Elsevier, 1967), pp. 479-490, where he lays down a postulate called "the stipulation of meaning," which requires that concepts and distinctions are only admissible to the extent that observable facts can be assigned to them without ambiguity. He considers this postulate to be of "fundamental importance" epistemologically.

[74] See his Herbert Spencer lecture of 1933, "On the Methods of Theoretical Physics," in *The World as I See It* (New York: Covici-Friede, 1934).

[75] Einstein, "Autobiographical Notes," pp. 11-13. Cf. his statement, "Concepts can only acquire content when they are connected, however indirectly, with sensible experience" (Albert Einstein, "The Problem of Space, Ether, and the Field in Physics," in *Ideas and Opinions*, trans. Sonja Bargmann [New York: Crown Publishers, 1954], p. 277).

[76] Ibid., p. 49.

[77] Frank, "Einstein, Mach, and Logical Positivism," p. 274. Frank concludes that the difference between Einstein and later logical positivists is purely verbal (Ibid., p. 282). Cf. Ernst Nagel, "Relativity and Twentieth-Century Intellectual Life," in *Some Strangeness in the Proportion*, ed. Harry Woolf (Reading, Mass.: Addison-Wesley Publishing Co., 1980), p. 41, who claims that logical positivists came to embrace a moderate operationalism like Einstein's, an operationalism which came to dominate the philosophy of science during the second half of the twentieth century.

[78] Albert Einstein, "Reply to Criticism," in *Einstein: Philosopher-Scientist*, p. 683.

the rigid body is a basis for space, so is the clock for time."[79] "Objective local time is based on the correlation of the temporal course of experience with the indications of a 'clock'.... Objective extended time is based upon synchronization of distant clocks by signals."[80] Lenzen goes on to contrast these concepts to the "hypostatization" of time and space in classical theory. These re-definitions of time and space in operational terms were and remain positivistic in character.

In any case, Einstein's later views remain somewhat beside the point for our purposes. What is relevant are his early epistemological views, and these are solidly Machist. In his "Autobiographical Notes," Einstein admits that his epistemological credo "actually evolved only much later and very slowly and does not correspond with the point of view I held in younger years."[81] In those years, it was Mach who held sway: "I see Mach's greatness in his incorruptible skepticism and independence; in my younger years, however, Mach's epistemological position also influenced me very greatly, a position which today appears to me to be essentially untenable."[82] Holton concludes, "Taking the early papers as a whole, and in the context of the physics of the day, we find that Einstein's philosophical pilgrimage did start on the historic ground of positivism. Moreover, Einstein thought so himself and confessed as much...."[83]

POSITIVIST RECEPTION OF SR

Positivistic philosophers and physicists were quick to recognize in SR a kindred spirit and embraced the theory eagerly. According to Holton,

> This was the kind of operationalist message which, for most of his readers, overshadowed all other philosophical arguments in Einstein's paper. His work was enthusiastically embraced by the groups who saw themselves as philosophical heirs of Mach, the Vienna Circle of neopositivists and its predecessors and related followers, providing a tremendous boost for the philosophy that had initially helped to nurture it.[84]

In fact Friedman believes that SR helped to spawn the logical positivism of 1925-35, particularly that school's theoretical/observational distinction.[85] The earliest adherents to the theory, such as Planck, Wien, and Minkowski, embraced SR because they found it "congenial," stressing that no empirical facts stood *against* it.[86] Planck noted in particular the "devastating results of [Einstein's] new interpretation of the time concept, which have repercussions on the whole of

---

[79] Victor F. Lenzen, "Einstein's Theory of Knowledge," in *Einstein: Philosopher-Scientist*, pp. 369-370.

[80] Ibid., p. 370.

[81] Einstein, "Autobiographical Notes," p. 11.

[82] Ibid., p. 21.

[83] Holton, "Search for Reality," p. 171. Einstein's pilgrimage began with "a philosophy of science in which positivism was the center" (Holton, "Where Is Reality?" p. 45).

[84] Holton, "Search for Reality," pp. 170-171.

[85] Friedman, *Foundations of Spacetime Theories*, pp. 24-25.

[86] Gerald Holton, "Einstein's Scientific Program: The Formative Years," in *Strangeness in the Proportion*, p. 58.

physics, and above all on mechanics and thence deep into epistemology."[87] Walter Kaufmann, who thought his experimental data falsified SR, maintained that the only advantage of Einstein's theory was "epistemological" in that it did not involve unknown velocities.[88] Max von Laue opined that one can hardly avoid preferring the relativity principle to the aether, since "it contradicts all epistemological principles to ascribe to a body physical reality when it can never be detected."[89] At the first meeting of the *Gesellschaft für positivistische Philosophie* in 1912, Petzoldt declared, "Lorentz's theory is, at its conceptual center, pure metaphysics, nothing else than Schelling or Hegel's philosophy of nature," but Einstein's theory represents "the victory over the metaphysics of absolutes in the conceptions of space and time....a mighty impulse for the development of the philosophical point of view in our time."[90] In an Appendix entitled "The Relationship of Mach's Worldview to Relativity Theory" to the eighth edition of Mach's *Mechanik*, Petzoldt points to Mach's section on Newton as responsible for creating the atmosphere "without which Einstein's relativity theory would not have been possible."[91] He dismisses Newton's absolute space and absolute motion as having "no physical meaning" and concludes that relativity theory "fits in fully with Mach's worldview" and is "a golden fruit of his deeply rooted and powerfully expansive tree of thought."[92]

P. W. Bridgeman, the father of operationalism, extolled Einstein's procedure:

> ...he recognized that the meaning of a term is to be sought in the operations employed in making application of the term. If the term is one which is applicable to concrete physical situations, as 'length' or 'simultaneity,' then the meaning is to be sought in the operations by which the length of concrete physical objects is determined, or in the operations by which one determines whether two concrete physical events are simultaneous or not.[93]

---

[87] Max Planck, cited in C. Seelig, *Albert Einstein: A Documentary Biography* (London: Staples Press, 1956), pp. 144-145.

[88] W. Kaufmann, "Über die Konstitution des Elektrons," *Annalen der Physik* 19 (1906): 487-553.

[89] M. Laue, *Das Relativitätsprinzip*, Sammlung Naturwissenschaftlicher und mathematischer Monographien 38 (Braunschweig: Friedrich Vieweg & Sohn, 1911), p. 33.

[90] Holton, *Search for Reality*, p. 171.

[91] Joseph Petzoldt, "Das Verhältnis der Machschen Gedankenwelt zur Relativitätstheorie," in *Die Mechanik in ihrer Entwicklung*, 8th ed., by Ernst Mach (Leipzig: F. A. Brockhaus, 1921), p. 494. Petzoldt records Einstein's agreement with this judgement.

[92] Ibid., pp. 495, 516-517.

[93] P. W. Bridgeman, "Einstein's Theories and the Operational Point of View," in *Einstein: Philosopher-Scientist*, p. 335. Gaston Bachelard adds, "This *operational* definition of simultaneity dissolves the notion of *absolute* time. Since simultaneity is linked to physical experiments which occur in space, the temporal contexture (*contexture*) is one with spatial contexture. Since there is no absolute space, there is no absolute time" (Gaston Bachelard, "The Philosophic Dialectic of the Concepts of Relativity," in *Einstein: Philosopher-Scientist*, p. 571). According to Prokhovnik, "Perhaps Einstein's most radical innovation was his introduction of precise operational definitions (or 'conventions') for synchronizing separated clocks and for determining the space and time co-ordinates of an event. By implication he was thereby asserting that there was no other meaning for these co-ordinates apart from their determination in terms of agreed measurements—he had no need for any abstract notions of space and time" (S. J. Prokhovnik, *Light in Einstein's Universe* [Dordrecht: D. Reidel, 1985], p. 33). Epstein observed, "The theory of relativity had an immense educational influence on physicists and other scientists by teaching them the approach to scientific concepts that has been more recently named *the operational point of view*;" the main philosophic value of relativity is that it has divested concepts of

Reichenbach recognized the positivism underlying such a procedure:

> ...he saw that certain physical problems could not be solved unless the solutions were preceded by a logical analysis of the fundamentals of space and time, and he saw that this analysis, in turn, presupposed a philosophic readjustment of certain familiar conceptions of knowledge. The physicist who wanted to understand the Michelson experiment had to commit himself to a philosophy for which the meaning of a statement is reducible to its verifiability, that is, he had to adopt the verifiability theory of meaning if he wanted to escape a maze of ambiguous questions and gratuitous complications. It is this positivist, or let me rather say, empiricist commitment which determines the philosophical position of Einstein.[94]

Ernst Cassirer, in his essay "Einstein's Theory of Relativity considered from the Epistemological Standpoint," clearly recognized the crucial role played by positivism in SR and GR. All historical examples of "the real inner connection between epistemological problems and physical problems," he says, are outdone by relativity theory; "from its very beginning" in its analysis of the concept of time, the theory extends into the field of epistemology.[95] SR's advantage over competing explanations, such as Lorentz's contraction hypothesis, "is based not so much on its empirical material as on its pure logical form, not so much on its physical as on its general *systematic* value."[96] He explains,

> The Lorentzian hypothesis...was sufficient to give a complete explanation of all known observations. An experimental decision between Lorentz and Einstein's theories was

---

space and time "of their former metaphysical connotations" (Paul S. Epstein, "The Time Concept in Restricted Relativity," *American Journal of Physics* 10 [1942]: 1).

[94] Hans Reichenbach, "The Philosophical Significance of the Theory of Relativity," in *Einstein: Philosopher-Scientist*, pp. 290-291. Norton observes that when in the 1905 paper Einstein stated that he had no need to introduce the ether, "Einstein was not consciously applying the verifiability principle when he wrote these words. The criterion was not formulated until over twenty years later. However, if we are to identify any justifiable, scientific application of the criterion, then this case surely would be included" (John D. Norton, "Philosophy of Space and Time," in *Introduction to the Philosophy of Science* [New Jersey: Prentice-Hall, 1992], p. 183). Since the state of absolute rest had no observable consequences, "We are thereby enjoined to despise the entity or state as an idle metaphysical conception and to banish it from our discourse" (Ibid.). Earlier Reichenbach had blasted neo-Kantian philosophers for failing to assimilate Einstein's revolution:

> "the attempts by some philosophers to relegate the achievements of the theory of relativity to physics show only that these philosophers are no match for the physicist Einstein in philosophical competence. Einstein's discovery is not a philosophical triviality.... It is a serious failing of post-Kantian philosophy to have left the development of the epistemological space-time problem to the mathematicians and physicists....it is high time that philosophy adjust itself to this development, and begin its work where the mathematicians left off instead of guarding itself with a *noli me tangere*" (Hans Reichenbach, "The Present State of the Discussion on Relativity," [*Logos* 10 (1921): 316-378] in *Modern Philosophy of Science*, trans. and ed. Maria Reichenbach [London: Routledge, Kegan & Paul, 1959], p.32).

Cf. Earman's more recent remark that Einstein's misguided epistemological arguments in favor of taking GR as an extension of the principle of relativity "should serve as a cautionary tale for philosophers of science who seek to draw wisdom from the philosophical pronouncements of scientists, even the greatest scientists" (John Earman, *World Enough and Space-Time* [Cambridge, Mass.: MIT Press, 1989], p. 105).

[95] Ernst Cassirer, *Substance and Function and Einstein's Theory of Relativity*, trans. W. C. Swabey and M. C. Swabey (1923; rep ed.: Dover, 1953), pp. 354-355.

[96] Ibid., p. 354.

> thus not possible; it was seen that between them there could fundamentally be no *experimentum crucis*. The advocates of the new doctrine accordingly had to appeal—an unusual spectacle in the history of physics—to general philosophical grounds, to the advantages over the assumption of Lorentz which the new doctrine possessed in a systematic and epistemological respect.[97]

Cassirer goes on to explain that Lorentz's approach was deemed "epistemologically unsatisfactory" because it ascribed to the aether definite effects which preclude the aether's ever being an object of possible observation. What in the last analysis counted against Lorentz's theory was not an empirical, but a methodological defect:

> It conflicted most sharply with a general principle, to which Leibniz has appealed in his struggle against the Newtonian concepts of absolute space and time, and which he formulated as the 'principle of observability' (*principe de l'observabilité.*) When Clarke, as the representative of Newton, referred to the possibility that the universe in its motion relatively to absolute space might undergo retardation or acceleration which would not be discoverable by our means of measurement, Leibniz answered that nothing fundamentally outside the sphere of observation possessed 'being' in the physical sense: *quand il n'y a point de changement observable, il n'y a point de changement du tout.* It is precisely this principle of 'observability' which Einstein applied at an important and decisive place in his theory, at the transition from the special to the general theory of relativity, and which he has attempted to give a necessary connection with the general principle of causality.[98]

From start to finish, then, relativity theory was seen to be based in a positivistic epistemology which won for it its principal advantage over its competitors.

POSITIVISM AND THE REJECTION OF METAPHYSICS

Under the influence of positivism and the verificationist criterion of meaning, physicists and philosophers of space and time during the first half of the twentieth century had an abhorrence for what was called "metaphysics." An instructive piece is a 1941 article by Henry Margenau ostensibly defending metaphysical elements in physics. Observing that "our time appears to be distinguished by its *taboos*, among which there is to be found the broad convention that the word *metaphysics* must never be used in polite scientific society," Margenau counters that there not only are, but ought to be, metaphysical elements in physical science.[99] But then he emasculates this bold contention by explaining that he means thereby that we must have "epistemology"; but as physicists "we reject ontology."[100] He reduces metaphysics to what he calls the methodology of science and insists that we must not relax our standards here, lest the "obnoxious ontological elements" find their

---

[97] Ibid., pp. 375-376.

[98] Ibid., pp. 376-377. Cassirer's analysis can mislead. The debate between Clarke and Leibniz concerned substantivalism vs. relationalism. Leibniz was just as non-relativistic as Clarke, and contemporary relativists are often substantivalists. Moreover, Einstein's positivism tended to deny that statements about an absolute frame had any meaning, so that his principle resembled a verification criterion of meaning in a way Leibniz's did not. Still, both critiques of the classical concept of time dispute that concept based on verificationist arguments.

[99] H. Margenau, "Metaphysical Elements in Physics," *Reviews of Modern Physics* 13 (1941): 176.

[100] Ibid., p. 177.

way back into science.[101] What these elements are he leaves in no doubt: the luminiferous aether and simultaneity in different Lorentz frames are classed along with the external world and the *Ding-an-sich* and dismissed as "ultra-perceptory and hence meaningless."[102] This judgement is based on the "positivistic criticism" that propositions not verifiable in principle are meaningless, a criterion which elicits Margenau's ringing endorsement: "this recognition should be one of the premisses of philosophy of science; it enjoys, indeed, almost universal consent."[103]

It is not difficult to find examples of this anti-metaphysical bent in treatments of relativity theory. Complaining that "it is hardly possible to open a textbook on the theory of relativity—even if written by an otherwise competent physicist—without coming upon sentences of an entirely metaphysical character," Frank asserted that such sentences are wholly meaningless in physics.[104] By a metaphysical sentence, he meant one which asserts a fact not verifiable by observation. All such sentences form a system of sentences which is coherent and isolated from the verifiable sentences. "One usually calls these isolated sentences 'metaphysical' sentences which state something about a real world. Being non-verifiable sentences they are meaningless as far as science is concerned."[105] Relativity theory, through its operational definitions of space and time, relegates sentences about metaphysical space and time to a state of "splendid isolation."[106]

Similarly, Eddington endorsed a verificationist analysis of meaning. He wrote, "If we happen to make a deduction which could not conceivably be corroborated or disproved by these diligent measures, there is no criterion of its truth or falsehood and it is thereby a meaningless deduction."[107] With the advent of relativity theory, he explains, concepts of space and time are defined by pointer readings rather than having the metaphysical significance we might have expected.[108] Thus, in an imaginary dialogue between a physicist and a relativist we find this exchange concerning time dilation:

> *Phys.* I think your unit of time would change according to the motion of your 'clock' through the aether.

---

[101] Ibid.

[102] Ibid., p. 178.

[103] Ibid.

[104] Frank, *Interpretations and Misinterpretations of Modern Physics*, Actualités Scientifiques et Industrielles 587: Exposés de Philosophie Scientifique 2 (Paris: Hermann & Cie, 1938), p. 34. Cf. his later remark, "It can be said without exaggeration that there is no philosophical congress, no philosophical textbook, not even an issue of a philosophical journal, where we do not encounter examples of attempts to draw arguments in favor of metaphysical opinions from the statements of the theory of relativity" (Ibid., p. 46). He gives as an example James Jeans's attempt to deduce fatalism from SR.

[105] Ibid., p. 38.

[106] Ibid., p. 39.

[107] Arthur Eddington, *The Nature of the Physical World*, with an Introductory Note by Sir Edmund Whittaker, Everyman's Library (1928; rep. ed.: London: J. M. Dent and Sons, 1964), p. 223.

[108] Ibid., p. 247.

> *Rel.* Then you are comparing it with some notion of absolute time. I have no notion of time except as the result of measurement with some kind of clock.[109]

Again, when discussing the possibility that the aether could serve to partition spacetime into privileged, successive three-dimensional slices, Eddington demurs, "It seems an abuse of language to speak of a division *existing*, when nothing has ever been found to pay any attention to the division."[110] Those who insist on the existence of a "true time," he believed, are holding on to the possibility that some day a discriminating test may be found. "But the off-chance that a future generation may discover a significance in our utterances is scarcely an excuse for making meaningless noises."[111]

Similarly, Max Born discarded absolute time with the words, "But what exists 'without reference to any external object whatsoever' is not ascertainable, and is not a fact" and absolute space with the same aplomb: "A conception has physical reality only when there is something ascertainable by measurement corresponding to it in the world of phenomena."[112] With respect to distant simultaneity he advises, "the quantitative physicist...sees no meaning in the statement that an event at A and an event at B are simultaneous, since he has no means of deciding the truth or the incorrectness of this assertion."[113] Since Einstein showed that there is no means of determining absolute time, "this signifies that absolute time has no physical reality."[114] Generalizing, Born draws attention to the particular emphasis of his exposition that "only ascertainable facts have physical reality."[115]

When the contemporary student of physics reads such anti-metaphysical polemics, he must feel as though he were peering into a different world. For it is now widely recognized that the boundaries of science are impossible to fix with precision, and during the last few decades theoretical physics has become characterized precisely by its metaphysical, speculative character. In various fields such as quantum mechanics, classical cosmology, and quantum cosmology, debates rage over overtly metaphysical issues. Take quantum physics, for example. In 1935 Born wrote, "For what lies within the limits is knowable, and will become known; it

---

[109] Arthur Eddington, *Space, Time and Gravitation*, Cambridge Science Classics (1920: rep. ed.: Cambridge: Cambridge University Press, 1987), p. 15.

[110] Ibid., p. 39.

[111] A. S. Eddington, *The Mathematical Theory of Relativity*, 2d ed. (Cambridge: Cambridge University Press, 1930), p. 8.

[112] Max Born, *Einstein's Theory of Relativity*, trans. Henry L. Brose (New York: E. P. Dutton, n.d.), pp. 51, 61.

[113] Ibid., p. 193. Cf. the strictures of d'Abro:

> "For we may speak of simultaneity throughout space, of an instantaneous space, as much as we please, but until we know enough about these evasive concepts to be able to distinguish events that are simultaneous from those which are not, we cannot claim to have any definite idea of what we are talking about....
>
> ...if two events take place in different places, we can no longer attribute any universal meaning to the opinion that these two events have taken place at one and the same instant of time" (A. d'Abro, *The Evolution of Scientific Thought*, [1927] 2d rev. ed. [n.p.: Dover Publications, 1950], pp. 186, 201).

[114] Born, *Einstein's Theory*, p. 210.

[115] Ibid., p. 252.

is the world of experience, wide, rich enough in changing hues and patterns to allure us to explore it in all directions. What lies beyond, the dry tracts of metaphysics, we willingly leave to speculative philosophy."[116] Commenting on this statement, Euan Squires marvels, "how far from the truth such an attitude really is."[117] In Squires's opinion, "In an effort to understand the quantum world, we are led beyond physics, certainly into philosophy and maybe even into cosmology, psychology and theology."[118] Is there a quantum world at all, or is there only, as at least one version of the Copenhagen Interpretation would have it, an abstract quantum physical description which does not at all purport to find out how nature really is? Or do quantum realities come into being, as other Copenhagen Interpreters maintain, by and upon the occasion of measurement with a classical apparatus? How and when does the wave function of a quantum entity collapse? Since any physical measuring apparatus can itself be described in quantum physical terms, are we forced to regard observation by a conscious being as the final link in the chain, thus resulting in an observer-dependent reality? But since human consciousnesses are linked to physical brains, which can be described quantum mechanically, would we not be led to posit some transcendent Consciousness, or God, who is finally responsible for the collapse of all wavefunctions in the physical universe? But perhaps the wavefunction of quantum entities never collapses. Perhaps we should adopt the Many Worlds Interpretation, according to which the universe splits at every quantum measurement into parallel worlds, each similar except for the differing value of the measurement in each respective world—Or, even more radically, the Many Minds Interpretation, which posits an infinite number of selves associated with any mind which respectively apprehend the different values of the wave function. Or perhaps Einstein was right after all: a non-local hidden variables theory along the lines of the de Broglie-Bohm pilot wave is mathematically consistent and complete and would explain all empirical observations. Then again, maybe the wave function is an *ens fictum*; maybe quantum physics describes only the behavior of ensembles of particles, not individual quantum entities. All of these theories take up metaphysical positions and are part and parcel of contemporary debate in physics.

Or take the field of classical cosmology. George Gale has observed that most cosmologists in the twentieth century have remarked with relief the disengagement of modern cosmology from its origin in philosophy. But such relief was short-lived. "Cosmology, even as practiced today," says Gale, "is science done at the limit: at the limit of our concepts, of our mathematical methods, of our instruments, indeed, of our very imaginations."[119] In a recent article in *Astronomy*, Rothman and Ellis pose the question, "Has astronomy become metaphysical?" and answer that it has.[120]

---

116 Max Born, *Atomic Physics*, 5th ed., trans. J. Douglas (London: Blackie, 1951), p. 306.

117 Euan Squires, *The Mystery of the Quantum World* (Bristol: Adam Hilger, 1986), p. 4.

118 Ibid.

119 George Gale, "Cosmos and Conflict," paper presented at the conference "The Origin of the Universe," Colorado State University, Ft. Collins, Colorado, 22-25 September, 1988.

120 Tony Rothman and George Ellis, "Has Astronomy become Metaphysical?" *Astronomy* (February 1987), 7.

Think, for example, about the origin of the universe. Einstein's field equations for GR (minus the cosmological constant) predicted the expansion of the universe, which was confirmed in 1927 by Hubble's observation of the galactic red-shift, which indicated a universal and isotropic expansion. In 1928 Eddington wrote of the beginning of the universe predicted by the Big Bang model, "It is one of those conclusions from which we can see no logical escape—only it suffers from the drawback that it is incredible. As a scientist I simply do not believe that the present order of things started off with a bang.... But I can make no suggestion to avoid the deadlock."[121] Since then the Steady State Theory has come and gone. Today, a standard text reports, "No problem of cosmology digs more deeply into the foundations of physics than the question of what 'preceded' the 'initial state' of infinite (or near infinite) density, pressure, and temperature."[122] Did matter and energy, space and time, all of physical reality come into being out of nothing at a point in the finite past? How can this be understood? Does it point, as Milne and Whittaker thought, to an extra-mundane Creator of the universe? Or should we prefer naturalistic, but equally metaphysical, explanations, such as an oscillating universe? —Or perhaps adjust the standard model by adding some sort of inflationary scenario, according to which the observable universe is but a bubble in a sea of foam? Then again, there is the staggeringly improbable complexity of our universe to be dealt with. The old design argument, thought to have been banished by Hume and Darwin, has come roaring back in the debate surrounding the Anthropic Principle. The delicate balance of physical constants and quantities in the initial conditions of the universe necessary for the universe to be life-permitting cries out for explanation. But what sort of explanation? A cosmic Intelligence, who designed the world, as John Leslie argues? Or are we forced to posit an infinite and exhaustively random World Ensemble, so as to give purchase to the Anthropic Principle? Or should we insist that no explanation is needed for the life-permitting conditions in the first place? These questions, hotly debated in physical journals, are metaphysical in character. Gale concludes, "Although both sides—Inflationists as well as Anthropic Big Bangers—accuse the other of being 'metaphysical,' and just insofar as that, unscientific, it is clear that metaphysics continues to play an honorable role in cosmology. And, to the extent that it is an honorable role, it is no dishonor to use metaphysics in one's cosmologizing."[123]

Physics becomes most metaphysical in the budding field of quantum cosmology. Relativistic physicists in the first half of this century exulted in the four-dimensional spacetime view of reality which relativity theory had given them. Now quantum theory threatens to undo the fabric of their world. At distances of the order of the Planck length, 1.6×10^{-33} cm and less, quantum fluctuations take place in the geometry of space. Quantum geometrodynamics, as this field of study is called, posits a superspace as the dynamical arena in which all three-dimensional geometries exist. The various three-geometries are assigned probability wave

---

[121] Eddington, *Nature of the Physical World*, pp. 90-91.

[122] C. Misner, K. S. Thorne, and J. A. Wheeler, *Gravitation* (San Francisco: W. H. Freeman, 1973), p. 769.

[123] Gale, "Cosmos and Conflict."

functions which make it possible only to approximate a classical spacetime as a leaf of superspace. "*No prediction of spacetime, therefore no meaning for spacetime*, is the verdict of the quantum principle. That object which is central to all of classical general relativity, the four-dimensional spacetime geometry, simply does not exist, except in a classical approximation."[124] If this vision of reality did not raise metaphysical questions enough, some theorists like Wheeler go further, proposing to analyze superspace in terms of a "pre-geometry," mathematical objects which are the constituents that make up superspace. Quantum geometrodynamics becomes relevant to cosmology in the on-going quest of a quantum theory of gravity. Proponents of quantum cosmology assign a wave function to the universe as a whole. But if one adopts some version of the Copenhagen Interpretation, then one is forced, it seems, to posit some sort of Ultimate Observer who transcends the universe and collapses its wave-function.[125] In order to avoid this implication, which smacks of theism, quantum cosmologists often embrace a Many Worlds Interpretation, the worlds being all the three-geometries in superspace. One can then try to employ quantum mechanical methods to calculate the probability of our universe. In the most celebrated of such models, the Hartle-Hawking model, the further metaphysical position must be adopted that time is imaginary (that is, the time coordinate is an imaginary number), if one is to avoid intractable infinities in performing the quantum calculations. But clearly, the notion that our universe exists in imaginary time raises profound metaphysical difficulties.

Metaphysical questions, hypotheses, and difficulties are abundant in these and other fields of modern physics—for example, particle physics, which Victor Weisskopf mischievously accuses of suffering from the defect of developing "too many new unproven ideas per unit time."[126] Like it or not, theoretical physics has become thoroughly impregnated with metaphysics.

---

[124] Misner, Thorne, and Wheeler, *Gravitation*, p. 1183.

[125] John Barrow writes that the program of quantum cosmology "immediately faces an impasse that can only be overcome by coming to grips with the meaning of quantum observership. If we only ascribe reality to what is observed, who observes the Universe?" Unless we adopt the Many Worlds Interpretation, "...we are left asking the question 'who or what collapses the wave function of the universe?'—some 'ultimate Observer' at the world's end, or outside the Universe of space and time altogether?" The Many Worlds Interpretation "is adopted by quantum cosmologists because it does not require the Universe to be observed" (John Barrow, *The World within the World* [Oxford: Clarendon Press, 1988], pp. 156, 363).

[126] Victor Weisskopf, comment on Steven Weinberg, "Elementary Particle Physics in the Very Early Universe," in *Astrophysical Cosmology*, ed. H. A. Brück, G. V. Coyne, and M. S. Longair, Pontificiae Academiae Scientiarvm Scripta Varia 48 (Vatican City: Pontificia Academia Scientiarvm, 1982), p. 527. He thinks that particle physics has become "even more uncertain and hypothetical than cosmology"! (Ibid., p. 528). For a discussion of the metaphysical issues raised by this field, see George Gale, "Some Metaphysical Perplexities in Contemporary Physics," paper presented at the 36th Annual Meeting of the Metaphysical Society of America, Vanderbilt University, March 14-16, 1985. He asserts that "...although quarks apparently satisfy the classical metaphysical desiderata of the atomistic research program, they do so at enormous cost: the sacrifice of the epistemological and methodological criteria which explicitly define modern science," since they are "*in principle* unobservable." Of course, the same may be said about string theory, in which the boundary between physics and metaphysics has been thoroughly erased. See Michio Kaku, *Introduction to Superstrings and M-Theory*, 2d ed., Graduate Texts in Contemporary Physics (New York: Springer, 1999) for an account of these 11-dimensional theories.

But we have forgotten one of the prime examples of the role of metaphysics in scientific theorizing: relativity theory itself! As Sklar points out, those who are realists about problems of space and time reject the positivist doctrine that observational equivalence is a sufficient condition of the full equivalence of two theories.[127] Realists often hold that two theories, in order to be equivalent, must not only be empirically indistinguishable but also have some structural isomorphism at the non-observational level. For example, in response to the sort of conventionalism propounded by Poincaré and Reichenbach, according to which there is no factual difference between a curved spacetime and a flat spacetime conjoined with compensating universal forces, the realist will insist, in Sklar's words, that

> Two theories might have all the same observational predictions but be so radically different in their structure at the theoretical level that one ought to take them as attributing (realistically) quite different explanatory structures to the world. Only commonality of structure at the level of the theoretical ontology introduced by the theories to explain the commonly predicted observational results is enough for us to say that the two accounts are genuinely, realistically, equivalent to one another.[128]

Sklar quite rightly observes that this response raises the further question as to what then justifies the choice of one theory as true over a non-equivalent, but empirically indistinguishable alternative. But notice that this is, indeed, a *further* question. At worst, we are left with a sort of watered-down, epistemic conventionalism, such as Clark Glymour espouses:

> Cases do arise, however, where it is unreasonable to think that two theories can both be true, yet we cannot decide between them on empirical grounds. It is with such cases that conventions in physics have a real importance, and that importance is epistemic. The significant cases of conventional topologies are those in which the world is one way or the other, not both, but things are so arranged that we cannot discover which way the world is.[129]

This epistemic conventionalism goes no distance toward justifying a positivistic, ontological conventionalism, which is the issue at hand. As Nerlich emphasizes, "Maybe we could never know, even in principle, whether the earth or our whole space was Euclidean or spherical or what. But we are not concerned with the problems of finding out about spaces. We are worried about the spaces themselves. Our question was whether the space was determinate in its global topology, not whether our knowledge was."[130]

Now with respect to SR, we have two theories which are on all hands admitted to be observationally equivalent, but which differ fundamentally in terms of their theoretical ontology, namely, the aether compensatory approach of a man like Lorentz and the relativistic approach of Einstein. The central difference between these two lay in the structures posited at the theoretical level, which yield quite

---

[127] Lawrence Sklar, *Philosophy and Spacetime Physics* (Berkeley: University of California Press, 1985), p. 6.

[128] Ibid., p. 56.

[129] Clark Glymour, "Topology, Cosmology, and Convention," in *Space, Time, and Geometry*, ed. Patrick Suppes, Synthèse Library (Dordrecht: D. Reidel, 1973), pp. 213-214.

[130] Graham Nerlich, *The Shape of Space* (Cambridge: Cambridge University Press, 1976), p. 146.

different accounts of the predicted phenomena. In Kilmister's words, what makes relativity theory interesting is the "major change in ontology" wrought by Einstein *vis à vis* nineteenth century aether compensatory theories.[131] René Dugas points out,

> Lorentz's theory already embodies the essential results of special relativity, namely, the transformation (T) of the coordinates of space and time, the law (C) of the transformation of electrical and magnetic fields, the law (M) of the variation of the mass with the velocity. To these results, Einstein's first paper was to add only the law (V) of the composition of the velocities and the formula relating the mass to the energy.
>
> But in the passage of Lorentz's theory to that of Einstein, what is essential is the novelty of the Einsteinian point of view. Lorentz's theory is not relativistic, in this sense that a system at absolute rest in relation to the ether continues to enjoy special properties there.[132]

In Lorentz's theory there is a preferred rest frame and temporal parameter and, hence, absolute length and absolute simultaneity. Relativistic phenomena are the effects of dynamic causes operative as a result of motion relative to the rest frame. Einstein denies all of this theoretical structure. Early interpreters of SR, no doubt under the influence of positivism, tended to paper over these differences, often referring to SR as the Lorentz-Einstein theory. But these theoretical differences are so stark that in time Einstein's relativistic interpretation came to be distinguished from and preferred to the aether compensatory approach. This involved positivist thinkers in a strange inconsistency: in order to reject Lorentz's theory in favor of Einstein's they appealed to a verifiability criterion of meaning to eliminate the aether frame, absolute simultaneity, and so on; but in so doing they undermined the thesis of the equivalence of observationally indistinguishable theories, thereby exposing the metaphysical elements in science. In any case, philosophers appealing to SR as justification for a B-Theory of time clearly see the metaphysical differences between the two approaches, since Lorentz's theory preserves the classical concept of time and so offers no obstacle to the objectivity to tense and temporal becoming. If we take a positivistic line, then we cannot say that SR is incompatible with an A-Theory of time, since it is equivalent to Lorentz's theory. Only by recognizing the metaphysical differences between the theories can we say that SR constitutes even *prima facie* support for a B-Theory of time.

---

131 C. W. Kilmister, "Why is Relativity Interesting?" *British Journal for the Philosophy of Science* 42 (1991): 414. Dieks makes the point well:

> "The philosophical difference between Lorentz's electron theory and the special theory of relativity is exactly the different ontology of the two theories. Whereas an adherent of the electron theory works with classical spacetime relations, and therefore feels obliged to *explain* deviations from these relations which occur if moving rods are used, the relativist will, following Einstein's postulates take it for granted that the world is relativistic, and not classical" (Dennis Dieks, "The 'Reality' of the Lorentz Contraction," *Zeitschrift für allgemeine Wissenschaftstheorie* 15/2 [1984]: 341).

According to Dieks, the Relativity Principle thus assumes a central explanatory role within SR which can be compared to the role of the classical ontology of space and time in the Lorentz theory.

132 René Dugas, *A History of Mechanics*, with a Foreword by Louis de Broglie, trans. J. R. Maddox (New York: Central Book Company, 1955), p. 489. Cf. the comment by David Bohm, *The Special Theory of Relativity*, Lecture Notes and Supplements in Physics (New York: W. A. Benjamin, 1965), pp. vii-viii: "Einstein's basic contribution was less in the proposal of new formulas than in the introduction of fundamental changes in our basic notions of space, time, matter, and movement." See further ibid., p. 71.

Moreover, we have seen that SR itself is patient of two radically different interpretations, namely, Einstein's original relativity interpretation and Minkowski's spacetime interpretation. Again, these two interpretations are observationally equivalent but are fundamentally different at the theoretical level. The spacetime realist posits a four-dimensional ontology radically different from the three-dimensional ontology (with its reference frames, light signals, and so forth) of the space and time relativist. The B-theorist has to recognize this metaphysical difference because the relativity interpretation is not incompatible with the A-Theory of time, as we have seen. Thus, unless we are to collapse the Lorentzian aether compensatory theory, the Einsteinian relativity interpretation of SR, and the Minkowskian spacetime interpretation of SR into the same theory, despite the structural differences between them at the theoretical level (in which case the theory is indeterminate with respect to the debate of the A- vs. B-Theory of time), then we have no choice but to recognize in SR itself the metaphysical currents which breach the dike of positivism.[133] John Earman concludes that when it comes to questions about the nature of space and time, there is simply no way to justify an empirical/philosophical dichotomy; the appropriate term for the study is the old one: Natural Philosophy.[134]

In a recent essay, "Is Physics at the Threshold of a New Stage of Evolution?" Rompe and Treder echo Planck's question of 1908 and answer in the affirmative: "For several decades physics '*wächst über sich selbst hinaus*' (increases beyond its own limits."[135] Gale, in surveying some of "the metaphysical perplexities abounding in today's physics," contends, "...we are entering a phase of scientific activity during which the physicist has out-run his philosophical base camp, and, finding himself cut off from conceptual supplies, he is ready and waiting for some

---

[133] The metaphysical element in relativity theory becomes most apparent when it is given a spacetime formulation, in which case it becomes hypocritical to reject Newton's absolute time and space on the grounds that they are metaphysical. Arthur comments,

> "Mach himself, of course, is no longer a force to be reckoned with in the philosophy of space and time, the positivist position which he inaugurated having come in for its own share of debunking in recent years. This applies particularly to his attack on Newton's conception of an enduring absolute space, a conception which the latest generation of spacetime theorists has done much to rehabilitate" (Richard T. W. Arthur, "Newton's Fluxions and Equably Flowing Time," *Studies in History and Philosophy of Science* 26 [1995]: 323).

The question remains whether the preference on the part of these theorists for a four-dimensional spacetime metaphysic over an absolute space enduring through time is not itself grounded in a positivist critique of the Newtonian concepts. See also Carl Hoefer and Nancy Cartwright, "Substantivalism and the Hole Argument," in *Philosophical Problems of the Internal and External Worlds*, ed. John Earman, Alan I. Janis, Gerald J. Massey, and Nicholas Rescher, Pittsburgh-Konstanz Series in the Philosophy and History of Science (Pittsburgh: University of Pittsburgh Press, 1993), pp. 23-43.

[134] John Earman, "Who's Afraid of Absolute Space?" *Australasian Journal of Philosophy* 48 (1970): 317.

[135] R. Rompe and H.-J. Treder, "Is Physics at the Threshold of a New Stage of Evolution?" in *Quantum, Space and Time—The Quest Continues*, ed. Asim O. Barut, Alwyn van der Merwe, and Jean-Pierre Vigier, Cambridge Monographs on Physics (Cambridge: Cambridge University Press, 1984), p. 608.

relief from his philosophical comrades-in-arms."[136] Noting that in recent years such "metaphysical conundrums" as *creatio ex nihilo* "have entered the mainstream of scientific discussions," John Barrow remarks, "Traditional dogmas as to what criteria must be met by a body of ideas for it to qualify as a 'science' now seem curiously inappropriate in the face of problems and studies far removed from the human enterprise."[137] The point is that the positivistic, anti-metaphysical view of physics which dominated the first two-thirds of the twentieth century is simply outmoded in light of contemporary theoretical physics.

In summary, we have seen that the key justificatory factor in Einstein's rejection of Newton's absolute or metaphysical time and space and consequently his reduction of time and space to measured or physical time and space was Machian positivism, which eschewed such metaphysical entities as absolute time and space as literally meaningless and which in Einstein's case issued in a verificationist analysis of time and space along operationalist lines. Despite the triumph of Einstein's Special Theory, contemporary physics has burst the old wineskins of positivism and verificationism, as metaphysical elements have increasingly entered the mainstream of modern physical discussions.

---

136 Gale, "Metaphysical Perplexities."

137 Barrow, *World within the World*, pp. 2, vii-viii.

CHAPTER 4

SR'S ELIMINATION OF METAPHYSICAL TIME

In light of what we have seen in our previous chapter, what can be said of SR's putative elimination of metaphysical time and space?

POSITIVISM'S ESSENTIAL ROLE IN SR

The first thing to be noticed is that the positivism which characterized the historical formulation of SR belongs essentially to the philosophical foundations of the theory. The relativity of length depends upon the relativity of simultaneity, which in turn rests upon Einstein's re-definition of simultaneity in terms of clock synchronization by light signals. But that re-definition assumes necessarily that the time light takes to travel between two relatively stationary observers *A* and *B* is the same from *A* to *B* as from *B* to *A*. That assumption presupposes that *A* and *B* are not both in absolute motion, or in other words that neither metaphysical space nor a privileged rest frame exists. The only justification for that assumption is that it is empirically impossible to distinguish uniform motion from rest relative to such a frame. But if metaphysical space and absolute motion or rest are undetectable empirically, therefore they do not exist (and may even be said to be meaningless). Such an inference is clearly verificationist, and therefore positivistic.

In a clear-sighted analysis of the epistemological foundations of SR, Lawrence Sklar underlines the essential role played by this verificationism:

> Certainly the original arguments in favor of the relativistic viewpoint were rife with verificationist presuppositions about meaning, etc. And despite Einstein's later disavowal of the verificationist point of view, no one to my knowledge has provided an adequate account of the foundations of relativity which isn't verificationist in essence.[1]

It would be desirable to do so, muses Sklar, but "what I don't know is either how to formulate a coherent underpinning for relativity which isn't verificationist to begin with, or how, once begun, to find a natural stopping point for verificationist claims of under-determination and conventionality."[2]

POSITIVISM'S UNTENABILITY

But if positivism belongs essentially to the foundations of SR, the next thing to be noted is that positivism has proved to be completely untenable. This fact is so

---

1 Lawrence Sklar, "Time, Reality and Relativity," in *Reduction, Time and Reality*, ed. Richard Healey (Cambridge: Cambridge University Press, 1981), p. 141.
2 Ibid.

universally acknowledged that it will not be necessary to rehearse the objections against positivism here.[3] In a recent review of the literature, Tyler Burge has remarked that "the central event" in philosophy during the last half century has been "the downfall of positivism and the re-opening of discussion of virtually all the traditional problems in philosophy."[4] This is the case not only in philosophy of language and of mind. Healey observes that "...positivism has come under such sustained attack that opposition to it has become almost orthodoxy in the philosophy of science."[5] The rug has thus been pulled from beneath the positivists' anti-metaphysical polemics. Richard Kitchener comments,

> As anyone familiar with the recent history of philosophy knows, the once popular view of logical empiricists that 'metaphysics is meaningless' is no longer tenable; indeed, in retrospect it seems clear that the logical empiricists had a metaphysics and that what they were opposed to was a metaphysics that was *transcendent*, one that made claims about a supernatural (or 'super-empirical') reality that could in no way be checked empirically.[6]

Positivism provides absolutely no justification for thinking that Newton erred, for example, in holding that God exists in a temporal series which transcends our physical measures of it and may or may not be accurately registered by them. It matters not a whit whether we finite creatures know what time it is in God's metaphysical time; God knows, and that is enough.

With the failure of positivism, the A-theorist, especially if he is a theist of Newtonian stripe, is quite free to make, with Newton, a distinction between physical time and space (clock and rod measurements) and metaphysical time and space (ontological time and space independent of physical measures thereof). SR is a theory about physical time and space and says nothing about the nature of metaphysical time and space. Questions dealing with the latter are philosophical in nature and must be dealt with as such. In an unusual appreciation of this point in relativity discussions, Max von Laue wrote in the *Physikalische Zeitschrift* of 1912:

> In addition it seems to me that the whole question of the existence of the aether and of absolute time could without harm be banned from physical discussion; so long as entirely new facts, like, say, experimental proof of the existence of superluminal velocities or a contradiction between the appearances of gravitation and the Relativity Principle, do not permit us to decide physically between both of the mentioned viewpoints. They deliver exactly the same statements about all the quantities susceptible to measurement. That is not to say that the question has no interest, on the contrary, it seems to me to be of great philosophical significance. But precisely because of that, it should remain reserved for treatment with philosophical methods.[7]

---

[3] See the excellent survey in Frederick Suppe, "The Search for Philosophic Understanding of Scientific Theories," in *The Structure of Scientific Theories*, 2d ed., ed. F. Suppe (Urbana, Ill.: University of Illinois Press, 1977), pp. 3-118.

[4] Tyler Burge, "Philosophy of Language and Mind," *Philosophical Review* 101 (1992): 49.

[5] Richard Healey, "Introduction," in *Reduction, Time and Reality*, p. vii.

[6] Richard F. Kitchener, "Introduction: The World View of Contemporary Physics: Does It Need a New Metaphysics?" in *The World View of Contemporary Physics*, ed. Richard F. Kitchener (Albany: State University of New York Press, 1988), p. 5.

[7] M. Laue, "Zwei Einwände gegen die Relativitätstheorie und ihre Widerlegung," *Physikalische Zeitschrift* 13 (1912): 120.

How tragic that this quite sensible advice was ignored by the succeeding generation of positivist philosophers and physicists, who misinterpreted the relativization of physical time as the destruction of metaphysical time! Richard Swinburne does not exaggerate when he complains that "scientific talk about space and time has been influenced by verificationist presuppositions for the past century, perhaps more than any other scientific talk."[8] In an interesting survey article, John Norton concurs that it is in the philosophy of space and time that one finds some of the "clearest applications" of the ideas of logical positivism, including

> 1. An application of the verifiability criterion in Einstein's SR, its target being the absolute state of rest of Newton's theory of space and time,
> 2. Anti-realist claims about the geometry of space and the simultaneity of distant events (conventionality claims),
> 3. A reduction of spatio-temporal relations to causal relations in the causal theory of time.[9]

What is striking about this list is that applications (2) and (3), namely, conventionalism and causal theories of time, are now widely rejected, whereas (1) continues to be just as widely accepted, as if immune to the collapse of the epistemology essential to its underpinning. This constitutes an untenable situation. In his presidential address to the Royal Astronomical Society, E. A. Milne declared, "It is a good rule that however well a theory appears to fit the facts, it cannot be accepted as satisfactory if it is not *philosophically satisfactory.*"[10] The contemporary A-theorist should insist that an exclusively physical methodology is simply inadequate to deal with the problems of time and space. P. J. Zwart concurs,

> According to modern physics,...the duration of a certain interval of time is dependent upon the state of motion of the observer or his clock. However, strictly speaking this is not a philosophical standpoint, but a purely scientific one, since it only relates to the *meaning* of time and not to its nature. Only when adopting the positivistic view that time is nothing more than what is measured by a clock, could the relativistic theory of time be considered a philosophical one.... However, when one is of the opinion that the scientific description of time as a quantity applied in certain formulas and measured by means of a clock is far from complete, one should restrict oneself to the statement that clocks (and processes in general) in systems in motion run more slowly than in systems at rest.[11]

---

[8] Richard Swinburne, "Verificationism and Theories of Spacetime," in *Space, Time, and Causality*, ed. Richard Swinburne, Synthèse Library 157 (Dordrecht: D. Reidel, 1983), p. 63. He aptly remarks, "A satisfactory science ought to reveal the extent of our ignorance, not pretend that what is not knowable is not true" (Ibid., p. 74). Stating that "...I know of no good argument for verificationism," Swinburne contends that there is "no need to follow Einstein in his verificationism" (Richard Swinburne, *Space and Time*, 2d ed. [London: Macmillan, 1981], pp. 6, 201).

[9] John D. Norton, "Philosophy of Space and Time," in *Introduction to the Philosophy of Science*, ed. Merilee Salmon (New Jersey: Prentice-Hall, 1992), p. 179.

[10] E. A. Milne, "Presidential Address to the Royal Astronomical Society," *Monthly Notices of the Royal Astronomical Society* 104 (1944): 121.

[11] P. J. Zwart, "The Flow of Time," *Synthèse* 24 (1972): 134. Cf. the statement of A. N. Prior: "the theory of relativity isn't about *real* space and time, in which the earlier-later relation is defined in terms of pastness, presentness, and futurity; the 'time' which enters into the so-called space-time of relativity theory isn't this, but is just part of an artificial framework which scientists have constructed to link

SR may provide a perfectly adequate account of physical phenomena without giving us any philosophical insight into the nature of time. Decrying the "illegitimate passage from the results of physics to metaphysics," Mary Cleugh asserts,

> ...physicists have a perfect right to use what concepts they please: and criticism can enter only when one of these is labeled 'Time'. Then it is fair to ask whether those characteristics of Time which are neglected and even implicitly negated are not important. If they are, it is reasonable to hold that, however useful '*t*' might be for physics, its *complete* identification with Time is fallacious.[12]

Take absolute simultaneity, for example. It is granted by all parties that we have an intuitive sense that on the planet Neptune, for example, there are events occurring right now, simply simultaneously with events here on Earth. To point out that observers in different inertial frames synchronizing their clocks by reflected light signals will calculate different events on Neptune to be simultaneous with events here is to raise an irrelevancy. We are not talking about light signals or clocks, we are talking about what is happening at this moment of time. It is only by a positivistic collapse of ontological time into measured time that absolute simultaneity is drawn into question. As Sir James Jeans put it, SR "deals with measures of things, and not with things themselves, and so can never tell us anything about the nature of the things with the measure of which it is concerned. In particular it can tell us nothing as to the nature of space and time."[13] Thus, when Ralph Baierlein asserts, "...the fault is not with our clocks. It is with our intuition. We expect all observers to measure the same time interval between a pair of events, but space and time are not that way,"[14] one is tempted to reply, "*Au contraire*, the fault is not with the clocks or with our intuition. It is with our definitions. We are told that the time interval between a pair of events is the same as what observers measure, but space and time are not that way." Mary Cleugh concludes, "Relativity physics, then, in raising the question as to the precise significance which temporal relations such as simultaneity have *for it*, is not necessarily saying anything which is relevant for the metaphysician, still less is it laying down dicta to which metaphysics must conform."[15] "It cannot be too often emphasized that physics is concerned with

---

together observed facts in the simplest way possible...." (A. N. Prior, "Some Free Thinking about Time," in *Logic and Reality*, ed. B. J. Copeland [Oxford: Clarendon Press, 1996], p. 51).

[12] Mary F. Cleugh, *Time and its Importance in Modern Thought*, with a Foreword by L. Susan Stebbing (London: Methuen, 1937), pp. 29-30, 61.

[13] James Jeans, *Physics and Philosophy* (Cambridge: University Press, 1942), p. 68; cf. p. 66. Some theorists have disputed the reduction of time to its measures but have gratuitously assumed that time is that quantity described by relativity theory. See, for example, the critique of operational definitions of time by Mario Bunge, "Physical Time: The Objective and Relational Theory," *Philosophy of Science* 35 (1968): 355-388, though Bunge's concept of time is still physical, not metaphysical, in that it is analyzed as a map from sets of events, reference frames, and chronometric scales to the real numbers; see also Henry Mehlberg, *Time, Causality, and the Quantum Theory*, 2 vols., ed. Robert S. Cohen, Boston Studies in the Philosophy of Science 19 (Dordrecht: D. Reidel, 1980), 1: 1, 189-190, 251, who thinks that physical time is but an aspect of universal time and yet seems to assume that universal time is relativistic.

[14] Ralph Baierlein, *Newton to Einstein: The Trail of Light* (Cambridge: Cambridge University Press, 1992), p. 215.

[15] Cleugh, *Time*, p. 61.

the measurement of time, rather than with the essentially metaphysical question as to its nature."[16]

What is astounding is the degree to which contemporary treatments of time, despite the demise of positivism, still cling to the old verificationist perspective. For example, Bondi declares,

> Time is that which is measured by a clock. This is a sound way of looking at things. A quantity like time, or any other physical measurement, does not exist in a completely abstract way. We find no sense in talking about something unless we specify how we measure it. It is the definition by the method of measuring a quantity that is the one sure way of avoiding talking nonsense about this kind of thing.[17]

A verificationist theory of meaning clearly underlies this thinking, and it never occurs to Bondi that collapsing time to measured time might be one sure way of distorting reality. Verificationism also underlies Møller's claim that "The concept of simultaneity between two events in different places obviously has no exact objective meaning at all, since we cannot give any experimental method by which this simultaneity could be ascertained."[18] Like Bondi, Pagels makes the simple inference: clocks run slow in a gravitational field; time is what clocks measure; so time runs slow![19] Concerning a related phenomenon, Clifford Will advises, "Both descriptions are physically equivalent.... The observable phenomenon is unambiguous: the received signal is blue-shifted. To ask for more is to ask questions without observational meaning. This is a key aspect of relativity, indeed of much of modern physics: we focus only on observable, operationally defined quantities, and avoid unanswerable questions."[20] Capek asserts that to admit the "intrinsic unobservability" of certain events in spacetime and yet still to insist on their existence "would contradict the most elementary rules of scientific methodology."[21] In line with this thinking, Barbour declares that because absolute motion in space is undetectable, therefore it does not exist.[22] Freundlich in his discussion of temporal asymmetries repeatedly asserts that in order for a statement about the physical world to be meaningful, it must make claims about the "ways we are appeared to"; two physical statements which do not make different claims about the ways we are appeared to do not differ in meaning.[23] This same sort of reasoning appears to underlie the claim of persons like Duffy that the Lorentzian and Einsteinian alternatives concerning relativity, since they are empirically equivalent,

---

16 Ibid., p. 51.

17 Herman Bondi, *Relativity and Common Sense* (New York: Dover Publications, 1964), p. 65.

18 C. Møller, *The Theory of Relativity*, 2d ed. (Oxford: Clarendon Press, 1972), p. 31.

19 Heinz Pagels, *The Cosmic Code* (London: Michael Joseph, 1982), p. 50.

20 Clifford M. Will, *Was Einstein Right?* (New York: Basic Books, 1986), pp. 49-50.

21 M. Capek, "The Inclusion of Becoming in the Physical World," in *The Concepts of Space and Time*, ed. Milic Capek, Boston Studies in the Philosophy of Science 22 (Dordrecht: D. Reidel, 1976), p. 520; cf. his verificationist remarks that "the succession of causally unrelated events" is devoid of physical meaning, so that the simultaneity of distant events and the succession of causally independent events *simply do not exist* (Ibid., pp. 514-515).

22 Julian B. Barbour, *Absolute or Relative Motion?*, vol. 1: *The Discovery of Dynamics* (Cambridge: Cambridge University Press, 1989), pp. 8-9.

23 Yehudah Freundlich, "'Becoming' and the Asymmetries of Time," *Philosophy of Science* 40 (1973): 497-498.

are in fact the *same* theory, interpreted in different ways.[24] According to d'Inverno, the FitzGerald-Lorentz hypothesis "had the philosophical defect that its fundamental assumptions were unverifiable," whereas Einstein's derivation of the Lorentz transformations "was physically meaningful."[25]

Sometimes the conflation of time with its measures even leads to faulty theological inferences. For example, P. C. W. Davies, observing that if a spacetime singularity did occur at the Big Bang, as predicted by Friedman models, then it will be impossible "to continue physics, or physical reasoning, through it to an earlier stage of the universe,"[26] goes on to conclude from this that it is "meaningless" to speak of God's creating the universe.[27] For a cause must precede its effect temporally, but there is no temporal moment before the Big Bang. Therefore, the Big Bang can have no cause.[28] But (leaving aside the faulty premiss that a cause must temporally precede its effect) if we draw a distinction between metaphysical time and physical time as Newton did, it is quite evident that a beginning of the latter does not imply a beginning of the former. God in metaphysical time could be quite active prior to creation (perhaps creating angelic realms) and could bring physical space and time into being after having existed without their being co-existent with Him.

In discussions of quantum theory, too, one frequently encounters extravagant metaphysical inferences based on positivistic analyses of time. For physical time disintegrates at the elementary level due to quantum fluctuations in the geometry of physical space. Explaining that we must forego the view that every event occupies a position in spacetime with the Einstein interval from each event to its neighbor eternally established, Wheeler roundly proclaims, "There is no spacetime, there is no time, there is no before, there is no after. The question of what happens 'next' is

---

[24] Michael C. Duffy, "The Modified Vortex Sponge: a Classical Analogue for General Relativity," paper delivered at the International Conference of the British Society for the Philosophy of Science, "Physical Interpretations of Relativity Theory," Imperial College of Science and Technology, London, 16-19 September, 1988. He quotes with approval H. P. Robertson's remark, "Ives had in fact set up a theory completely equivalent in substance to the special theory of relativity....but I was never able to convince him that since what he had was in fact indistinguishable in its predictions from the relativity theory within the domain of physics, it was...the same theory...." Ives was apparently the keener epistemologist of the two.

[25] Ray d'Inverno, *Introducing Einstein's Relativity* (Oxford: Clarendon Press, 1992), p. 16.

[26] P. C. W. Davies, *Space and Time in the Modern Universe* (Cambridge: Cambridge University Press, 1977), p. 160.

[27] P. C. W. Davies, *God and the New Physics* (New York: Simon & Schuster, 1983), pp. 38-39.

[28] He asks,

> "But what does it mean to say that God *caused* time to come into existence, when by our usual understanding of causation a cause must precede its effect? Causation is a temporal activity. Time must already exist before anything can be caused. The naive image of God existing *before* the universe is clearly absurd if time did not exist—if there was no 'before'"(Ibid., p. 44).

More recently, in his acceptance speech of the Templeton Prize for Progress in Religion, Davies enunciates his belief that the quantum origin of spacetime "has illuminated the ancient theological debate" over divine eternity because the contingency of time reveals that the ultimate Being must be atemporal (Paul Davies, "The Acceptance Speech of Professor Paul Davies," Westminster Abbey, London, 3 May 1995). Davies naively applies physical time concepts to a metaphysical question.

without meaning."[29] Such inferences are justified only if one is presupposing a positivistic reductionism of the time concept. When one connects quantum geometrodynamics with cosmology, these fluctuations also become important because in the earliest stages of the universe prior to the Planck time the observable universe becomes so small that its scale factor is of sub-atomic proportions, where quantum physics predominates. The uncertainty this introduces into the time coordinate is exploited by Banks to draw a marvelous metaphysical inference:

> ...as we enter this regime [prior to the Planck time] the intuitive concept of time loses all meaning. There is no content in the question of what happened before the big bang, not because the universe becomes singular, but because quantum fluctuations invalidate the notion of 'absolute time....'*
>
> ---
>
> * I. Newton, *Philosophiae naturalis principia mathematica* (1687)....[30]

It is not explained, of course, how the indeterminacy of physical time is supposed to invalidate Newton's absolute time, which, based in God's eternity, ought, to recall Newton's words, "to be distinguished from what are only sensible measures thereof." Sometimes the metaphysical conclusions proclaimed on the basis the positivistic analysis of time in quantum theory can be quite ludicrous. For example, appealing to the invariance of quantum field theories under consecutive reversals of time, charge, and space (TCP invariance), Henryck Mehlberg states,

> If all natural laws are time reversed invariant and no irreversible processes occur in the physical universe, then there is no inherent, intrinsically meaningful difference between past and future—just as there is no such difference between 'to the left of' and 'to the right of.' If this is actually the case, then all mankind's major religions which preach a creation of the universe (by a supernatural agency) and imply, accordingly, a differentiation between the past and the future, *i.e.*, an intrinsic difference between both, would have to make an appropriate readjustment of man's major religious and 'creationist' creeds and the scientific findings.[31]

This solemn and ridiculous pronouncement clearly rests on the identification of God's time with physical time, a reduction which is positivistic in character. In general, the whole debate surrounding TCP invariance and the asymmetry or anisotropy of time is predicated on the verificationist assumption that time is identical with physical time, so that the direction or arrow(s) of time must be found exclusively in physical processes like entropy increase, the expansion of space, or other irreversible processes. But if we reject the verificationist identification of time with its physical measures, then these irreversible processes are merely *evidence* of the asymmetry of time, not *constitutive* of it. The A-theorist sees the distinction between the past and future as rooted in the fact of absolute (ontological, mind-

---

[29] C. Misner, K. S. Thorne, and J. A. Wheeler, *Gravitation* (San Francisco: W. H. Freeman, 1973), p. 1183. Cf. Wheeler's declaration, "*There is no such thing as spacetime in the real world of quantum physics*....superspace leaves us space but not spacetime and therefore not time. With time gone the very ideas of 'before' and 'after' also lose their meaning" (J. A. Wheeler, "From Relativity to Mutability," in *The Physicist's Conception of Nature*, ed. J. Mehra [Dordrecht: D. Reidel, 1973], p. 227).

[30] T. Banks, "*TCP*, Quantum Gravity, the Cosmological Constant, and All That...," *Nuclear Physics B* 249 (1985): 340.

[31] Henryk Mehlberg, "Philosophical Aspects of Physical Time," *Monist* 53 (1969): 363.

independent) becoming.[32] Events of which any observer in the metaphysical present can have traces have occurred and are therefore past, events in the actual world (tenselessly speaking of the possible world that obtains) of which no observer in the metaphysical present can have traces have not yet happened and are therefore future. Since becoming is real, metaphysical time is intrinsically anisotropic.

In fact, when one reflects on some of the counter-intuitive situations into which one is forced by the reduction of time to physical time, it becomes all the more evident that the two are to be distinguished. The dissolution of time in quantum physics would be just one example.[33] Another example would be GR's permitting time travel in universes having a global rotation of matter.[34] By following a wide enough curve, a spaceship can travel into its own past (or future) and meet up with itself before its journey (Figure 4.1).

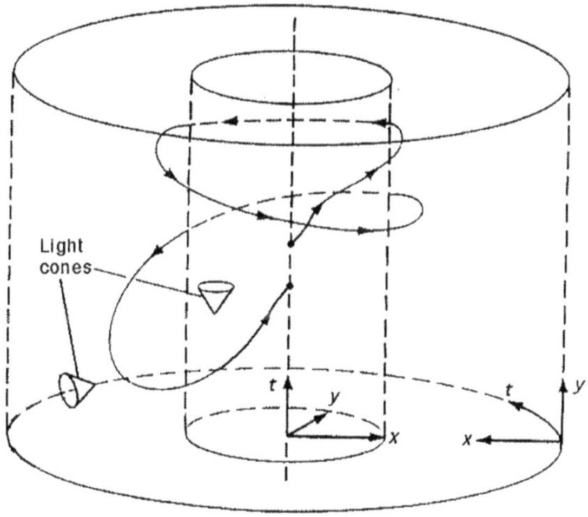

[32] See my *The Tensed Theory of Time: a Critical Examination*, Synthèse Library (Dordrecht: Kluwer Academic Publishers, forthcoming), chap. 7.

[33] To quote Wheeler again:

> "'Before' and 'after' don't rule everywhere, as witness quantum fluctuations in the geometry of space at the scale of the Planck distance. Therefore, 'before' and 'after' cannot legalistically rule anywhere. Even at the classical level, Einstein's standard closed-space cosmology denies all meaning to 'before the big bang' and 'after the big crunch.' Time cannot be an ultimate category in the description of nature. We cannot expect to understand genesis until we rise to an outlook that transcends time" (J. A. Wheeler, *Frontiers of Time* [Amsterdam: North-Holland, 1979], p. 20).

Wheeler should have said, "transcends *physical* time," *i.e.*, to God's outlook in metaphysical time in which "before" and "after" exist regardless of any breakdown in physical time concepts.

[34] On such models see Kurt Gödel, "A Remark About the Relationship Between Relativity Theory and Idealistic Philosophy," in *Albert Einstein: Philosopher-Scientist*, ed. P. A. Schilpp, Library of Living Philosophers 7 (LaSalle, Ill.: Open Court, 1949), pp. 555-562. Gödel also recognized that such models were incompatible with objective becoming, but he opted for the static models, since "...no reason can be given why an objective lapse of time should be assumed at all" (Ibid., p. 561). He hoped to forestall the paradoxes of time travel by practical considerations like excessive fuel requirements. For discussion see

Figure 4.1. Time travel in a Gödel space-time. Space-time is divided into two parts, an internal cylinder with a vertical temporal axis and the surrounding space-time, whose temporal dimension is circular. The world-line of the rocket ship begins in the central axis, moves upward and outward through the cylinder boundary, spirals down outside the cylinder, re-enters it, and then moves back toward the center, ending at a point earlier than its origin.

The only restrictions on such scenarios' coming to pass are technological: excessive fuel requirements, for example, for journeys into the recent past. But with the right fuel, GR permits such a trip. This, however, generates all sorts of difficulties. For example, suppose the space traveler as a child gets the formula for the super-efficient, light-weight fuel from his future self, while the future self is on his journey into the past. As he grows up, he becomes a space engineer and directs the construction of a spacecraft designed to use the fuel specified in the secret formula. He then embarks on his journey through space and time and delivers the formula to his earlier self. So where does the knowledge for the formula come from? Such circular causation seems evidently absurd, and the best way to break it is to deny that time and physical time are identical. In metaphysical time such a journey is impossible, given an A-Theory, because what is metaphysically present is not affected by motion through space, and, therefore, model universes permitting time-travel are engaging mathematical constructs which cannot be descriptive of reality.

Or consider what happens to physical time in connection with an object undergoing gravitational self-collapse. When the object collapses to its so-called Schwarzschild radius at which $r = 2M$, where r is its gravitational radius and M its mass,—a process which elapses in a finite proper time but takes infinite coordinate time as reckoned by an external observer—, then *the time and space coordinates switch*, that is to say, the spacelike coordinate r becomes at that point the temporal coordinate and the time coordinate t becomes spatial. The further decrease of r to zero represents the passage of time. But unless we distinguish metaphysical from physical time, so that the same metaphysical quantity is being represented by different coordinates, then this seems to be a blatant example of spatializing time, indeed, of actually turning time into space at a certain juncture. This is surely absurd, and their failure to distinguish metaphysical and physical time lead Misner, Thorne, and Wheeler into the following *mélange:*

> The unseen power of the world which drags everyone forward willy-nilly from age twenty to forty and from forty to eighty also drags the rocket in from time coordinate $r = 2M$ to the later value of the time coordinate $r = 0$. No human act of will, no engine, no rocket, no force...can make time stand still.[35]

In this description of a rocket's being sucked into a black hole from its Schwarzschild radius, we have a colorful mix of physical and metaphysical time. The flow of Newtonian absolute time is represented as a power pulling people along which nothing can arrest. But then this is identified as the same power that pulls the

---

Paul Horwich, *Asymmetries in Time* (Cambridge, Mass.: MIT Press, 1987), chap. 7 and William Lane Craig, *Divine Foreknowledge and Human Freedom* (Leiden: E. J. Brill, 1990), chap. 6.

[35] Misner, Thorne, and Wheeler, *Gravitation*, p. 823.

rocket into the black hole, namely, gravitation or spacetime curvature! Finally, the radius of the black hole is identified with time, so that one's powerlessness to halt the gravitational collapse of the black hole is identified with one's inability to stop the flow of time! Surely this is an egregious example of the confusion which inevitably results when time is identified with its measures.

Or consider the attempt on the part of Hartle and Hawking to avoid the initial cosmological singularity by having the time co-ordinate assume imaginary values at an early epoch of the universe, which again has the effect of converting time into space.[36] According to Hawking, "...there is no difference between the time direction and directions in space...time is imaginary and is indistinguishable from directions in space."[37] As Isham points out, the Hartle-Hawking model depends crucially on (i) the legitimacy of defining "time" phenomenologically in terms of the gravitational and/or material content of the universe (for example, its radius) and (ii) employing what in special relativistic contexts is a mere mathematical trick, namely, the introduction of imaginary values for the time co-ordinate, in general relativistic contexts, where its use entails radical differences for the nature of spacetime.[38] Taken literally, the Hartle-Hawking model is metaphysically absurd, since time is distinct from space, so that it must be regarded as the manipulation of physical time and its quantities only, not as a temporal ontology of the universe.

A final *Gedankenexperiment* suggested by Grünbaum will illustrate my point.[39] He invites us to imagine a particle moving circularly on a platform without observer and with no light. Each time the particle "returns" to point *A*, it is in an identical state. Therefore, given the Identity of Indiscernibles, it would really be the *same* time. There would be no basis in the particle's attributes or in the relation of events for saying that the same kind of event is recurring eternally. Rather we should have a closed time. If the motion of the particle were pendulum-like, then termini could be established which are different from points on the path, and so time would be open. But on a circular path, in the absence of light or contemplation, the particle would exist in closed time. If it is said that time is by nature open, then the same thing could be said about space, which would be false, as non-Euclidean geometry shows.

Now if we reduce time to its empirical measures, then Grünbaum is quite correct. But the conclusion is obviously wrong: a particle circling in the dark would return to the same point again and again, regardless of physical observation of it. The fallacy lies in the positivistic interpretation of the Identity of Indiscernibles. For the particle's being at *A* obtains at different moments of metaphysical time, so that these successive states are discernible, having different temporal properties, even if

---

[36] James Hartle and Stephen W. Hawking, "Wave Function of the Universe," *Physical Review D* 28 (1983): 2960-2975.

[37] Stephen W. Hawking, *A Brief History of Time: From the Big Bang to Black Holes*, with an Introduction by Carl Sagan (New York: Bantam Books, 1988), pp. 14,35.

[38] C. J. Isham, "Creation of the Universe as a Quantum Process," in *Physics, Philosophy, and Theology: A Common Quest for Understanding*, ed. R. J. Russell, W. R. Stoeger, and G. V. Coyne (Vatican City: Vatican Observatory, 1988), pp. 376, 391, 399, 402-403.

[39] Adolf Grünbaum, *Philosophical Problems of Space and Time*, 2d ed., Boston Studies in the Philosophy of Science 12 (Dordrecht: D. Reidel, 1973), pp. 197-201.

they are empirically indistinguishable. Arthur Pap asserts that the proposition "No event precedes itself" is a synthetic *a priori* truth;[40] becoming is intrinsic to metaphysical time. Grünbaum's analogy to non-Euclidean space evinces the tendency of B-theorists to spatialize time; as Capek notes, if we believe that time is adequately represented by a geometrical line, then there is no reason why this line should not be curved or closed.[41] But the A-theorist considers such a representation inadequate. Even if no basis exists for distinguishing physical times, it is evident in this case that different metaphysical times are involved.

In a fascinating review of the time concept in various fields of physics alone, Carlo Rovelli has emphasized how unlike the intuitive notion of time physical time concepts are and how diverse they are when compared among themselves.[42] He lists eight characteristics commonly associated with time:

1. One dimensional: time can be thought of as a collection of instants which can be arranged in a one-dimensional line.
2. Metric: time intervals can be measured such that two intervals can be said to have equal duration.
3. Temporally global: the real variable *t* which we use to denote the measure of time goes through every real value from –infinity to +infinity.
4. Spatially global: the time variable can be uniquely defined at all space points.
5. External: the flow of time is independent of the specific dynamics of the objects moving in time.
6. Unique: there are not many times, but just the time.
7. Directional: it is possible to distinguish the past from the future direction of the time-line.
8. Present: there always exists an ontologically preferred instant of time, the Now.

Rovelli then provides the following chart to illustrate the diversity of the physical time concept (Figure 4.2).

---

40 Arthur Pap, *Introduction to the Philosophy of Science* (London: Eyre & Spottiswoode, 1963), pp. 97-98.

41 M. Capek, "Introduction," in *Concepts of Space and Time*, p. xxx.

42 Carlo Rovelli, "What Does Present Days [*sic*] Physics Tell Us about Time and Space?", lecture presented at the Annual Series Lectures of the Center for Philosophy of Science of the University of Pittsburgh, 17 September 1993. Elsewhere he writes, "We lack a general coherent picture of the physical world capable of embracing what, or at least most of what, we have learned about it. The 'fundamental scientific view of the world' of the present time is characterized by an astonishing amount of perplexity, and disagreement, about what time, space, matter, and causality are" (Carlo Rovelli, "Halfway through the Woods: Contemporary Research on Space and Time," in *The Cosmos of Science*, ed. J. Norton and John Earman [Pittsburgh: University of Pittsburgh Press, 1998], p. 180).

The Notion of Time Used in:	*has properties:*
Natural language	1, 2, 3, 4, 5, 6, 7, 8
Thermodynamics	1, 2, 3, 4, 5, 6, 7
Newtonian mechanics	1, 2, 3, 4, 5, 6
SR	1, 2, 3, 4, 5
Cosmology	1, 2, 3, 4
GR - proper time	1, 2, 3, 5
GR - coordinate time	1, 3, 4
GR - clock times	1, 2
Quantum gravity	none

Figure 4.2. The concept of time in various fields of physics in comparison with the customary concept.

Even if one is disposed to dispute some of the specifics, there is, I think, no gainsaying Rovelli's point that physical time is both very different from our intuitive notion of time and, moreover, that because time is differently defined in different fields there is no unitary notion of physical time. It is difficult to resist the conclusion that all of these operationally defined "times" are not really time at all, but just various measures of time suitable for their respective fields of inquiry. The *reductio ad absurdum* of the positivistic reduction of time to its measures is surely the conclusion of someone like Barbour that because time in quantum gravity disappears, therefore time does not exist![43]

In sum, I think that in light of the collapse of positivism a re-appraisal of the time concept(s) in physics and in relativity theory in particular is long overdue. The demise of positivism should be neither ignored nor mourned; on the contrary, its lingering shadow over certain discussions of the time concept in contemporary physics has obfuscated the philosophical issues involved and sometimes led to quite unjustified metaphysical and theological conclusions. The above examples well bear out the verdict of Jonathan Powers: "In the end positivism itself passed into history: its ambitions had proved to be a mirage. But in the slogans of the New Physics, its ghost goes marching on."[44]

Certain thinkers have, we may be thankful, been more careful and discriminating than their above cited colleagues in their discussions of the concept of time. Such discrimination is already evident in the title of Peter Kroes's *Time: Its Structure and Role in Physical Theories*, for example. In his preface, Kroes makes it quite clear that the concept handled in his book is the concept of physical time, that is, time as used by physicists in their theories. He explicitly limits his discussion so as not to pronounce upon metaphysical time, commenting, "nor is it my intention to examine the metaphysical status of time. I shall restrict myself to an analysis of the structure

---

[43] Julian B. Barbour, "The Timelessness of Quantum Gravity: I & II," *Classical and Quantum Gravity* 11 (1994): 2853-2873, 2875-2897. Barbour holds that instants exist, but these are analyzed as three-dimensional relative configurations of the universe in superspace.

[44] Jonathan Powers, *Philosophy and the New Physics*, Ideas (London: Methuen, 1982), p. 12.

of physical time and of the role time plays in physical theories."[45] That he takes this distinction seriously is evident when one comes to his discussion of the objectivity of temporal becoming. He construes the controversy to concern whether it is possible to introduce the notion of the flow of time into the physicist's conception of nature.[46] He concludes, "Whether or not it is in principle impossible for physics to incorporate the flow of time in its description of physical reality, is still an open question. Up to the present, all attempts to capture this mysterious *but essential aspect of time* in the language of physics have failed."[47] Kroes carefully differentiates between physical time, into which the notions of past, present, and future have yet in his opinion to be successfully introduced, and what he calls "real time," for which these notions are essential. Positivistic philosophers would have concluded from the absence of becoming in physical time that objective becoming and temporal transience are therefore unreal. But with the failure of positivism, so rash an inference without further justification would only betray a lack of care.[48]

It must be admitted, too, that even some of the positivist thinkers I have cited above employed a verificationist analysis of time with respect to natural science only, but showed themselves open to a more full-orbed concept of time when discussing issues of a metaphysical character. For example, Margenau, after dismissing absolute simultaneity and other notions as ultra-perceptory and, hence, meaningless, adds the qualifier that he is not saying that these notions are meaningless in the sense intended by the logical positivist, but only that "...they cease to be the physicist's professional business."[49] Similarly, Frank's claim concerning metaphysical sentences was that "...they are meaningless *as far as science is concerned.*"[50] But when we begin to ask questions that are broader in scope, the meaningfulness of such sentences may change considerably:

> Our judgement about the usefulness of such expressions may change considerably if we consider not only the realm of physical facts in the narrower sense (*e.g.*, the motion of planets) but ask also for a general picture of the world and include the phenomena of human behavior as facts to be represented.[51]

Frank even opines that here religious ideas may enter the picture. Finally, Eddington also limited his verificationism to physical science and seemed open to

---

45 Peter Kroes, *Time: Its Structure and Role in Physical Theories*, Synthèse Library 179 (Dordrecht: D. Reidel, 1985), p. xiii; cf. p. viii.

46 Ibid., p. 196.

47 Ibid., p. xxiv [my emphasis]; cf. p. 209.

48 See also Herbert Dingle, "Time in Philosophy and Physics," *Philosophy* 54 (1979): 99-104. Though overly generous to Einstein in stating that "'time,' the ever-rolling stream, never entered his thoughts" in the 1905 paper, Dingle is correct that Einstein's theory had only to do with clock-times and proved nothing about the nature of time itself. Cf. idem, *Science at the Crossroads* (London: Martin Brian & O'Keefe, 1972), p. 137.

49 H. Margenau, "Metaphysical Elements in Physics," *Reviews of Modern Physics* 13 (1941): 183.

50 Philipp Frank, *Interpretations and Misinterpretations of Modern Physics*, Actualités Scientifiques et Industrielles 587: Exposés de Philosophie Scientifique 2 (Paris: Hermann & Cie., 1938), p. 38 [my emphasis].

51 Philipp Frank, *Philosophy of Science* (Englewood Cliffs, N. J.: Prentice-Hall, 1957), p. 144.

wider realms of reality. In his imaginary dialogue between a physicist and a relativist we find the following exchange:

> *Phys.* What you have shown is that we have not sufficient knowledge to determine in practice which are simultaneous events on the Earth and Arturus. It does not follow that there is no definite simultaneity.
>
> *Rel.* That is true, but it is at least possible that the reason why we are unable to determine simultaneity in practice...in spite of many brilliant attempts, is that there is no such thing as absolute simultaneity of distant events. It is better therefore not to base our physics on this notion of absolute simultaneity, which may turn out not to exist, and is in any case out of reach at present.[52]

Here the relativist admits that the relativity of simultaneity may be merely an epistemic matter; but because we cannot establish absolute simultaneity and because it may not exist, we ought not *to base our physics* on it. Such a contention, even if correct, does absolutely nothing to undermine Newton's metaphysical time. In another place, Eddington distinguishes two questions that may be asked concerning time: (i) what is the true nature of time, and (ii) what is the nature of that quantity which has under the name of time become a fundamental part of the structure of classical physics? Einstein's theory, says Eddington, cleared up the *second* of these questions.[53] But since this distinction is precisely that drawn by Newton between absolute and relative time, Eddington in effect agrees that the contribution made by SR is the relativization of relative or physical time and that the theory does not even speak to the question of the nature of absolute or metaphysical time.

These thinkers erred in believing that positivism could be successful even as a philosophy of science, but at least they did not use verificationism as a proofstone for all conceptions of reality. Their restricted positivism was quite compatible with the reality of metaphysical time. Unless one is a full-blooded positivist, then, one cannot support the claim that SR abolished notions of absolute simultaneity, absolute length, and so on. And full-blown positivism is widely recognized today as simply indefensible.

ARBITRARINESS OF SR'S SIMULTANEITY CONCEPT

The third thing to be noticed about SR's time concept is that it is predicated upon a definition of simultaneity which we are under no obligation to adopt. Contrary to the widespread, popular impression that Einstein proved the relativity of simultaneity, in fact all SR does is re-define simultaneity. The operational definitions Einstein gives to define simultaneity, and, hence, time, are completely arbitrary for someone who has not yet accepted the theory. Dingle remarks, "It is

---

52 Arthur Eddington, *Space, Time and Gravitation*, Cambridge Science Classics (1920; rep. ed.: Cambridge: Cambridge University Press, 1987), p. 17.

53 Arthur Eddington, *The Nature of the Physical World*, with an Introductory Note by Sir Edmund Whittaker, Everyman's Library (1928; rep. ed.: London: J. M. Dent & Sons, 1964), p.47; so also S. C. Tiwari, "Fresh Look on Relativistic Time and Lifetime of an Unstable Particle," paper presented at "Physical Interpretations of Relativity Theory," International Conference of the British Society for the Philosophy of Science, Imperial College of Science and Technology, London, 16-19 September, 1988.

necessary to emphasize that this definition is entirely arbitrary.... Prejudices die hard, and there are still physicists who have not entirely rid their minds of the idea that Einstein was *discovering* the right way of timing a distant event, instead of *freely inventing* something to which he gave that name."[54] Why should we regard two separated events as simultaneous, just because a light signal is inferred to reach the reflection event at the time of an event midway between the signal's emission and reception? Why is this what we mean by "simultaneity?" As Cleugh points out, "It is one thing to say that we cannot make judgments of simultaneity with regard to events at some distance from each other without the help of light signals: it is quite another to *define* simultaneity as depending on light signals."[55]

Einstein himself freely confessed that his operational definitions were conventional. He commented,

> The theory of relativity is often criticized for giving, without justification, a central theoretical role to the propagation of light, in that it founds the concept of time upon the law of the propagation of light. The situation, however, is somewhat as follows. In order to give physical significance to the concept of time, processes of some kind are required which enable relations to be established between different places. It is immaterial what kind of processes one chooses for such a definition of time.[56]

Two things are evident in these remarks: (i) Einstein's concern was to define time in such a way as to give it physical significance, and (ii) the choice of the processes used to establish temporal relations is arbitrary. One who has not already adopted the theory is thus under no obligation at all to regard as simultaneous events calculated to be simultaneous by Einstein's clock synchronization prescription. One could have used bullets instead of light signals to synchronize distant clocks. One could with equal justification (or lack thereof) have defined simultaneity by an exchange of signals in which the outbound signal is a light beam and the return signal is sound waves. For in the absence of any absolute time we do not know how distant simultaneity is to be defined.

What we mean intuitively by simultaneity is "the property of occurring at the same moment of time." If we feel bound to give operational definitions, then they ought to preserve the intuitive content of the concept or else be used to define a new and different concept. We ought to ask at the outset whether there may not be some operational method of defining distant simultaneity which is in tune with our intuitive beliefs. Continuity with the intuitive concept of simultaneity is preserved by T. Sjödin's proposal to define simultaneity operationally by using the concept of infinitely fast signals.[57] He emphasizes that "The *concept* 'infinitely fast signal'

---

[54] Herbert Dingle, "Time in Relativity Theory: Measurement or Coordinate?" in *The Voices of Time*, 2d ed., ed. with a new Introduction by J. T. Fraser (Amherst, Mass.: University of Massachusetts Press, 1981), p. 462.

[55] Cleugh, *Time*, p. 57.

[56] Albert Einstein, *The Meaning of Relativity*, 6th ed., (1922; rep. ed.: London: Chapman & Hall, 1967), p. 27.

[57] T. Sjödin, "On the One-Way Velocity of Light and its Possible Measurability," paper delivered at "Physical Interpretations of Relativity Theory." See also Richard Gale, "Human Time: Introduction," in *The Philosophy of Time: a Collection of Essays*, ed. R. M. Gale (New York: Humanities Press, 1968), p. 299; M. Capek, *The Philosophical Impact of Contemporary Physics* (Princeton: D. Van Nostrand, 1961),

exists even if there is an upper limit to the velocity of causal propagations in our actual world. Consequently, also the *concept* 'simultaneity' exists even if there is no possibility for man to know if two events are simultaneous, or not."[58] He offers the following definition of simultaneity:

> *Definition of simultaneity:* Two events E_1 *and* E_2 at the points P_1 *and* P_2 are *simultaneous*, if no signal, independently how fast, may depart from P_1 after the event E_i and arrive at P_j before the event E_j (i, j = 1, 2).

While intuitively on target, this definition needs to be tightened up a bit. In the first place, it is really a proposed definition of distant simultaneity, not simultaneity *simpliciter*. Moreover, it states *prima facie* only a sufficient condition of distant simultaneity and so cannot constitute a definition. Finally, it allows that a signal might depart from P_1 after E_1 and arrive at P_2 at E_2 or that a signal might depart from P_1 at E_1 and arrive at P_2 before E_2, so that it fails to give even a sufficient condition of distant simultaneity. We may revise Sjödin's definition accordingly:

> *Definition of distant simultaneity:* Two events E_1 and E_2 at the points P_1 and P_2 are simultaneous iff no signal, independently how fast, may depart from P_1 after E_1 and arrive at P_2 at or before E_2.

In other words, the time of E_1 is the latest that a signal, no matter how fast, can depart from P_1 and arrive at P_2 by the time of E_2. If signals travel infinitely fast, their moment of reception at a distance is the same as the moment of their emission. E_1 and E_2 must therefore occur at the same moment, since they are connectable only by an infinitely fast signal.

Sjödin's definition of simultaneity makes no use of clocks. But on the basis of his definition, he is prepared to define absolute synchronization as well:

> *Definition of absolute synchronisation*: The synchronisation of two identical clocks at rest in some inertial frame is *absolute* if they show the same time simultaneously (according to the definition above).

Sjödin recognizes that because we are dependent on finite signals, we are obliged to choose from among the infinitely many possibilities of synchronizing clocks that all give different (but equivalent) descriptions of the physical world. Of all these different choices the standard Einstein synchronization procedure is of special practical importance, but it enjoys no conceptual importance in defining simultaneity.

Sjödin's definitions are so perspicuous and in harmony with the intuitive notion of simultaneity that they seem vastly preferable to Einstein's definitions (if one feels the need for operational definitions at all). What is especially intriguing about these definitions is that relativity theory does not prohibit the existence of superluminal velocities, even arbitrarily high velocities, so that Sjödin's prescription describes a

---

p. 172. It is noteworthy that the Galilean transformations follow automatically from the Lorentz equations if we substitute for c an infinite value.

[58] Sjödin, "One Way Velocity of Light."

nomologically possible procedure.[59] There is an intriguing body of literature surrounding the nature of and the search for particles that travel at superluminal velocities (dubbed "tachyons").[60] Grünbaum, who touts the conventionality of simultaneity, acknowledges that if we admit superluminal velocities then relativistic time order relations are not "ontologically sacrosanct."[61] Referring to a universal tachyonic background as constitutive of a special reference frame, Grünbaum poses the provocative question, "Perhaps this is tantamount to an ether reference frame whose time qualifies as physically preferred?" This inference is usually disputed on the grounds that infinite velocity tachyons, while sufficing to eliminate the conventionality of simultaneity within a single reference frame, would not serve to establish absolute simultaneity unless their velocity were also invariant.[62] For since E_1 and E_2, being connected by an infinite velocity tachyon beam in frame R, are spacelike separated, SR requires that in some other frame R', E_2 occurs later than E_1. Thus, while a tachyon beam connecting E_1 and E_2 would possess an infinite velocity in R, it would possess only a finite velocity in another frame R'. In R' there is no reason to take E_1 and E_2 as simultaneous, since they are connected by a merely finite velocity signal, and faster tachyons in R' could serve to establish the non-simultaneity of E_1 and E_2. Worse, SR requires that in some frame R^* E_1 occurs later than E_2, so that the tachyon beam literally travels backward in time (and possesses negative energy). Now this bizarre result might provide just the motivation which is needed for regarding R as the preferred reference frame and for treating solutions in which E_1 and E_2 are not simultaneous as purely mathematical and non-physical.[63] Bilaniuk and Sudarshan, two pioneers of tachyon research, insist that this would not serve to overthrow SR. They ask,

> After all, is not the exclusion of a preferred frame what relativity is all about? *No, it is not.* The postulates of special relativity require the laws of physics, including the speed of light, to be the same in all inertial frames. They do not preclude the existence of cosmological boundary conditions that permit us to single out a particular local frame as a preferred reference system.[64]

Bilaniuk and Sudarshan are being somewhat disingenuous here, since the admission of a preferred frame would require us to adopt a neo-Lorentzian rather than Einsteinian relativity theory. Loathe to admit such a privileged frame, relativity theorists frequently invoke instead a so-called Reinterpretation Principle in order to

---

59 SR prohibits the acceleration of particles from subluminal to luminal or superluminal speeds, not particles that always travel at such speeds. SR may also be incompatible with superluminal *signals* (controllable causal connections), but nothing in Sjödin's definition depends on the use of signals as opposed to raw tachyon beams occurring naturally.

60 See the discussion and literature listed in Craig, *Divine Foreknowledge and Human Freedom.*

61 Grünbaum, *Space and Time*, p. 827.

62 Michael Friedman, *Foundations of Spacetime Theories* (Princeton: Princeton University Press, 1983), p. 167.

63 See Wesley C. Salmon, *Space, Time, and Motion* (Encino, Calif.: Dickenson Publishing, 1975), pp. 119-122.

64 Olexa-Myron Bilaniuk, *et al.*, "More about Tachyons," *Physics Today* (December 1969), p. 52.

rule out backward-travelling tachyons.[65] According to this principle, a negative energy tachyon traveling backward in time may be reinterpreted as a positive energy tachyon traveling forward in time. Thus, what in one frame is the emission of a tachyon beam at E_1 and its reception at E_2 is in another frame the emission of the beam at E_2 and its reception at E_1. This Reinterpretation Principle, however, seems clearly to be just one more vestige within the philosophy of space and time of precisely that defunct positivism which we have been decrying. Because observers cannot tell the difference between a negative energy particle traveling backward in time and a positive energy particle travelling forward in time, it is declared that there is no objective, factual difference between the two. Apart from other objections,[66] the untenability of the Reinterpretation Principle is evident from the simple consideration that the observer associated with E_2 might not even have a tachyon transmitter but only a receiver, so that E_2 could not possibly be reinterpreted as the emission of a tachyon beam. Thus, resort to the Reinterpretation Principle to preclude backward-traveling tachyons is a desperate and ultimately futile expedient. Rather what needs to be done, in Tim Maudlin's words, is to posit the extra structure to spacetime requisite to preclude them, structure "which looks suspiciously like an absolute notion of simultaneity."[67] If there is a privileged frame in which E_1 and E_2 are simultaneous in virtue of their being connected by an infinite velocity tachyon beam, then E_1 and E_2 are non-conventionally and absolutely simultaneous.

In any case, Sjödin's definitions are unproblematic because they do not define simultaneity in terms of connectability by infinite velocity signals. His definitions are independent of reference frames and rule out backward-traveling signals. Since his operational definitions are intuitively continuous with the common notion of simultaneity, they are preferable to Einstein's definitions.

Of course, one might complain that an operational definition presupposing the possibility of infinitely fast tachyonic signals is of no practical significance, since we have no such signals. But, in the first place, it should be understood that Einstein's own definitions are impracticable as well. Holton underscores this point:

> That Newton's absolute space and absolute time were not meaningful concepts in the sense of laboratory operations was, of course, not the original discovery of Mach; rather it was freely acknowledged by Newton himself. But Einstein was also quite explicit that in replacing absolute Newtonian space and time with an infinite ensemble of rigid meter sticks and ideal clocks he was not proposing a laboratory-operational definition. He stated it could be realized only to some degree, 'not even with arbitrary approximation' and that the fundamental role of the whole conception, both on factual and logical grounds, 'can be attacked with a certain right.'*

---

65 See, for example, Michael Redhead, "Nonlocality and Peaceful Coexistence," in *Space, Time and Causality*, ed. Richard Swinburne, Synthèse Library 157 (Dordrecht: D. Reidel, 1983), pp. 167-179; idem, "The Conventionality of Simultaneity," in *Philosophical Problems of the Internal and External Worlds*, ed. John Earman, Alan I. Janis, Gerald J. Massey, and Nicholas Rescher, Pittsburgh-Konstanz Series in the Philosophy and History of Science (Pittsburgh: University of Pittsburgh Press, 1993), pp. 116-125; Graham Nerlich, *What Spacetime Explains* (Cambridge: Cambridge University Press, 1994), pp. 68-71.

66 See Craig, *Divine Foreknowledge and Human Freedom*, pp. 122-127.

67 Tim Maudlin, *Quantum Non-Locality and Relativity*, Aristotelian Society Series 13 (Oxford: Blackwell, 1994), pp. 115-116; cf. pp. 73-75.

* Albert Einstein, *Les prix Nobels en 1921-1922* (Stockholm, 1923), p. 2.[68]

It is at this juncture in his article that Holton suggests that Einstein has merely substituted an abstract *Gedankenexperimenter* for God. There seems to be no in principle reason why our abstract *Gedankenexperimenter* should not employ tachyon beams rather than light rays in his synchronization procedure. In any case, issues of practicality are irrelevant for purposes of *defining* the concept of simultaneity. Even if Einstein's procedure were completely practicable, that does nothing to commend it as defining the concept of simultaneity. On the contrary, Einstein's prescription seems manifestly counter-intuitive in comparison with Sjödin's suggestion.

The arbitrariness of Einstein's operational definitions and their lack of conceptual continuity with the intuitive (and classical) conception of simultaneity provokes a critical problem with regard to the claim that relativity theory has superseded Newtonian theory, for it appears that the two theories are not even talking about the same thing. Cleugh cautions, "'Simultaneity' is now a dangerously ambiguous word, standing for at least two quite different notions."[69] On this account, the Newtonian physicist and the relativist do not mean the same thing in asserting, "Events e_1 and e_2 are simultaneous in frame *F*." Therefore, their theories do not contradict each other, since the conceptions of simultaneity are diverse. The Newtonian would concede quite happily that two events which are simultaneous in absolute time are not simultaneous according to Einstein's clock synchronization procedure.

In analyzing this problem, Friedman asserts that if we view the referents of theoretical terms as determined by the theoretical principles within which they occur, then we cannot say Newtonian mechanics represents a false view of time and simultaneity, for its terms have no referents, and so the theoretical principles involving those terms are not false, but truth-valueless.[70] But this verdict seems to presuppose the legitimacy of Einstein's operational definitions; if we choose Sjödin's, then the Newtonian concept of simultaneity does have a referent independent of the behavior of clocks. In this case, the Newtonian assertion that "Events e_1 and e_2 are simultaneous in frame *F*" and the Einsteinian assertion "Events e_1 and e_2 are not simultaneous in frame *F*" can both be true because the operational definitions of what it is to be "simultaneous" are different. Friedman proposes to untangle the muddle by adopting a causal theory of reference instead, according to

---

68 Gerald Holton, "On the Origin of the Special Theory of Relativity," in *Thematic Origins of Scientific Thought*, by G. Holton (Cambridge, Mass.: Harvard Univ. Press, 1973), pp. 170-171. Dugas comments, "We shall not go as far to complain, as certain critics do, that Einstein made use of magical clocks and enchanted measuring rods in order to return, in the last analysis, to Lorentz's transformation (T). But there is no doubt that those ideal clocks and rods can only be used in idealised experiments" (René Dugas, *A History of Mechanics*, trans. J. R. Maddox [N. Y.: Central Book Co., 1955], p. 490).

69 Cleugh, *Time*, p. 59. She asserts, "...the simultaneity with which Einstein deals is only a very distant cousin of the simultaneity with which the plain man and the metaphysician are concerned...," and, she might have added, the classical physicist as well (p. 58).

70 Friedman, "Simultaneity in Newtonian Mechanics and Special Relativity," in *Foundations of Spacetime Theories*, ed. John Earman, *et al.* (Minneapolis: Univ. of Minnesota Press, 1977), pp. 411-415.

which what a theoretical term refers to is a matter of which actual entities have the right sort of historical connection with the use of the term. The use of "time" and "simultaneous" in Newtonian mechanics provides Friedman with an illustration of the proper procedure in determining the meaning of theoretical terms. Given the entities postulated by our best current theory, we look for some among these which: (i) give a plausible distribution of truth values for the sentences involving "time" and "simultaneous" used by Newtonian physicists, (ii) are actually responsible for the phenomena explained by Newtonian mechanics, and (iii) are actually measured by the measuring procedures used to test Newtonian mechanics. One then shows how Newtonian physicists were referring to time and simultaneity as these concepts play a role in SR or else that these terms had no referents.

Unfortunately, Friedman still leaves us with a disjunction: either the Newtonian referent of simultaneity was the same as in SR or there was no such referent. Since different operational definitions could provide such a referent, and since such a referent's theoretical term could with equal, if not greater, plausibility also be claimed to stand in historical continuity with the use of "time" and "simultaneity" in Newtonian theory (after all, Poincaré's use of the light signal synchronization procedure was only valid in the rest frame of the aether, and he remained uncompromising on this in the face of SR), one reaches the conclusion that the terms "time" and "simultaneity" in Newtonian theory were referring either to the referents mentioned in Einstein's definitions or in, say, Sjödin's definitions. But since these are mutually incompatible, the referents of Newtonian theory cannot be identical with the referents of both.

Perhaps the difficulty here only serves to reveal the ineptness of operational definitions.[71] On an operationalist view, what is intuitively a single concept can be operationally defined in a multiplicity of ways, so that the resulting referents of the terms so defined are not after all identical. In the case at hand, it seems evident that "simultaneity" is a unitary, intuitive concept and that what ought to be at issue in the operational procedures is not *defining* simultaneity, but *determining* or *discovering* it. All of the definitions discussed share the common, intuitive core meaning of "simultaneous" as "occurring at the same time," but they differ in how to determine or discover physically which events occur at the same time. Einstein adopts the convention of using light signals and clocks to determine which events are simultaneous. Adopting this prescription leads in turn to the relativity of simultaneity. But another prescription is possible, whose adoption does not lead to the conventionality or relativity of simultaneity. The proponent of this other prescription will agree that the behavior of light rays, clocks, and measuring rods is

---

[71] As Suppe explains,

> "Often there is more than one experimental procedure used to determine, for example, an object's mass. Since concepts or properties are identified with unique combinations of operations, each different experimental procedure defines a distinct concept, and so there are as many distinct concepts of *mass* as there are procedures for determining it. In actual scientific practice, however, these different procedures are taken as measuring the same thing, *mass*; thus operational definition is unsatisfactory as an analysis of the meaning (hence the cognitive significance) of theoretical terms" (Suppe, "Search for Philosophic Understanding," p. 19).

such that "simultaneity" according to the Einsteinian prescription is, indeed, relative; but he will regard this as a pseudo-simultaneity, not a genuine simultaneity. Genuine simultaneity is discovered, not constituted, by the relevant operations and is therefore non-conventional and absolute.

Einstein's definitions, then, are admittedly conventional and, arguably, not very plausible intuitively. It is interesting to read the defense of one relativist of Einstein's re-definition of time:

> ...the primitive concept of absolute time, which has been discarded, is so deep and has been so successful for so long that it has become part of our language. It is a concept so much a part of our thinking that it is difficult to change—to accept a definition of time that is different for different observers. And the mind always struggles, the question is always raised, 'Is it necessary? Is this what the experiments tell us?' The answer is 'No, it is not necessary.' And the fact that it is not necessary makes Einstein's achievement the more brilliant. For he has not shown us truth. Better, he has shown us a way to regard our experience as so perfect that, having seen his canvas, we find it difficult to look at the world another way.[72]

According to this most peculiar and interesting analysis, Einstein's theory is not required by experimental evidence but is based upon arbitrary and counter-intuitive re-definitions, so that what the theory asserts cannot even be claimed to be true. Nevertheless, it gives us a view of the world which is so compelling that we prefer it to alternative viewpoints! Engaging as the relativistic view of time may be, however, the empirical content of the theory is not incompatible with absolute simultaneity, so that the discarding of absolute time is unwarranted. Geoffery Builder concludes, "...the measurement procedures prescribed by the restricted theory are conventional; failure to recognize this has led many exponents of the restricted theory to assert, without sufficient justification, that these procedures demand a complete revision of the older concepts of space and time."[73] Revisions in Newtonian mechanics are necessary, of course; Newton did not realize that physical time—the time kept by clocks—needed to be relativized. But, as Frank explains, "The 'relativization' of space and time consists actually in the introduction of new operational definitions that are better adapted to the actual needs of the scientist. This 'relativization' of space and time is an advance in semantics and not, as has been frequently said, an advance in metaphysics or ontology."[74]

---

72 Leon N. Cooper, *An Introduction to the Meaning and Structure of Physics* (New York: Harper & Row, 1968), p. 401. Cf. the comment by Henry Margenau and Richard A. Mould, "Relativity: an Epistemological Appraisal," *Philosophy of Science* 24 (1957): 303:

> "...it does not make sense to speak of settling the matter between Einstein and Lorentz by experiment. What is significant about Einstein's principle is that it *can* be maintained, not that it *must* be maintained. In rejecting it, Lorentz did not commit any technical error, but he did forego its conceptual and heuristic advantages."

But we have already seen that Einstein's theory is conceptually hobbled in comparison with Lorentz's, being based on an arbitrary and counter-intuitive re-definition of distant simultaneity motivated by an untenable epistemology; as for heuristic advantages, see discussion in the sequel.

73 G. Builder, "The Constancy of the Velocity of Light," *Australian Journal of Physics* 11 (1958): 457-480, rep. in *Speculations in Science and Technology* 2 (1979): 421.

74 Frank, *Philosophy of Science*, p. 143.

SUMMARY

We have thus seen that positivism belongs essentially to the philosophical foundations of SR, that such an epistemological outlook has been justifiably and universally rejected as untenable, and that SR is predicated upon arbitrary and counter-intuitive re-definitions of the key concepts of time and distant simultaneity. Lawrence Sklar concludes,

> The original Einstein papers on special relativity are founded, as is well known, on a verificationist critique of earlier theories.... Now it might be argued that Einstein's verificationism was a misfortune, to be encountered not with a rejection of special relativity, but with an acceptance of the theory now to be understood on better epistemological grounds....
>
> But I don't think a position of this kind will work in the present case. I can see no way of rejecting the old aether-compensatory theories, originally invoked to explain the Michelson-Morley results, without invoking a verificationist critique of some kind or other. And I know of no way to defend the move to a relativized notion of simultaneity, so essential to special relativity, without first offering a critique, in the same vein as Einstein's, of the pre-relativistic absolutist notion, and then continuing to observe that even the relativistic replacement for this older notion is itself, insofar as it outruns the 'hard data' of experiment, infected with a high degree of conventionality.75

It seems to me, therefore, that SR's elimination of absolute time is a miscarriage, resting as it does on defective epistemological foundations. The A-theorist both can and should distinguish between metaphysical time and physical time and maintain that the former is characterized by a universal and objective present and, hence, relations of absolute simultaneity. The fatal shortcoming in Putnam's argument against the objectivity of temporal becoming[76] lies in his uncritical acceptance of Einstein's positivistic elimination of metaphysical time and, consequently, his misdirected search for the foundation of temporal becoming in physical, measured time rather than in time itself—hence, his remark that it is physics, not philosophy, which has solved the problem of the reality of future events and that, therefore, there are no longer any philosophical problems about time, a remark which drew Sklar's riposte, "I think that such a naive view is as wrong as can be."[77] Once we examine critically the philosophical foundations of Einstein's Special Theory, we discover that there is nothing there that warrants the conclusion that metaphysical time, objective becoming, and absolute simultaneity do not exist. It is difficult, indeed, to understand how the authors quoted earlier can therefore speak of SR's "forcing" us to abandon the classical concepts of space and time or of its "destruction" of Newtonian absolute time. Sklar asserts, "One thing is certain. Acceptance of relativity cannot force one into the acceptance or rejection of any of the traditional views about the reality of the past and future."[78] Of course, we are still left with D.

---

75 Sklar, "Time, Reality, and Relativity," p. 132.

76 Recall discussion in chap. 1, pp. 1-5.

77 Sklar, "Time, Reality, and Relativity," p. 131. He adds, "One can extract only so much metaphysics from a physical theory as one puts in."

78 Ibid., p. 140. In Gödel's words,

> "the observational results by themselves really do not force us to abandon Newtonian time and space as objective realities, but only the observational results together with

H. Mellor's challenge[79] of exactly how the relativity of physical time, which has been well-established empirically, is to be reconciled with the objectivity of temporal becoming and the absolute simultaneity that the A-theorist postulates of metaphysical time. It is to that task that we now turn.

---

certain general principles, *e.g.*, the principle that two states of affairs which cannot be distinguished by observers are also objectively equal" (unpublished *Nachlass* cited by Palle Yourgrau, *The Disappearance of Time* [Cambridge: Cambridge University Press, 1991], p. 54).

79 See chap. 1, pp. 7-9.

CHAPTER 5

THE VINDICATION OF LORENTZ

NEO-LORENTZIAN RELATIVITY

In Newton's view absolute time, God's time, is the seat of temporal becoming and objective tense. The A-theorist may concur with Newton on this point, seeing SR as nothing more than a theory about relative time, that is, our physical measures of time. But how, then, does A-theoretic time connect with physical time? Given our rejection of the relativity interpretation of SR, it follows from the reality of tense that a Lorentzian theory of relativity is correct after all. The A-theorist may plausibly contend that that reference frame whose associated time coincides with and thus constitutes the measure of metaphysical time is thereby privileged and that the familiar relativistic phenomena are the result of motion relative to this frame.

Lorentz's understanding of relativity theory is well-suited to an A-Theory of time. First, it is one of the marks of Lorentz's genius that he stood out virtually alone among the major figures of early twentieth century physics in resisting the pervasive positivistic epistemology which lay at the root of Einstein's SR. Gracious almost to a fault, Lorentz always spoke appreciatively of Einstein's alternate approach and lectured sympathetically on both SR and GR, while remaining finally unconvinced that Einstein had abolished the classical conceptions of time and space. A major reason that Lorentz remained unconvinced is that he was not a positivist. In 1913 he wrote,

> According to Einstein it has no meaning to speak of motion relative to the aether. He likewise denies the existence of absolute simultaneity.
>
> It is certainly remarkable that these relativity concepts, also those concerning time, have found such a rapid acceptance.
>
> The acceptance of these concepts belongs mainly to epistemology.... It is certain, however, that it depends to a large extent on the way one is accustomed to think whether one is attracted to one or another interpretation. As far as this lecturer is concerned, he finds a certain satisfaction in the older interpretations, according to which the aether possesses at least some substantiality, space and time can be sharply separated, and simultaneity without further specification can be spoken of. In regard to this last point, one may perhaps appeal to our ability of imagining arbitrarily large velocities. In that way, one comes very close to the concept of absolute simultaneity.
>
> Finally, it should be noted that the daring assertion that one can never observe velocities larger than the velocity of light contains a hypothetical restriction of what is accessible to us, [a restriction] which cannot be accepted without some reservation.[1]

[1] In H. A. Lorentz, A. Einstein, H. Minkowski, *Das Relativitätsprinzip*, Fortschritte der mathematischen Wissenschaften 2, mit Anmerkungen von A. Sommerfeld und Vorwort von O. Blumenthal (Leipzig: B. G. Teubner, 1913), p. 23 (Pais translation).

Here Lorentz clearly discerns the crucial role played by Einstein's verificationist theory of meaning and rejects it. In defense of absolute simultaneity, he appeals to the use of arbitrarily fast signals, even though they are not presently observable. He quite rightly disregards the assumption that it is meaningless to speak of such unobservables. Elsewhere Lorentz affirms that it makes sense, if there is an aether, to speak of motion relative to it even if observers could not detect such motion.[2] He writes,

> But it needs to be clearly recognized that A could never assure himself of the immobility in the ether which we have attributed to him by supposition and that physicist B could with the same right, or rather with the same absence of right, claim that it is he who finds himself in these privileged circumstances. This incertitude, this impossibility of even disclosing a movement in relation to the ether, led Einstein and numerous other modern physicists to abandon completely the notion of an aether.
>
> There, it seems to me, is a question toward which each physicist must take a position which best accords with the manner of thinking to which he is accustomed.[3]

Lorentz clearly felt no obligation to abandon what seemed to him meaningful notions out of deference to a verificationist epistemology. The subsequent collapse of positivism has vindicated his feeling and left his approach to problems of space and time, in contrast to Einstein's, unshaken by this epistemological revolution.

Second, Lorentz's conception of the aether was virtually equivalent to space itself. His aether was so dematerialized that Einstein, lecturing at the University of Leiden in 1920, could tease the Dutch physicist by declaring, "As regards the mechanical nature of Lorentz's aether one might say of it, with a touch of humor, that immobility was the only mechanical property which H. A. Lorentz left it."[4] Later Einstein would speak of Lorentz's "great discovery," namely, that all the phenomena of electromagnetism then known could be explained on the basis of two assumptions: that the aether is firmly fixed in space—that is to say, unable to move at all—and that electricity is firmly lodged in the mobile elementary particles.[5] "Today his discovery may be expressed as follows: physical space and the ether are only different terms for the same thing: fields are physical states of space."[6] For

---

2 H. A. Lorentz, "Het relativiteitsbeginsel. Voordrachten in Teyler's Stichtung," *Archives Musée Teyler* 2 (1914): 26. See the outstanding historical study by Jozsef Illy, "Einstein Teaches Lorentz, Lorentz Teaches Einstein. Their Collaboration in General Relativity, 1913-1920," *Archive for History of Exact Sciences* 39 (1989): 272, who comments, "Lorentz distinguishes the existence of a state of motion from its observability—a distinction void of meaning according to Einstein."

3 H. A. Lorentz, "Considérations élémentaires sur le principe de relativité," *Revue générale des Sciences* 25 (1914): 179ff, in H. A. Lorentz, *Collected Papers*, ed. P. Zeeman and A. D. Fokker (The Hague: Martinus Nijhoff, 1934), 7: 165.

4 Albert Einstein, *Äether und Relativitätstheorie* (Berlin: Julius Springer Verlag, 1920), p. 7.

5 Albert Einstein, "The Problem of Space, Ether, and the Field in Physics" [*Mein Weltbild*, 1934], in *Ideas and Opinions*, trans. Sonja Bargmann (New York: Crown Publishers, 1954), p. 281. So also Tetu Hirosige, "Origins of Lorentz' Theory of Electrons and the Concept of the Electromagnetic Field," *Historical Studies in the Physical Sciences* 1 (1969): 197; idem, "The Ether Problem, the Mechanistic Worldview, and the Origins of the Theory of Relativity," *Historical Studies in the Physical Sciences* 7 (1976): 69.

6 Einstein, "Space, Ether, and the Field," p. 281. I am skeptical that Lorentz would have agreed that fields are physical states of space rather than that physical fields exist in space. Lorentz held to the classical conceptions of time and space.

Lorentz the aether just is the privileged spatial frame. He thus accepts a 3+1 ontology of spatial objects enduring through time, a metaphysic which is well-suited to an A-Theory of time.

Third, Lorentz distinguished true time, which was measured by a clock at rest in the aether frame, from merely local time and ascribed to events absolute simultaneity in true time. Writing in 1910, he contrasted his view with Einstein's:

> Assume there were an aether; then there would be among all systems *x, y, z, t* one singled out in that the coordinate axes as well as the clock is at rest in the aether. If one conjoins with this the idea...that space and time are something wholly different and that there is a 'true time' (simultaneity would then exist independently of location, in accord with the circumstance that it is possible for us to conceive of infinitely great velocities), then one easily sees that this true time would have to be indicated just by clocks which are at rest in the aether. If, then, the principle of relativity were generally valid in nature, then one would not be in a position to determine whether the coordinate system employed is that distinguished one. One thus comes to the same results as when one in agreement with Einstein and Minkowski denies the existence of the aether and the true time and treats all coordinate systems as equivalent. Which of the two modes of thought one may agree with is best left to the individual.[7]

Lorentz, realizing that his aether compensatory theory is empirically equivalent to the Einstein-Minkowski theory, leaves it up to the individual to choose which he shall adopt. But Lorentz preferred the classical conceptions of time and space on metaphysically intuitive grounds, as he made clear in his 1922 lectures at Cal Tech:

> All our theories help us form pictures, or images, of the world around us, and we try to do this in such a way that the phenomena may be coordinated as well as possible, and that we may see clearly the way in which they are connected. Now in forming these images we can use the notions of space and time that have always been familiar to us, and which I, for my part, consider as perfectly clear and, moreover, as distinct from one another. My notion of time is so definite that I clearly distinguish in my picture what is simultaneous and what is not.[8]

---

[7] H. A. Lorentz, "Alte und neue Fragen der Physik," *Physikalische Zeitschrift* 11 (1910): 1234ff, in *Collected Papers*, 7: 211. Notice that Lorentz eschews operational definitions of time and does not equate time with clock readings. This is particularly evident in his failure to ascribe at first physical reality to his local times. He recalled,

> "I had the idea that there was an essential difference between the systems *x, y z, t* and *x' y', z', t'*. In one one uses—such was my thought—coordinate axes which have a fixed position in the aether and which one can call the 'true' time; in the other system, on the other hand, it would be a matter of mere auxiliary quantities whose introduction is nothing but a mathematical artifice. In particular, the variable *t'* could not be called 'time' in the same sense as the variable *t*" (H. A. Lorentz, "Deux mémoires de Henri Poincaré sur la physique mathématique," *Acta mathematica* 38 [1914]: 293ff, in *Collected Papers*, 7: 262).

Eventually Lorentz came to see the reality of clock retardation. But even the readings of a clock at rest in the aether are but a sensible measure of true time, not definitive of true time. Illy comments, "When he says that 'we make the readings of...clocks match our fundamental conception of time' he implies that time is not to be identified with the mere reading of clocks, in the way that Einstein defined time in his relativistic paper of 1905; consequently measurements cannot be interpreted without some previous agreement upon fundamental concepts; measurements may not be used for defining time (and space)" (Illy, "Einstein Teaches Lorentz," p. 275).

[8] H. A. Lorentz, *Problems of Modern Physics*, ed. H. Bateman (Boston: Ginn, 1927), p. 221. Here Lorentz identifies himself as a physicist "of the old school" who says, "'I prefer the time that is measured

Here Lorentz quite rightly refuses to jettison the intuitively obvious reality of absolute simultaneity among events in the world just because one cannot determine which spatially separated events are simultaneous or because Einstein's operationally re-defined notion of simultaneity is relative to reference frames. Moreover, he sees no good reason to scrap the intuitive distinctness of space and time in favor of Minkowski's unified reality, spacetime. It seems to me that Lorentz was quite justified in sticking with his intuitions. In any case, the ease with which his approach can be assimilated with a tensed theory of time is obvious.

Finally, Lorentz himself hinted at a theological application of his ideas. In January of 1915 Lorentz, convinced that Einstein's work on GR was quite compatible with the notion of a privileged aether frame, wrote to Einstein in response to the latter's paper "The Formal Foundations of the General Theory of Relativity."[9] In a passage redolent of the General Scholium and *Opticks* of Newton, Lorentz broached considerations whereby "I cross the borderland of physics":

> A 'World Spirit' who, not being bound to a specific place, permeated the entire system under consideration or 'in whom' this system existed and who could 'feel' immediately all events would naturally distinguish at once one of the systems *U, U', etc.* above the others.[10]

Such a being could "directly verify simultaneity."[11] Here we see from Lorentz's own hand the integratability of his approach to relativity theory with a theological perspective such as Newton advocated.

I do not mean to suggest, of course, that we should return to Lorentz's own, defunct electron theory. As Zahar has shown,[12] that theory was doomed by the advent of quantum physics. But Lorentz's own attempted explanation of the phenomenon of length contraction is incidental to his basic approach. By 1908 Lorentz had already realized the incompatibility of his electron theory with Planck's quantum hypothesis, and by the 1911 Solvay Congress there was a general sense that the electron theory would have to be radically reformed in light of the advent of quantum physics.[13] Nonetheless, Lorentz continued to adhere to an approach to relativity theory which preserved the classical notions of space and time. A theory may be classified as Lorentzian just in case it affirms (i) the round trip vacuum propagation of light is isotropic in a preferred (absolute) reference frame R_0 (with

---

by a clock that is stationary in the ether, and I consider this as the true time, though I admit that I cannot make out which of the two times is the right one, that of A or that of B'" (Ibid.).

9 Albert Einstein, "Die formale Grundlage der allgemeinen Relativitätstheorie," *Physikalische Zeitschrift* 14 (1913): 1249-1262.

10 H. A. Lorentz to A. Einstein, January, 1915, Boerhaave Museum, cited in Illy, "Einstein Teaches Lorentz," p. 274. The text reads, "Ein 'Weltgeist,' der ohne an einem bestimmten Ort gebunden zu sein, das ganze betrachte System durchdränge, oder 'in dem' dieses System bestände, und der unmittelbar alle Ereignisse 'fühlen' könnte, würde natürlich sofort eins der Systeme *U, U'*, u.s.w. vor den anderen auszeichnen." See comment by A. J. Knox, "Hendrik Antoon Lorentz, the Ether, and the General Theory of Relativity," *Archive for History of Exact Sciences* 38 (1988): 75.

11 Lorentz to Einstein.

12 Elie Zahar, "Why Did Einstein's Programme Supersede Lorentz's? (I & II)," *British Journal for the Philosophy of Science* 24 [1973]: 95-123; 223-262.

13 See Russell McCormmach, "H. A. Lorentz and the Electromagnetic View of Nature," *Isis* 61 (1970): 486-488.

speed $c=1$) and independent of the velocity of the source, and (ii) lengths contract and time rates dilate in the customary special relativistic way only for systems in motion with respect to R_0.[14] In light of the reality of tense and temporal becoming, any metaphysically adequate theory of space and time must be Lorentzian in this sense. In order to avoid confusion with Lorentz's original electron theory, we may call such theories neo-Lorentzian.

Lucas and Hodgson emphasize that "although the Special Theory cannot single out any one of a set of frames of reference that are equivalent to one another under the Lorentz transformation as being at absolute rest, there is no reason why the Special Theory plus some other theory or some other consideration should not do so."[15] In fact they argue that considerations from both quantum theory and cosmology do imply the existence of a preferred frame. Metaphysical and theological considerations must also be allowed here.

EINSTEINIAN VS. NEO-LORENTZIAN RELATIVITY

Although in semi-popular expositions of SR scorn is usually poured upon a neo-Lorentzian interpretation, particularly the FitzGerald-Lorentz contraction, the fact is that a small minority of physicists, including such notable figures as H. E. Ives, Geoffery Builder, and S. J. Prokhovnik, have continued the Lorentzian research programme down to the present day, so that the correct physical interpretation of relativity theory remains a matter of debate.[16] In fact, that debate has intensified in recent years, due to what one participant has called a "sea change" in the foundations of physics community concerning the viability of a Bohmian interpretation of quantum mechanics, which plausibly requires absolute simultaneity and, hence, absolute time.[17] It is acknowledged on all hands that the neo-Lorentzian

---

14 See A. K. A. Maciel and J. Tiomno, "Analysis of Absolute Spacetime Lorentz Theories," *Foundations of Physics* 19 (1989): 507-508.

15 John R. Lucas and Peter E. Hodgson, *Spacetime and Electromagnetism* (Oxford: Clarendon Press, 1990), pp. 118-119. Cf. Lucas's earlier statement: "Although we need to say that so far as electromagnetism is concerned, all inertial frames are on a par, we should not be flying in the face of empirical facts if we were led by theological or metaphysical considerations or by considerations from some other part of physics to prefer one frame above all others" (J. R. Lucas, *The Future: an Essay on God, Temporality and Truth* [Oxford: Basil Blackwell, 1989], p. 219).

16 See the proceedings of the biennial conferences sponsored by the British Society for the Philosophy of Science, "Physical Interpretations of Relativity Theory," Imperial College of Science and Technology, London. It is noteworthy that in Zahar's analysis, SR did not by itself empirically supersede Lorentz's programme. Zahar can claim that Einstein's research programme superseded Lorentz's only by taking GR to be continuous with SR and by then appealing to the empirical confirmation of GR. (Elie Zahar, "Why Did Einstein's Programme Supersede Lorentz's? (I & II)." But as we shall see below, GR is definitely not continuous with SR; it is a theory of gravitation, not a theory of relativity. It therefore follows that Einstein's programme never in fact did supersede Lorentz's, in the sense of demonstrating its superiority over Lorentz's.

17 John Kennedy in a paper presented at the Central Division Meeting of the American Philosophical Association, April 23-26, 1997, Pittsburgh, Pennsylvania. On Bohm's interpretation see Sheldon Goldstein, "Review Essay: Bohmian Mechanics and the Quantum Revolution," *Synthèse* 107 (1996): 145-165; James T. Cushing, Arthur Fine, and Sheldon Goldstein, eds., *Bohmian Mechanics and Qunatum Theory: An Appraisal*, Boston Studies in the Philosophy of Science 184 (Dordrecht: Kluwer Academic

and the received, Einsteinian versions of SR are empirically equivalent, so that the decision between the two must be made on the basis of non-empirical considerations.

It is often asserted that the received version of the theory is simpler and therefore to be preferred. However, as is well-known, one must be very cautious about the connection between the simplicity of a theory and its *truth*, as opposed to its *utility*. What right do we have to infer that if the received version is simpler, it is therefore true, especially in light of the extraordinary metaphysical commitments it involves (no absolute space, no absolute time, but, plausibly, absolute spacetime)? Moreover, that simplicity is achieved by means of conventional definitions, which makes it difficult to regard the theory as being true rather than merely useful or expedient. One could also argue that counterbalancing the simplicity of the received version is the heuristic superiority of the neo-Lorentzian version.[18] Besides all this, the fact remains that the A-theorist has good reasons for believing that a neo-Lorentzian theory is correct, namely, the arguments for an A-Theory of time and the deficiencies of the relativity interpretation imply it, so that concerns about which version is simpler become of little moment.

In any case, the claim that the received theory is simpler is simply wrong. Although Lorentz's own theory was more complicated than Einstein's, H. E. Ives was able to derive the Lorentz transformation equations from (i) the laws of conservation of energy and momentum and (ii) the laws of transmission of radiant energy. He showed that there is an apparent discrepancy in the equations for a particle governed by these laws which demands that the particle's mass vary with velocity. He then derived from these variations of dimensions the Lorentz transformations. "The space and time concepts of Newton and Maxwell are retained without alteration," he wrote, "It is the dimensions of the material instruments for measuring space and time that change, not space and time that are distorted."[19] On Ives's achievement, Martin Ruderfer comments that he succeeded in elevating Lorentz's *ad hoc* theory to an equal status with SR and did so with the same number of basic assumptions as Einstein, so that his theory has the same "beauty." "The Ives and Einstein interpretations represent two different, but equally valid, views of

---

Publishers, 1996). Bohm's interpretation, in the words of James Cushing, demotes SR from "the status of an absolutely universal foundational theory and instead demands relativistic invariance only of the 'observational' content of a physical theory"—precisely a Lorentzian approach (James T. Cushing, *Philosophical Concepts in Physics* [Cambridge: Cambridge University Press, 1998], p. 337). It is instructive that was precisely the same positivist epistemology that underlay Einstein's SR that prevented previous generations' seriously considering Bohm's interpretation, an obstacle that has now collapsed (Mara Beller, "Bohm and the 'Inevitability' of Acausality," in *Bohmian Mechanics and Quantum Theory: An Appraisal*, ed. James T. Chushing, Arthur Fine, and Sheldon Goldstein, Boston Studies in the Philosophy of Science 184 [Dordrecht: Kluwer Academic Publishers, 1996], pp. 211-221).

18 See J. S. Bell, "How to Teach Special Relativity," in *Speakable and Unspeakable in Quantum Mechanics* (Cambridge: Cambridge University Press, 1987), pp. 66-80; see also S. J. Prokhovnik, "An Introduction to the Neo-Lorentzian Relativity of Builder," *Speculations in Science and Technology* 2 (1979): 226; G. Builder, "Ether and Relativity," *Australian Journal of Physics* 11 (1958): 279-297, reprinted in *Speculations in Science and Technology* 2 (1979): 239-241.

19 Hubert E. Ives, "Derivation of the Lorentz Transformations," *Philosophical Magazine* 36 (1945): 392-401, reprinted in *Speculations in Science and Technology* 2 (1979): 247, 255.

the same set of observations."[20] Hence, assertions that Einstein's version is simpler than a neo-Lorentzian theory are incorrect.

Neo-Lorentzian theorists, on the other hand, have complained that the received theory leaves us with physical effects without any causal explanation for them. Thus, for example, Builder asserts,

> The relative retardation of clocks, predicted by the restricted theory of relativity, demands our recognition of the causal significance of absolute velocities.... The observable effects of absolute accelerations and of absolute velocities must be ascribed to the interaction of bodies and physical systems with some absolute inertial system. We have no alternative but to identify this absolute system with the universe.[21]

Builder argues that the universe as a whole constitutes a unique, absolute inertial system, which he calls "the ether," that affects the masses and dimensions of bodies, as well as the running of clocks, in accordance with their absolute speed. He contrasts this interpretation with the received view:

> The conceptual difficulties associated with the restricted theory all arise out of the denial that these absolute concepts are permissible, and out of the consequent attempts to avoid them in the presentation of the theory. It is frequently maintained that the theory has forced us to discard entirely the old-fashioned commonsense notions of time and space, but nothing comprehensible or definable has been offered in their place. Moreover, any questions as to what causes the relativity of simultaneity, the measured constancy of the velocity of light in all inertial reference systems, or the reciprocity of relativistic variations of length, of mass, and of clock rates, are avoided by vague references to the principle of relativity, to the four-dimensional character of spacetime, and so on.
>
> On the other hand, the presentation of the restricted theory in terms of the absolute concepts (following generally the lines of its development by Poincaré and Lorentz) involves no conceptual difficulties. The relativity of simultaneity, the reciprocity of relativistic variations, and the constancy of the measured velocity of light, then all appear simply as comprehensible effects of the motions, relative to the ether, of the bodies of observers and of the measuring instruments used.[22]

Developing Builder's approach, Prokhovnik interprets length contraction and time dilation as retarded potential effects. He explains that Builder's notion of a fundamental inertial frame I(F) implies that energy propagation is strictly isotropic with respect to this frame. A body moving with respect to this frame will drag its gravitational field with it. The theory of retarded potentials applies to the gravitational or electromagnetic field of a body moving relative to I(F). For such a body energy propagation is not isotropic and its co-moving fields are no longer symmetric, which leads to a whole chain of interacting anisotropy effects. If the field of a moving electron is compressed in the direction of motion, for example, then its surface must be similarly compressed to maintain an equipotential equilibrium state, and the moving electron takes the shape of an ellipsoid. A system of particles would contract in the direction of motion in order to sustain the interparticle equilibrium of the system. No substantial aether is required for the

---

20 Martin Ruderfer, "Introduction to Ives' 'Derivation of the Lorentz Transformations'," *Speculations in Science and Technology* 2 (1979): 243.

21 Builder, "Ether and Relativity," p. 230.

22 Ibid., p. 240.

production of such effects; simply the existence of I(F) itself suffices, as Prokhovnik explains,

> Fitzgerald and Lorentz considered the contraction as an absolute effect due to movement relative to a unique reference frame associated with an aether. It is clear that the contraction can be even more satisfactorily considered as a secondary effect resulting from movement relative to a unique cosmologically-based fundamental reference frame, I(F), defined in terms of our greatly expanded view of the universe since 1930. The retarded potential field effect is a primary consequence of movement relative to I(F) and, as seen in the basis and derivation of this effect, it emerges essentially in consequence of the limiting velocity restriction on the transmission of energy; no aether or any other property of 'physical space' is required for the emergence of this effect.[23]

Length contraction can be considered as the reaction of a moving system to maintain its stationary-in-I(F) equilibrium state in the circumstance of an asymmetric gravitational field resulting from its motion in I(F). This phenomenon is thus as universal and general as the gravitational property of matter.

Following Builder, Prokhovnik sees time dilation as the consequence of the fact that a moving light clock has a longer unit of time and hence runs more slowly than a similar, stationary clock in I(F) (Figure 5.1).

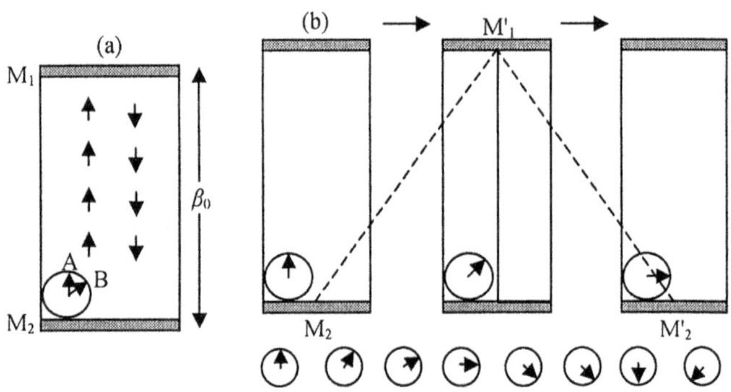

Figure 5.1. The light-pulse clock. (a) When at rest a light pulse is reflected back and forth from the mirrors M_1 and M_2; each return of the pulse to one end counts as a unit of time. The dial hand moves from A to B in that time. (b) When the clock is in motion the light travels the distance $M_2M'_1M'_2$ between ticks, so causing it to go slow. The dials underneath the moving clock are those stationary clocks distributed in space and used for comparison with the moving clock times.

This effect can also be generalized, according to Prokhovnik: "The effect will apply to all phenomena involving electromagnetic impulses and energy-exchanges; so it can be considered to apply to all clocks and physical phenomena and, by extension,

[23] S. J. Prokhovnik, *Light in Einstein's Universe* (Dordrecht, Holland: D. Reidel, 1985), pp. 84-85.

also to biological phenomena."[24] Thus, clock retardation, the Twin Paradox, and asymmetrical aging become comprehensible phenomena.

The adequacy of such causal explanations is not the primary point here. Rather the salient point is that a neo-Lorentzian theory involves appeal to some sort of causal explanation for the physical effects predicted by SR. By contrast, in the received theory, these effects simply follow from Einstein's denial of the existence of a fundamental frame and his two postulates. Given his postulates, a new kinematics follows according to which reciprocal length contraction and clock retardation occur, but without any dynamical explanation.

As a theory of principle rather than a constructive theory, Einstein's SR is based on postulates which are characterized by their very non-empirical character. As Holton and Goldberg have emphasized, although Einstein insisted early on that his SR was empirically grounded, in fact the two postulates on which the theory is based "were postulates for which there was and can be no direct empirical confirmation"; their status is that of "a non-verifiable and non-falsifiable presupposition."[25] The only version of SR which is experimentally verifiable, according to Builder, "is the theory that the spatial and temporal coordinates of events, measured in any one inertial reference system, are related to the spatial and temporal coordinates of the same events, as measured in any other inertial reference system, by the Lorentz transformations."[26] But this verifiable statement only concerns measurements made in inertial reference systems and is neutral with regard to the Lorentzian and Einsteinian interpretations. As a constructive theory, the neo-Lorentzian approach promises to enrich our understanding of the causal structure of the world in a way that Einstein's cannot.

Now, as we have seen, the spacetime interpretation of the received theory does furnish explanations of time dilation and length contraction in terms of the spacetime distances involved. Therefore, on a spacetime interpretation Einstein's SR does explain why these relativistic phenomena occur. But here it needs to be seriously called into question whether any such metaphysical reality as spacetime actually exists. In the first place, while a spacetime interpretation of SR is arguably superior to a relativity interpretation of SR[27], there are no compelling reasons to prefer a spacetime ontology over a classical ontology of space and time as the latter

---

24 Ibid., p. 89. See also S. J. Prokhovnik, "The Logic of the Clock Paradox," *Physical Interpretations of Relativity Theory I*, Proceedings of the International Conference of the British Society for the Philosophy of Science, September 16-19, 1988, London.

25 See Gerald Holton, "Where Is Reality? The Answers of Einstein," in *Science and Synthesis*, ed. UNESCO (Berlin: Springer Verlag, 1971), pp. 52-64; idem, "Mach, Einstein, and the Search for Reality," in *Ernst Mach: Physicist and Philosopher*, ed. Robert S. Cohen and Raymond J. Seeger, Boston Studies in the Philosophy of Science 6 (Dordrecht: D. Reidel, 1970), p. 178; idem, "Poincaré and Relativity," in *Thematic Origins of Scientific Thought: Kepler to Einstein* (Cambridge, Mass.: Harvard University Press, 1973), p. 190; see also Stanley Goldberg, *Understanding Relativity: Origin and Impact of a Scientific Revolution* (Boston: Birkhäuser, 1984), p. 105; idem, "Putting New Wine in Old Bottles: The Assimilation of Relativity in America," in *The Comparative Reception of Relativity*, ed. Thomas F. Glick, Boston Studies in the Philosophy of Science 103 (Dordrecht: D. Reidel, 1987), pp. 15-21.

26 G. Builder, "The Constancy of the Velocity of Light," *Australian Journal of Physics* 11 (1958): 457-480; reprinted in *Speculations in Science and Technology* 2 (1971): 422.

27 Recall our discussion in chap. 1.

comes to expression in a neo-Lorentzian theory of relativity. Spacetime realists are self-confessed, post-positivistic metaphysicians, and so cannot complain that the classical concepts of space and time are obnoxiously metaphysical. As Earman and Friedman write, "Newton's theory, and especially his conceptions of space and time, have often been criticized on operational and positivistic grounds; but it is rarely pointed out that relativity theory, which is often taken as vindicating Newton's philosophical critics, is not less subject to the same criticisms...."[28] When spacetime realists champion a spacetime ontology over a space and time ontology, virtually all their arguments are aimed at demonstrating the superiority of the spacetime interpretation of SR over the relativity interpretation of that theory. The Lorentz-Poincaré approach is simply passed over in benign neglect, under the assumption that any such theory is no longer a viable option. Spacetime realists are all of them Einsteinians and so take no notice of contemporary neo-Lorentzians. The irony of this situation is that the self-restriction of spacetime realists to interpretations of Einsteinian relativity theory is rooted, as we have seen, in the very positivism and verificationism which would be inimical to their own approach to questions of space and time. Having freed themselves from the metaphysical straitjacket of positivism, spacetime realists now need to ask why they should continue to philosophize only within the familiar walls of Einsteinian relativity and should not rather begin to explore alternatives in previously proscribed realms.

Occasionally arguments are offered on behalf of a spacetime interpretation of relativity theory which are relevant to a neo-Lorentzian approach, and these merit our attention. For example, oftentimes space and time theorists have accused spacetime realists of a gratuitous hypostatization of what is only a geometrical representation of space and time. There is no reason to think that because a thing's spatio-temporal location can be plotted on a Minkowski diagram, therefore an entity *spacetime* exists anymore than because temperature and pressure can be similarly related in a diagrammatic way, therefore an entity *temperaturepressure* exists. In response to this charge d'Abro points to the four-dimensional metric of Minkowski spacetime as its distinguishing feature, in virtue of which it is not like the artificial temperaturepressure continuum.[29] The temperaturepressure continuum can be separated into two distinct elements, but Minkowski spacetime cannot.

> In short, the reality of spacetime arises from Minkowski's discovery that it was possible to discover an invariant distance between two points in spacetime, holding for all observers; and that it was impossible to define any such invariant distance in space alone or in time alone, showing that space and time by themselves were phantoms. These last two concepts must henceforth be considered jointly, and no longer as separate entities.[30]

One can, of course, adopt a spacetime approach to Newtonian theory, as Friedman has done. But d'Abro's point is that there is no special four-dimensional metric for

---

[28] John Earman and Michael Friedman, "The Meaning and Status of Newton's Law of Inertia and the Nature of Gravitational Forces," *Philosophy of Science* 40 (1973): 341.

[29] A. d'Abro, *The Evolution of Scientific Thought*, 2d rev. ed. (1927; rep. ed.: n.p.: Dover Publications, 1950), p. 347.

[30] Ibid., p. 447.

such a continuum; it reduces to a mere geometric representation of the Galilean transformations.[31] Like the temperaturepressure continuum, no real unification takes place.

This is an odd argument on behalf of realism, since it treats Minkowski spacetime preferentially over Newtonian spacetime. The would-be Newtonian spacetime realist is dismissed along with the temperaturepressure realist, even though he holds to a spacetime realism, too. His spacetime is a mathematical fiction, we are told, whereas Minkowski spacetime is not, since the latter alone has a four-dimensional metric. But the reason why it has such a four-dimensional metric is precisely because of its preclusion of relations of absolute simultaneity. In the limit case where $c \rightarrow \infty$, the elsewhere region in spacetime is squeezed out and the Newtonian dichotomy between a universal, absolute past and future is recovered.[32] Thus the very existence of a spacetime metric and a lightcone structure in spacetime depends upon there not existing such a global, absolute separation of past and future. The Minkowskian spacetime realist can hardly justify relativistic spacetime over the Newtonian realists' classical spacetime on the grounds that only the former's has a spacetime metric, since it is a necessary condition of relations of absolute simultaneity (which are the distinguishing mark of classical spacetime) that no such metric exist in Newtonian spacetime. But if the presence of such a metric serves merely to differentiate Minkowskian from Newtonian spacetime, rather than to justify realism about the former over realism about the latter, then neither can it justify Minkowskian spacetime realism over neo-Lorentzian space and time realism, since the latter does not differ relevantly from Newtonian spacetime realism.

Probably at the root of many physicists' rejection of a neo-Lorentzian approach to relativity theory is the deep-seated conviction that comes to expression in Einstein's aphorism: "Subtle is the Lord, but malicious He is not."[33] That is to say, if there exists a fundamental asymmetry in nature, then nature will not conspire to conceal it from us by precisely countervailing effects. D'Abro expresses this sentiment clearly:

> If Nature was blind, by what marvelous coincidence had all things been so adjusted as to conceal a velocity through the ether? And if Nature was wise, she had surely other subjects to attend to, more worthy of her consideration, and would scarcely be interested in hampering our feeble attempts to philosophize. In Lorentz's theory, Nature, when we read into her system all these extraordinary adjustments *ad hoc*, is made to appear mischievous; it was exceedingly difficult to reconcile one's self to finding such human traits in the universal plan.[34]

---

[31] Ibid., pp. 472-473.

[32] Olivier Costa de Beauregard, *Time, the Physical Magnitude*, Boston Studies in the Philosophy of Science 99 (Dordrecht: D. Reidel, 1987), pp. 59-60. Cf. Lucas and Hodgson, *Spacetime and Electromagnetism*, pp. 228-238.

[33] "*Raffiniert ist der Herr Gott, aber boshaft ist er nicht*," remark of Albert Einstein during a visit to Princeton, upon being informed that D. C. Miller, a former colleague of Michelson, had claimed to have detected the ether wind (cited in Abraham Pais, '*Subtle is the Lord...*': *The Science and Life of Albert Einstein* [Oxford: Oxford University Press, 1982], pp. 113-114).

[34] D'Abro, *Evolution of Scientific Thought*, p. 138.

One difficulty with this objection is that it seems to be guilty of greatly over-exaggerating the extent of the alleged conspiracy. After all, SR is a restricted theory of relativity: it is only uniform motion relative to the privileged frame that fails to manifest itself. But in all other cases of motion, the absolute character of that motion is disclosed. This is not to say that acceleration or rotation proves the reality of privileged space, but it is to say that, given the classical concepts of time and space, nature does not at all conspire to conceal either absolute motion or the privileged space from us. Moreover, there are modern equivalents of the classical aether, such as the microwave background radiation or the quantum mechanical vacuum, which serve to disclose a privileged frame. Indeed, when Einsteinians complain that no evidence of a privileged space and time exist, one wonders what it would take to convince them of the contrary. If no empirical evidence of a fundamental frame is incapable of being explained away, then the supposed failure of nature to disclose such a frame to us becomes trivial. The more difficult it is for nature to provide evidence of the existence of a privileged frame, the less compelling the charge that she is conspiring against us to conceal it.

But even considered in abstraction from these wider considerations, why accept the assumption that fundamental asymmetries in nature must disclose themselves to us? This assumption is by no means obvious, as Martin Carrier explains:

> Science would be an easy matter if the fundamental states of nature expressed themselves candidly and frankly in experience. In that case we could simply collect the truths lying ready before our eyes. In fact, however, nature is more reserved and shy, and its fundamental states often appear in masquerade. Put less metaphorically, there is no straightforward one-to-one correspondence between a theoretical and an empirical state. One of the reasons for the lack of such a tight connection is that distortions may enter into the relation between theory and evidence, and these distortions may alter the empirical manifestation of a theoretical state. As a result, it is in general a nontrivial task to excavate the underlying state from distorted evidence.[35]

On a neo-Lorentzian theory, Carrier's general remarks on distortions of a theoretical state in its empirical manifestation are quite literally true, for the result of uniform motion relative to privileged space is length contraction and clock retardation, phenomena which it will be recalled, are every bit as real under the Einsteinian theory.[36] If it is in general difficult to excavate the underlying state of nature from distorted evidence, if nature's fundamental states often appear in masquerade, then

---

[35] Martin Carrier, "Physical Force or Geometrical Curvature?" in *Philosophical Problems of the Internal and External Worlds*, ed. John Earman, Allen I. Janis, Gerald J. Massey, and Nicholas Rescher (Pittsburgh: University of Pittsburgh Press, 1993), p. 3. Tim Maudlin has emphasized and illustrated Carrier's point in his *Quantum Non-Locality and Relativity*, Aristotelian Society Series 13 (Oxford: Blackwell, 1994). He surveys what he characterizes as the "teratological collection" of theories attempting to explain the Bell Inequalities and integrate the EPR results with relativity theory and concludes, "One way or another God has played us a nasty trick" (Ibid., p. 241). One cannot dismiss neo-Lorentzian relativity on the grounds that it would be deceptive, since partisans of each theory could say the same of rival positions. "…the real challenge falls to the theologians of physics, who must justify the ways of a Deity who is, if not evil, at least extremely mischievous" (Ibid., p. 242).

[36] It is perhaps noteworthy that d'Abro holds that "The Fitzgerald contraction is no longer a real physical contraction, as it was assumed to be in Lorentz's theory" (d'Abro, *Evolution of Scientific Thought*, p. 151).

why are the relativistic phenomena which mask the privileged frame improbable on a classical ontology?[37] One is tempted to quote in this connection another of Einstein's aphorisms: "God does not wear His heart on His sleeve."[38]

As for d'Abro's concern with finding "human traits in the universal plan," the neo-Lorentzian might plausibly appeal to the Anthropic Principle in response.[39] According to that principle, features of the universe can only be judged in their correct perspective when due allowance has been made for the fact that certain features of the universe are necessary if it is to contain observers like ourselves.[40] Since our existence depends on the maintenance of equilibrium states within us, the FitzGerald-Lorentz contraction and clock retardation are necessary pre-requisites of our existence as observers. Thus, nature's alleged conspiracy, when seen in anthropic perspective, seems much less mischievous. Unless it can be shown that length contraction and clock retardation are extremely improbable (and thus evidence of the universe's fine-tuning for intelligent life), then we should hardly be surprised at nature's "conspiracy." In any case, given Newton's theistic perspective, even fine-tuning should not surprise us. Given that God designed the universe to contain intelligent beings like us, we should expect that He will have chosen laws of nature that serve to maintain the equilibrium essential to our existence. Even if, as d'Abro says, Nature is blind, God is not; and if Nature is not wise, God is. It is not Nature, then, who is concerned with our feeble selves, who deems us worthy subjects to attend to, but the Creator and Sustainer of the universe who is mindful of man. *Raffiniert ist der Herr Gott, barmherzig ist er auch.*

Earman and Friedman, in defense of spacetime realism over the classical ontology, have asserted that "neglect of spacetime structure in the classical context has led to a number of philosophical errors and oversights."[41] Presumably, they have reference, for example, to classical theorists' overlooking Galilean spacetime, which preserves absolute time without absolute space. But such examples serve to show only the heuristic utility of a spacetime methodological approach to questions of space and time. In the same way that possible worlds semantics is an illuminating and perhaps indispensable tool for exploring questions of modality but does not commit one *ipso facto* to modal realism, so also spacetime methodology

---

37 Tooley argues that the received view posits a conspiracy of nature of its own to conceal from us the one-way velocity of light.

> "a conspiracy of silence with respect to absolute rest and motion can obtain only if, and precisely because, there is a conspiracy within nature with respect to the measurement of the one-way speed of light. So any conspiracy of silence within nature that one must countenance, on the modified [neo-Lorentzian] theory, is either identical with, or derives from, a conspiracy of silence that one must countenance according to any version of the Special Theory of Relativity...given the supposition that there is no experimental way of determining the one-way speed of light" (Michael Tooley, *Time, Tense, and Causation* [Oxford: Clarendon Press, 1997], p. 367).

38 Cited in T. F. Torrance, *Transformation and Convergence* (Grand Rapids: Eerdmans, 1984): 253.

39 I owe this point to Robin Collins.

40 John Barrow, *et al.*, *The Anthropic Cosmological Principle* (Oxford: Clarendon Press, 1986), p. 15.

41 Earman and Friedman, "Meaning and Status," p. 329; cf. John Earman, "Spacetime, or How to Solve Philosophical Problems and Dissolve Philosophical Muddles without Really Trying," *Journal of Philosophy* 67 (1970): 259-277.

can enlighten without implying a spacetime ontology. The same point may be made concerning the avoidance of "philosophical errors"—though here one must add that a spacetime approach might also lead one to commit philosophical errors which a space and time approach would have avoided. For example, attempts to relativize acceleration and rotation by appeal to four-dimensional structures come to mind, representations which may be of heuristic value but have also led some thinkers to conclude erroneously that absolute motions such as Newton pointed to have been eliminated. Or again, due to the isomorphism of the space and time dimensions in spacetime, which differ only in sign, erroneous philosophical conclusions about the interchangeability of time and space, as in the Schwarzschild metric or the Hartle-Hawking cosmology, might be drawn, which a space and time approach would perhaps prevent. To return to "oversights", the neo-Lorentzian will charge that a perspectival, four-dimensional approach to relativistic phenomena has resulted in the neglect of the search for the dynamic causes of time dilation and length contraction, a major oversight. The obvious difficulty here is that one man's insights are another man's oversights, and it is difficult to tell which one is correct without independent justification of one point of view.

Earman elsewhere argues that even if spacetime can be ignored in the context of classical physics, one cannot neglect it in the context of relativistic physics, since spacetime is not uniquely separable into space and time.[42] In fact, he points out, there are relativistic spacetimes which resist every effort to separate them into a three-dimensional space enduring through a one-dimensional time. But Earman is speaking here of GR spacetimes, not SR spacetime; we may reserve a discussion of that subject for elsewhere.[43] So far as SR is concerned, Einstein's original formulation of the theory shows that a space and time approach is necessarily equivalent to a spacetime approach. Moreover, one cannot automatically assume that just because a particular spacetime can be modeled that such a spacetime represents a realistic possibility (think, for example, of imaginary spacetimes). One might well question whether models such as Gödel's, alluded to by Earman, which permit pathological results like closed timelike loops, are anything more than mathematical curiosities. In any case, we must not forget that relativity theory concerns physical time only, not metaphysical time. In a universe such as Gödel's the impossibility of delineating a one-parameter family of spacelike hypersurfaces covering spacetime would only show that there exists no physical measure of metaphysical time in such a universe. There would be an absolute present and absolute simultaneity, even if it were physically impossible to gather present events into a single hypersurface. Hence, the possibility of such relativistic spacetimes does not suffice to establish the ontological reality of spacetime.

It has been suggested that a neo-Lorentzian interpretation of SR is disfavored because of the excessive spacetime structure it posits with respect to the laws of motion.[44] Recognizing the failure of verificationist critiques of absolute space on

---

42 Earman, "Spacetime," pp. 259-260; cf. Earman and Friedman, "Meaning and Status," p. 352.

43 See my *Time and the Metaphysics of Relativity* (Dordrecht: Kluwer Academic Publishers, forthcoming).

44 This has been suggested to me by Yuri Balashov.

the basis of its unobservability, Earman, for example, suggests that "this objection of unobservability is more accurately stated as an objection based on Occam's razor."[45] What Earman has in mind are criteria which would serve to establish what sort of structure spacetime is endowed with. He proposes two "symmetry principles" which he presents as conditions of adequacy on theories of motion:

> SP1: Any dynamical symmetry of a theory T is a space-time symmetry of T.
>
> SP2: Any space-time symmetry of a theory T is a dynamical symmetry of T.

What justification is there for these criteria? Earman explains, "Behind both principles lies the realization that laws of motion cannot be written on thin air alone, but require the support of various space-time structures. The symmetry principles then provide standards for judging when the laws and the space-time structure are appropriate to one another."[46] With respect to (SP1), which will be crucial in the case at hand, Earman writes,

> The motivation for (SP1) derives from combining a particular conception of the main function of laws of motion with an argument that makes use of Occam's razor. Laws of motion, at least in so far as they relate to particles, serve to pick out a class of allowable or dynamically possible trajectories. If (SP1) fails, the same set of trajectories can be picked out by the laws working in the setting of a weaker space-time structure. The theory that fails (SP1) is thus using more space-time structure than is needed to support the laws, and slicing away this superfluous structure serves to restore (SP1).[47]

A prime example of (SP1) at work is the demonstration that Newtonian mechanics does not, in fact, require Newtonian spacetime, but only Galilean spacetime. While the spacetime symmetries of the theory are Newtonian, the dynamic symmetries are Galilean, a "clear violation" of (SP1).[48] By excising absolute space, one produces a theory whose spacetime symmetries and dynamic symmetries are both Galilean, in compliance with (SP1). Now it might similarly be argued—though Earman does not do so—that a neo-Lorentzian theory also violates (SP1) in that it posits more structure to spacetime, such as hyperplanes of absolute simultaneity, than is necessary to explain the symmetries required by relativistic laws of motion, that is, the symmetries of the Lorentz-Poincaré group. Therefore, such structures should be excised by Ockham's razor.

Now Earman is undoubtedly on to something important here. One should like to have some sort of constraint upon the postulation of gratuitous spacetime structures. But the difficulty with (SP1) and (SP2) is that they are too restricted, or to put the point another way, they can be overridden by considerations broader than the laws of motion. Indeed, one of the central contentions of our present study is that questions of time and space are metaphysical in character, so that such

---

45 John Earman, *World Enough and Space-Time* (Cambridge, Mass.: MIT Press, 1989), p. 48.
46 Ibid., p. 486.
47 Ibid., pp. 46-47.
48 Ibid., p. 48.

considerations cannot be ignored. As we have seen, the foundations of Newtonian time and space were laid in Newton's theism, as Earman acknowledges, but Earman simply ignores such theological considerations, choosing instead to put a "modern gloss" on Newton's views.[49] Newton's postulation of absolute space was not gratuitous but based, rightly or wrongly, in his conception of divine omnipresence. I think that Newton's conception of God's immensity was flawed; but it was not irrelevant, and I have argued at length elsewhere that his position on divine temporality and, hence, absolute simultaneity, is eminently defensible.[50] Such metaphysical considerations qualify any force which Earman's symmetry principles might possess.

In any case, it needs to be understood that Earman's principles *presuppose* a spacetime ontology and therefore cannot be employed to justify a spacetime interpretation of SR over a neo-Lorentzian space and time interpretation. Earman employs a spacetime approach to all the theories he considers, speaking not only of Newtonian spacetime but even of Machian spacetime. This involves him in frequent anachronisms: he speaks of Newton's "insistence that the structure of space-time is immutable;" he says that "If we take Newton at his word in the Scholium, the space-time setting for the theory of motion and gravitation of the *Principia* is supposed to be a full Newtonian space-time...;" he explains that for Newton "the space-time is given once and for all as an emanative effect of God...."[51] Of course, Newton himself insisted on and believed none of this, since his was a 3+1 ontology involving physical objects enduring through time. His laws of motion are not supported by various spacetime structures, as Earman assumes, for there are on Newton's view no such structures. Time and space have intrinsic structure, and motion is to be explained, not in terms of the geometrical structure of spacetime, but in terms of forces acting on bodies in space enduring through time. Earman's principles, then, take for granted a spacetime realism. At best, therefore, with respect to SR, Earman's symmetry principles could only serve to justify a Minkowskian spacetime realism over a neo-Lorentzian spacetime realism. But if only a single slice of spacetime actually exists, then the question becomes whether the relativity interpretation is preferable to the neo-Lorentzian interpretation. And neo-Lorentzians and spacetime realists concur that the relativity interpretation is deficient. In short, one's ontology may warrant postulating more structure to space and time than symmetry principles alone would allow.

GR raises a further ostensible advantage of spacetime realism: in GR gravitation is understood in terms of the geometry of spacetime. As Nerlich urges, "...not only is spacetime curvature the fundamental explanatory concept of the theory, but the idea of spacetime geometry is actually used to reduce causal explanation by gravitational force in space during time."[52] Nerlich gives the example of orbital planetary motion, which on the space and time ontology, would have to be regarded as motion of an object under a gravitational force. But on a spacetime ontology we

---

49 Ibid., p. 10.

50 See my *God, Time, and Eternity* (Dordrecht: Kluwer Academic Publishers, forthcoming).

51 Ibid., pp. 35-36, 45, 48.

52 Graham Nerlich, *What Spacetime Explains* (Cambridge: Cambridge University Press, 1994), p. 181.

can still apply Newton's first law to it in spacetime, seeing the object's trajectory as the geodesic path of a force-free body, thus reducing gravitation to spacetime curvature.

The question raised by the geometrization of gravitation in GR is whether this is to be understood instrumentally or realistically. For what it is worth, most physicists are apparently content to take the theory instrumentally, interpreting spacetime curvature as a geometrical representation of gravitation. According to Arthur Fine, few working, knowledgeable scientists give credence to the realist existence claims for entities like the four-dimensional manifold and associated tensor fields in GR; rather GR is seen as "a magnificent organizing tool" for dealing with certain gravitational problems in astrophysics and cosmology: "most who actually use it think of the theory as a powerful instrument, rather than as expressing a 'big truth'."[53] I think we can safely say that no disadvantage arises from treating the geometrization of gravitation as a heuristic device only.

Moreover, the fact is that a gravitational field cannot be analyzed as identical to inertial forces, as the Principle of Equivalence, which lies at the foundation of GR, asserts. Their equivalence is strictly local, as Roger Angel explains,

> It does not follow that any real gravitational field can be duplicated by a field of inertia and therefore 'transformed away'. In other words, it would not be possible to duplicate any arbitrary gravitational field by an appropriate acceleration of a frame of reference. For example, the gravitational field of the earth is radial and falls off with the square of the distance. Such a field could not be produced in its entirety by an accelerating frame.... In short, the equivalence of gravitational and inertial fields is only a *local* one. From an ontological standpoint, this restriction of the principle of equivalence is important. However, since the fundamental laws of general relativity are partial differential field equations, they tell us what is taking place in the neighborhood of a point. Accordingly, the local equivalence of gravity and inertia is still of great methodological significance.[54]

---

[53] Arthur Fine, *The Shaky Game: Einstein, Realism, and the Quantum Theory* (Chicago: University of Chicago Press, 1986), p. 123. See the statement of Steven Weinberg, *Gravitation and Cosmology: Principles and Explications of the General Theory of Relativity* (New York: John Wiley & Sons, 1972), p. 147:

> "...the nonvanishing of the tensor $R_{\lambda\mu\nu\kappa}$ is the true expression of the presence of a gravitational field.... Riemann...introduced the curvature tensor $R_{\lambda\mu\nu\kappa}$ to generalize the concept of curvature to three or more dimensions. It is therefore not surprising that Einstein and his successors have regarded the effects of a gravitational field as producing a change in the geometry of space and time. At one time it was even hoped that the rest of physics could be brought into a geometric formulation, but this hope has met with disappointment, and the geometric interpretation of the theory of gravitation has dwindled to a mere analogy, which lingers in our language in terms like 'metric,' 'affine connection,' and 'curvature,' but is not otherwise very useful. The important thing is to be able to make predictions about the images on the astronomers' photographic plates, frequencies of spectral lines, and so on, and it simply doesn't matter whether we ascribe these predictions to the physical effect of gravitational fields on the motions of planets and photons or to a curvature of space and time."

[54] Roger B. Angel, *Relativity: The Theory and Its Philosophy*, Foundations and Philosophy of Science and Technology (Oxford: Pergamon Press, 1980), p. 184; cf. pp. 206, 248-249, where he argues that gravitational time dilation should not be understood in terms of geometry, but of the effect of the physical field on the behavior of clocks.

Since our concern is precisely with ontology rather than methodology, the failure of the Principle of Equivalence as an ontological principle calls into question the spacetime realist's reduction of gravitational fields to pure geometry.

Finally, it can be argued that a realist interpretation of spacetime actually obscures our physical understanding of nature by substituting geometry for a physical force, thereby impeding progress in connecting gravitational theory to particle physics. In the preface to his text *Gravitation and Cosmology* Steven Weinberg reflects,

> There was another, more personal reason for my writing this book. In learning general relativity, and then in teaching it to classes at Berkeley and M.I.T., I became dissatisfied with what seemed to be the usual approach to the subject. I found that in most textbooks geometric ideas were given a starring role, so that a student who asked why the gravitational field is represented by a metric tensor, or why freely falling particles move on geodesics, or why the field equations are generally covariant would come away with an impression that this had something to do with the fact that spacetime is a Riemannian manifold.
>
> Of course, this was Einstein's point of view, and his preeminent genius necessarily shapes our understanding of the theory he created. However, I believe that the geometrical approach has driven a wedge between general relativity and the theory of elementary particles. As long as it could be hoped, as Einstein did hope, that matter would eventually be understood in geometrical terms, it made sense to give Riemannian geometry a primary role in describing the theory of gravitation. But now the passage of time has taught us not to expect that the strong, weak, and electromagnetic interactions can be understood in geometrical terms, and too great an emphasis on geometry can only obscure the deep connections between gravitation and the rest of physics.[55]

Weinberg believes that the theory of gravitational radiation provides "a crucial link" between GR and the microscopic frontier of physics, since radiative solutions of Einstein's equations lead to the notion of a particle of gravitational radiation, the so-called graviton.[56] The geometrical approach of spacetime realism only impedes our gaining a more integral understanding of physics. Riemannian geometry, in Weinberg's view, should be understood "only as a mathematical tool" and "not as a fundamental basis for the theory of gravitation."[57]

The quest for a unified field theory underlines the need for construing gravitation in terms of a physical force, rather than as a geometrical structure. Carlo Rovelli observes that Einstein's identification of the gravitational field and spacetime geometry can be interpreted either as the claim that the gravitational field is nothing but a distortion of spacetime geometry or as the claim that spacetime geometry is nothing but a manifestation of a particular physical field and that while the choice between these two "is a matter of taste," Rovelli himself believes that the former view "tends to obscure, rather than enlighten, the profound shift in the view of spacetime produced by general relativity."[58] In Rovelli's view the search for a

---

[55] Weinberg, *Gravitation and Cosmology*, p. vii.

[56] Ibid., p. 251.

[57] Ibid., p. viii.

[58] Carlo Rovelli, "Halfway through the Woods: Contemporary Research on Space and Time," in *The Cosmos of Science*, ed. J. Norton and J. Earman (Pittsburgh: University of Pittsburgh Press, 1998), pp. 193-194.

quantum theory of gravity, whether in the form of string theory or nonperturbative quantum gravity, will be abetted by interpreting gravitation as a dynamic field, akin to electromagnetic and other fields. He concludes, "The recent efforts to construct a quantum theory of gravity, or a unified theory of all interactions (for instance, via string theory), or a thermodynamics of the gravitational field, and other efforts, suggest that the second point of view, namely viewing the metric/gravitational field as a field like any other, is the fruitful one."[59]

In sum, while we may agree that a spacetime interpretation of SR is in some respects, at least, superior to a relativity interpretation of SR, there do not seem to be comparably good reasons to prefer a spacetime interpretation of SR over a neo-Lorentzian conception of space and time.

On the other hand, I think we do have good reasons for rejecting spacetime realism and, therefore, a spacetime interpretation of SR. Inherent to the concept of spacetime is the indissoluble unification of space and time into a four-dimensional continuum. Hence, Minkowski's pronouncement that "Henceforth, space by itself, and time by itself, are doomed to fade away into mere shadows, and only a kind of union of the two will preserve an independent reality."[60] It seems to me, however, that we have powerful metaphysical grounds for believing that time can exist independently of space. In making his pronouncement, Minkowski forgot Kant's insight that time is a form applicable not only to the external world, but to consciousness as well.[61] Concerning Minkowski spacetime, Wenzl cautions:

> From the standpoint of the physicist, this is a thoroughly consistent solution. But the physicist will (doubtless) understand the objection, raised by philosophy, that time is by no means a merely physical matter. Time is, as Kant put it, the form not merely of our outer sense but also of our inner sense.... Should our experiences of successiveness and of memory be mere illusion...?[62]

A series of mental events alone is sufficient to set up a temporal sequence.[63] Thus, if we imagine Newton's God counting down to the moment of creation: "..., 3, 2, 1, *Fiat lux*!" then the beginning of spacetime would be preceded by a metaphysical time associated with the mental events of counting which would be wholly independent of space. Whether such a state ever existed in the world is beside the

---

59 Ibid., p. 219.

60 H. Minkowski, "Space and Time," in *The Principle of Relativity*, by A. Einstein, *et al.*, with Notes by A. Sommerfeld, trans. W. Perrett and G. B. Jeffery (1923; rep. ed.: New York: Dover Publications, 1952), p. 75.

61 Immanuel Kant, *Immanuel Kant's "Critique of Pure Reason,"* trans. Norman Kemp Smith (London: Macmillan, 1970), A34; B51 (p. 77).

62 A. Wenzl, "Einstein's Theory of Relativity, Viewed from the Standpoint of Critical Realism, and its Significance for Philosophy," in *Albert Einstein: Philosopher-Scientist*, ed. P. A. Schilpp, Library of Living Philosophers 7 (LaSalle, Ill.: Open Court, 1949), pp. 587-588.

63 A point emphasized by Reichenbach:

> "An act of thought is an event and therefore defines a position in time. If my experiences are always produced within the framework of a 'now,' that means that each act of thought defines a point of reference. We cannot escape the 'now' because the attempt to escape signifies an act of thought and therefore defines a 'now'" (Hans Reichenbach, "Les fondements logiques de la mécanique des quanta," *Annales de l'Institut Henri Poincaré* 13 [1952]: 157).

point; such a scenario is coherent and thus it is evident that physical events having spatial co-ordinates are not a necessary condition of temporal events. If God experiences a succession of contents of consciousness, He is clearly in metaphysical time even if space does not exist. Pointing out that there are two percepts which give birth to the concept of time, one in the physical realm and one in the mental, E. A. Ramige contends,

> The spacetime concept is found to be ever more of a necessity in the theories of modern physics. However, necessary as it is in certain realms, that does not thereby rule out the separate concepts of space and time, any more than the concept of the area of a figure rules out the separate concepts of length and breadth. In the non-physical realm which we shall go into in taking up the concept of God it is time rather than space that is important. Descartes had stressed that thought is a separate attribute of reality from extension. 'What is conscious is not spatial and what is spatial is not conscious.' It is still held that in the mental realm space plays a less important role than time.[64]

Indeed, time is more fundamental than space, as Lucas is wont to emphasize.[65] For while a series of mental events is a sufficient condition of time, it is not sufficient to constitute space. Since God is incorporeal, He would not be spatial prior to the creation of the universe even though He were temporal. Thus, on the orthodox view of God, time is more fundamental than space, since God's mere thinking discursively is sufficient for the existence of time, but not of space.

Reinforcing one's skepticism about spacetime realism is that fact that in Minkowski spacetime the spacetime interval between timelike separated events takes imaginary values. Christensen comments,

> This is disturbing in the first place because imaginary numbers seem like a paradigm of a convenient conceptual fiction. They were originally invented to make it the case that all quadratic equations, not just some, would have two roots. The serious question, however, is not whether there are imaginary numbers—numbers being a peculiar lot of entities to begin with—but whether there could really be such a thing as mathematically imaginary *physical quantities*.... No one doubts the convenience of computing with such quantities, even in an eminently practical field such as electrical engineering; but in the end the imaginary parts of the result are discarded. Could it be otherwise in SR, rationally?[66]

On a relativity interpretation of SR, it is easy to dismiss this feature of Minkowski spacetime as a mathematical property which does not correspond to any physical property in the world. But such a recourse is not open to the spacetime realist, who prides himself precisely on the interval as a luminous indication of space and time's unification. The spacetime interval is as real a magnitude—nay, more real, if Minkowski is to be believed—than the distance from New York to Los Angeles. But what sense is there in affirming that the spacetime separation of, say, my having lunch and my having dinner is imaginary?

In addition to these difficulties, spacetime realism raises a host of problems due to its entailment of the doctrine of perdurance, or the existence of four-dimensional

---

64 Eldon Albert Ramige, *Contemporary Concepts of Time and the Idea of God* (Boston: Stratford, 1935), pp. iii-iv.

65 John R. Lucas, *A Treatise on Time and Space* (London: Methuen, 1973), p. 3.

66 F. M. Christensen, *Space-like Time* (Toronto: University of Toronto Press, 1993), pp. 260-261.

objects comprised of temporal parts. Since these also constitute independent objections to a B-Theory of time, we may reserve discussion of these until Part II.

Thus, with little to commend spacetime realism over a neo-Lorentzian conception of space and time and with powerful objections lodged against it, we may conclude that there is no reason to adopt a spacetime interpretation of SR rather than a neo-Lorentzian approach to the problems of relativity theory. I can only agree with David Bohm when he complains that a Minkowski diagram is "misleading" in showing the past and future as existent[67] and with Max Black when he writes,

> ...this picture of a 'block universe,' composed of a timeless web of 'world-lines' in a four-dimensional space, however strongly suggested by the theory of relativity, is a piece of gratuitous metaphysics.... Here, as so often in the philosophy of science, a useful limitation in the form of representation is mistaken for a deficiency in the universe.[68]

Minkowski's four-dimensional, mathematical space serves as a convenient calculational and diagrammatical aid, a mathematical *Hilfsmittel*, but says absolutely nothing about ontology. As Arzeliès states, "The four-dimensional continuum should therefore be regarded as a useful tool, and not as a physical 'reality'."[69] Since time can exist independently of space, Minkowski's spacetime is at best a representation of the manifold of physical time and space as described by the equations of SR and cannot pretend to imply a four-dimensional ontology. Minkowski spacetime is a mathematical space in which the Lorentz transformation of coordinates can be diagrammatically displayed. Since it is analytically connected with the propositions of the theory it displays, it cannot *explain* anything in the theory, but only exhibit the theory's content graphically. Diagrams are useful in mathematics, as Plantinga says, because they enable us to see connections, entertain propositions, and resolve questions that would otherwise be seen, entertained, or resolved only with the greatest difficulty; but they do not serve as explanations.[70] In the same way, spacetime diagrams are indispensable aids in providing a visualization of the propositional content of the theory without serving as explanations of why time dilation and length contraction occur.[71] Sellars, noting that Minkowski spacetime is simply a metrical picture of the location of events, not a depiction of things which come to be and cease to be, concludes, "It has too often

---

[67] David Bohm, *The Special Theory of Relativity*, Lecture Notes and Supplements in Physics (New York: W. A. Benjamin, 1965), pp. 174-175.

[68] Max Black, review of *The Natural Philosophy of Time*, by G. J. Whitrow, *Scientific American* 206 (April 1962), 181-182.

[69] Henri Arzeliès, *Relativistic Kinematics*, rev. ed. (Oxford: Pergamon Press, 1966), p. 258; cf. A. S. Eddington, *The Mathematical Theory of Relativity*, 2d ed. (Cambridge: Cambridge University Press, 1930), p. 196; Mary F. Cleugh, *Time and Its Importance in Modern Thought*, with a Forward by L. Susan Stebbing (London: Methuen, 1937), pp. 66-67; Herbert Dingle, *The Special Theory of Relativity* (London: Methuen, 1940), pp. 43-44.

[70] Alvin Plantinga, "Reply to Robert Adams," in *Alvin Plantinga*, ed. James Tomberlin and Peter Van Inwagen, Profiles 5 (Dordrecht, Holland: D. Reidel, 1985), p. 378.

[71] For a graphic explanation of how non-explanatory such accounts are, see C. Misner, K. S. Thorne, and J. A. Wheeler, *Gravitation* (San Francisco: W. H. Freeman, 1973), pp. 32-33. See also J. D. North, "The Time Coordinate in Einstein's Restricted Theory of Relativity," *Studium Generale* 23 (1970): 222.

been supposed by philosophers (and physicists) who discuss the Special Theory of Relativity that the Minkowski mathematical apparatus automatically carries with it a commitment to an ontology of 'events.' This...is simply a mistake."[72]

CONCLUSION

It seems to me, therefore, that despite the widespread aversion to a neo-Lorentzian interpretation of relativity theory, such antipathy is really quite unjustified. Admitted on all sides to be empirically equivalent to the Einsteinian interpretation, the neo-Lorentzian interpretation is neither *ad hoc* nor more complicated than its rival. The physical effects it posits are no less real in the received version, only there they appear as axiomatic deductions lacking causal explanations. Indeed, its fecundity in opening the question about physical causes is an important advantage of the neo-Lorentzian interpretation. Therefore, our philosophically grounded preference for the neo-Lorentzian interpretation, however distasteful to the majority of physicists, cannot count against the truth of the A-Theory of time.

In summary, arguments appealing to the Special Theory of Relativity, such as Putnam's argument exposited in chapter 1, in order to justify a B-Theory of time are at best inconclusive. For these arguments must presuppose a spacetime interpretation of SR, if they are to succeed in showing the incompatibility of SR with the objective reality of tense and temporal becoming. Both the relativity interpretation and the neo-Lorentzian interpretation of SR are compatible with an A-theory of time. Even if the spacetime interpretation is in some respects superior to the relativity interpretation, the same cannot be said of it in comparison to the neo-Lorentzian interpretation. Therefore, Putnam's argument fails.

---

[72] Wilfrid Sellars, "Time and the World Order," in *Scientific Explanation, Space, and Time*, ed. Herbert Feigl and Grover Maxwell, Minnesota Studies in the Philosophy of Science 3 (Minneapolis: University of Minnesota Press, 1962), p. 578. Robert Geroch says, "If you like, 'four dimensions' is just a convenient way of describing the world and thinking about the world, nothing more....these pictures are merely a convenient way of looking at (not a necessary ingredient of) spacetime, while spacetime is merely a convenient way of looking at (not a necessary ingredient of) nature" (Robert Geroch, *General Relativity from A to B* [Chicago: University of Chicago Press, 1978], pp. 12-13). Dugas concludes,

> "Therefore it seems that Minkowski, carried away by a very natural enthusiasm for the remarkable geometrical synthesis that he had discovered, had to some extent gone beyond the relativistic doctrine, which does not in any way forbid that an observer should reason and calculate, as in daily life, in terms of space and time" (Dugas, *History of Mechanics*, p. 501).

SECTION 2: THE MIND-DEPENDENCE OF TEMPORAL BECOMING

CHAPTER 6

THREE ARGUMENTS FOR THE MIND-DEPENDENCE OF BECOMING

Apart from the support allegedly lent to the B-Theory of time by SR, there are few arguments of a positive nature in support of the B-Theory. In his oft-reprinted defense of the mind-dependence of temporal becoming,[1] Adolf Grünbaum briefly exposits three such arguments. According to Grünbaum's analysis, what qualifies a physical event at a time *t* as occurring now is that some mind-possessing organism *M* experiences the event at *t* in such a way that at *t M* is conceptually aware that his having the experience of the event is simultaneous with an awareness of the fact that he has that experience at all.[2] For example, in order for a gunshot at *t* to qualify as occurring now, *M* must be judgmentally aware at *t* (i) that he hears a gunshot and (ii) that that awareness of his hearing the gunshot is simultaneous with his hearing the gunshot. Note that *M*'s awareness of (i) involves a present-tense belief, for otherwise any event of which *M* ever has an experience could qualify as now. But since tense is an expression of what is the case in relation to the present or now, Grünbaum's characterization of nowness is circular. In a revised version of his article, he freely admits this; the nowness of an event *E* is made to depend on someone's knowing at *t* that he is experiencing (present tense) *E*, which is tantamount to someone's judging at *t*, "I am experiencing *E now*."[3] But Grünbaum insists that his characterization of nowness is non-viciously circular because he is not attempting to define "now" in such a way as to eliminate nowness in favor of tenseless temporal attributes or relations but rather is merely articulating the mind-dependence of nowness. He is, in effect, saying that in order for a physical event at *t* to qualify as occurring now, it must at *t* be consciously experienced as occurring now by some person.

---

1 The same piece has appeared with only minor alterations as "The Status of Temporal Becoming," in *Modern Science and Zeno's Paradoxes* (Middletown, Conn.: Wesleyan University Press, 1967), pp. 7-36; "The Meaning of Time," in *Basic Issues in the Philosophy of Time*, ed. E. Freeman and W. Sellars (LaSalle, Ill.: Open Court, 1971), pp. 196-227; "Is There A 'Flow' of Time or Temporal 'Becoming'?" in idem, *Philosophical Problems of Space and Time*, 2d ed., Boston Studies for the Philosophy of Science 12 (Dordrecht: D. Reidel, 1973), chap. 10, and has been often reprinted in various anthologies.

2 Grünbaum, "Status of Temporal Becoming," p. 17.

3 Grünbaum, "The Exclusion of Becoming from the Physical World," rep. in *The Concepts of Space and Time*, ed. Milic Capek, Boston Studies in the Philosophy of Science 22 (Dordrecht: D. Reidel, 1976), pp. 480-481.

According to Grünbaum, "...what confers nowness at a time *t* on either a physical or mental event is that the *experience* of the event satisfies the specified requirements. And by satisfying these requirements, the experience of a physical event qualifies at the time *t* as occurring *now*."[4] L. R. Baker draws attention to the fact that Grünbaum's requirements specify sufficient conditions only for one's *experience* of a physical event's occurring now; but they constitute merely necessary conditions of the physical event's itself occurring now.[5] The physical event itself could have occurred long before one's experience of the event. Hence, in order to qualify as occurring now, a physical event must not only meet the specified conditions but also meet some unspecified condition(s).

Being experienced, either directly or indirectly by being simultaneous with a directly experienced event, is thus essential to an event's occurring now and, hence, to temporal becoming: "...independently of being perceived, physical events themselves qualify at no time as occurring now and hence as such do not become."[6] "Becoming is mind-dependent because it is not an attribute of physical events per se, but requires the occurrence of states of *conceptualized awareness*."[7]

Why should we accept the above characterization of nowness? Grünbaum answers that the mind-dependence of the nowness of an event emerges from a consideration of the kind of information which the statement "It is 3 p.m. now" is supposed to convey. Such a statement is clearly informative. But if the word "now" does not designate a particular content of conceptualized awareness (or the linguistic utterance expressing it), then there seems to be nothing left for it to designate other than the events already identified as occurring at 3 p.m. But then the initially informative temporal statement "It is 3 p.m. now" turns into the utter triviality that the events of 3 p.m. occur at 3 p.m.! The implication is that "now" refers only to the conceptualized awareness of the event's occurrence, nothing more.

But Grünbaum appears to have overlooked the tense of the verb once more in this analysis of the content of "now." Since the "is" in "It is 3 p.m. now" is present-tense, the "now" is simply a flourish of ordinary language which serves to reinforce the presentness of the tense, but is, philosophically speaking, redundant.[8] The judgement remains, however, informative because the tense of the verb prevents the reduction of its meaning to "It is 3 p.m. at 3 p.m.," which fails to inform us of the *present* time.[9] From the point of view of the A-theorist, "now" normally serves to

---

4 Grünbaum, "Status of Temporal Becoming," p. 18. Because he overlooks Grünbaum's use of the present-tense, McGilvray is led to complain that the stipulated necessary conditions for nowness, if true at some time, are always true and that the notion of temporal becoming needs to be introduced in order to make the conditions sufficient for an event's occurring now (James A. McGilvray, "A Defense of Physical Becoming," *Erkenntnis* 14 [1979]: 293).

5 Lynn Rudder Baker, "Temporal Becoming: The Argument from Physics," *Philosophical Forum* 6 (1975): 220.

6 Grünbaum, "Status of Temporal Becoming," p. 19.

7 Ibid., p. 8.

8 See also Arthur Prior, "On Spurious Egocentricity," in *Papers on Time and Tense* (Oxford: Clarendon Press, 1968), pp. 20-21.

9 See our discussion of the ineliminability of tense in *The Tensed Theory of Time: a Critical Examination*, Synthèse Library (Dordrecht: Kluwer Academic Publishers, forthcoming), chaps. 2-3.

pick out a time and ascribe present tense to it. The A-theorist may recognize a certain ego-centricity in the notion of something's happening now without compromising this dual function of the term "now."[10]

In discussing the notion of "nowness," it is critical that we understand that on the A-theorist's view, "now" is neither a tenseless indexical term nor a relational predicate, but an irreducibly tensed expression.[11] The word "now" is frequently used in philosophical literature as a temporal indexical term, analogous to the spatial indexical term "here," which serves to pick out a tenseless date. So employed, "now" is not a tensed expression and is perfectly consonant with a B-Theory of time. If one says at t_1, "It is now raining" and then at t_2, "It was raining then," one has not, if "now" merely picks out a tenseless temporal location, reported any different information content by means of the two expressions. Whether "now" is being used as a tenseless temporal indexical in any particular case can be ambiguous and lead to confusion.[12] A similar situation holds for terms like "yesterday," "today," and "tomorrow." It needs to be emphatically said that as the A-theorist normally employs such terms, they are not tenseless temporal indexicals, but irreducibly tensed expressions.

In the same way, neither are terms like "now" or "past," "present," and "future" relational predicates. Very frequently these words are used as synonyms for the relational predicates "earlier than," "simultaneous with," or "later than" respectively.[13] One says, for example, that "God foreknows future contingents," meaning that He knows what events occur later than the moment of His cognizing them, or again that "A retro-cause produces a past effect," meaning that the effect was earlier than the operation of the cause. Employed in this way, "past," "present," and "future" do not relate to the moment which is now or express any tensed fact. Again, for the A-theorist this usage does not exhaust the legitimate function of these terms, which may also be employed in a non-relational way to indicate tense.

Nor, finally, should the A-theorist want, I think, to claim that pastness, presentness, and futurity are absolute, monadic properties of events/things, which

---

10 See discussion in *Tensed Theory of Time*, chap. 4.

11 This underappreciated distinction is neatly drawn by Paul Horwich, *Asymmetries in Time* (Cambridge, Mass.: MIT Press, 1987), pp. 20-21. Ironically, for an example of indexical confusion see E. J. Lowe, "The Indexical Fallacy in McTaggart's Proof of the Unreality of Time," *Mind* 96 (1987): 62-70. For a very stimulating discussion of the different ways in which "now" may be used indexically to refer to past and future times, imaginary times, and non-temporal items, see Quentin Smith, "The Multiple Uses of Indexicals," *Synthèse* 78 (1989): 167-191. But it is noteworthy that all his other uses of "now," though grammatically present-tense, in fact concern tenseless uses of "now."

12 A prime example would be Paul Helm's misconceived attempt to prove, on the basis of the analogy between the spatial indexical "here" and the temporal indexical "now," that arguments against divine timelessness from considerations of tense have parallel arguments that would also serve to undermine divine spacelessness (Paul Helm, *Eternal God* [Oxford: Clarendon Press, 1988], pp. 41-55; see especially p. 44, where he admits the absence of spatial tenses, but thinks this to be irrelevant, and p. 47 where he explicitly states that the spatial analogy to past/present/future is here/there or before/behind).

13 See Nelson Goodman, "Time and Language, and the Passage of Time," in *The Structure of Appearance* (Cambridge, Mass.: Harvard University Press, 1951), chap. 11, §§ 2, 3, reprinted in *Problems of Space and Time*, ed. J. J. C. Smart, Problems of Philosophy (New York: Macmillan Co., 1964), p. 363. Goodman's example of a purely relational use of the terms in question is "We can know at a given time only what is past at that time or present at that time, not what is future at that time."

exist tenselessly and which successively exchange the property of futurity for that of presentness, then presentness for that of pastness. This sort of hybrid A-B-Theory, such as is defended by Smith[14] and has been aptly characterized by Fitzgerald[15] as a "property acquisition model" of temporal becoming, even if it evades McTaggart's Paradox, seems to fall prey to Grünbaum's objection that the property of nowness or presentness becomes altogether vacuous. For given that all events/things exist, it is unclear exactly what the property of presentness amounts to. Moreover, while we can perhaps make sense of, say, "The presidency of Bill Clinton presently has presentness," it does not seem intelligible to assert, "The Battle of Waterloo presently has pastness," since the Battle of Waterloo does not exist or occur at the present time so as to possess any properties. Yet Smith's view that the pastness or futurity of an event presently inheres in that event seems to imply such a conclusion. So long as event *e* is present, we can make sense of Smith's contention that the futurity of *e*'s pastness presently inheres in *e* and likewise that the pastness of *e*'s futurity presently inheres in *e*. But if *e* is a past or future event, it seems unintelligible to claim that presentness inheres in the inherence of pastness/futurity in *e*, since *e* does not presently exist so as to sustain the inherence of any properties in it.[16] Perhaps Smith would prefer to say of a past event that pastness presently inheres in the inherence of presentness in *e*. But then events never lose or acquire their property of presentness; what changes is the tense of presentness's inherence in *e*. But that interpretation seems as unintelligible as the first interpretation; for how can pastness presently inhere in *e*'s presentness, if *e* is a past event, since *e*'s presentness does not exist at the present time?

The most consistent A-theoretical employment of these temporal terms, I think, would be to regard nowness or, better, presentness as an absolute mode of existence or absolute property of existents and "pastness" and "futurity" as relational terms anchored in what is absolutely present. Such an understanding would be rooted in what Freddoso has called "the metaphysical primacy of the pure present."[17]

---

[14] Quentin Smith, "The Infinite Regress of Temporal Attributions," *Southern Journal of Philosophy* 24 (1986): 383-396; idem, "The Logical Structure of the Debate about McTaggart's Paradox," *Philosophy Research Archives* 14 (1988-89): 371-379, the latter being a reply to L. Nathan Oaklander, "McTaggart's Paradox and the Infinite Regress of Temporal Attributions: a Reply to Smith," *Southern Journal of Philosophy* 25 (1987): 425-431.

[15] Paul Fitzgerald, "Nowness and the Understanding of Time," in *PSA 1972*, Boston Studies in the Philosophy of Science 20, ed. Kenneth F. Schaffner and Robert S. Cohen (Dordrecht: D. Reidel, 1974), p. 261.

[16] See remarks of Ferrel Christensen, "The Source of the River of Time," *Ratio* 18 (1976): 137; Arnold B. Levison, "Events and Time's Flow," *Mind* 96 (1987): 346-350; also my "In Defense of Presentism: A response to Smith's paper 'Reference to the Past and Future'," *Time, Tense, and Reference*, ed. Quentin Smith and Alexander Jokic (forthcoming).

[17] See Alfred J. Freddoso, "Accidental Necessity and Logical Determinism," *Journal of Philosophy* 80 (1983): 264. He writes,

> "I take the claim that the pure present is metaphysically primary to be tantamount to the assertion that for any moment *t* and any logically possible world *w* there is a set *k* of purely present-tense propositions such that (a) each member of *k* is true at *t* in *w*, and (b) *k* determines what is true at *t* in *w* in a temporally independent way, i.e., in a way which does not temporally depend on what has been or will be true at moments of *w* other than *t*" (Ibid., p. 266).

According to this notion, the antecedent or subsequent truth of future- and past-tense propositions is parasitic on what present-tense propositions are true, which depends in turn on what states of affairs are actual. "Pastness" and "futurity" thus need not be regarded as absolute or even as properties of events or things; when we say, "The Battle of Waterloo is a past event," this is to be understood as the assertion "The Battle of Waterloo occurred," and this statement is true because the present-tense proposition "The Battle of Waterloo is occurring" was true in 1815.[18] Furthermore, as E. J. Lowe admonishes us, rather than saying of some event *e* that *e* is "present in the past" or that *e* is "present in the future," we should at most say that the sentence "*e* is present" either was true or will be true.[19] On the A-Theory, then, tense is ultimately non-relational in that it is anchored in the pure present.

There is an interesting and helpful analogy to presentness in the notion of actuality in possible worlds semantics.[20] We can characterize presentism by allowing tensed states of affairs to be constituents of possible worlds. A tensed possible world is then a maximal possible state of affairs at some time *t*. Tensed possible words which did, do, or will obtain are tensed actual worlds. The tensed actual world at *t* will be the tensed actual world which obtains when *t's being present* obtains, or when *t* is present. The tensed history of any possible world *W* will be all the tensed possible worlds constituted by the states of affairs entailed by *W* and each successive *t*'s being present in *W*. To say that a temporal entity *x* exists in a tensed possible world W^t is to say that if W^t were actual, then *x* would exist (present tense). To say that *x* exists in a tensed actual world α^t is to say that when α^t becomes actual, then *x* exists (present-tense). Each tensed possible world exists in each such world. The tensed actual world *ν* is the maximal state of affairs that

---

That this implies a sort of power over the past (as well as the future) is explained by idem, "Accidental Necessity and Power over the Past," *Pacific Philosophical Quarterly* 63 (1982): 54-68. For a discussion see my *Divine Foreknowledge and Human Freedom: The Coherence of Theism I: Omniscience*, Brill's Studies in Intellectual History 16 (Leiden: E. J. Brill, 1991), pp. 83-90, 179-190. If one does not accept that propositions are tensed, then one could reformulate the metaphysical primacy of the pure present in terms of non-propositional, present-tensed features of reality (*e.g.*, God's *de se* knowledge of what He is creating).

[18] See Levison, "Time's Flow," pp. 349-350. Horwich objects that statements like "The present-tense proposition 'The Battle of Waterloo is occurring' is true in 1815" fail to express any fact in the portion of the sentence enclosed in single quotation marks. He says that the term "fact" should not be used for those aspects of reality whose explicit descriptions are true only relative to some context (*e.g.*, "It is raining") but should be reserved for those aspects of reality whose explicit descriptions are sentences that are true *simpliciter* (*e.g.*, "It is raining in Manchester"). Hence, we should not countenance a sentence like "*E* is past" as a fact but reject it as an idiosyncratic and unmotivated conception of "fact." The real facts are described by sentences of the form "*E* is past at *t*," in which "past" has become a relational predicate meaning "earlier than" (Horwich, *Asymmetry of Time*, pp. 22-23). All this amounts to, however, is the mere rejection of tensed facts on Horwich's part. He provides no argument as to why no such absolute, tensed facts obtain. Ironically, his example of a genuine fact is not, on his own account, a fact at all, since "It is raining in Manchester" is not true *simpliciter*, but only relative to a time. If we adopt his canonical form, what is a fact is "It *rains* in Manchester at *t*"—but then we no longer know whether it is *now* raining in Manchester. The A-theoretic understanding of tensed facts is no more idiosyncratic than the B-theorist's interpretation of the same and is far from unmotivated, seeking to account coherently for the objectivity of temporal becoming.

[19] Lowe, "Indexical Fallacy," pp. 64-65.

[20] See discussion in *Tensed Theory of Time*, chap. 6.

obtains (present-tense). Were some other tensed possible world actual, then ν would not obtain but it would still exist as a tensed possible state of affairs. When some other tensed actual world obtains (present-tense), then ν either does not yet or no longer obtains, but ν nonetheless exists as a tensed state of affairs which was or will be actualized. Since ν alone is (present-tense) actual, none of the other tensed actual worlds (not to speak of tensed merely possible worlds) is (present-tense) actual, though they either were or will be actual. Thus, ν is uniquely distinguished as *the* tensed actual world, the one tensed possible world which obtains (present-tense).[21]

From the foregoing, I think that we can see that presentness is a restricted form of actuality, a tensed form, to be more precise. "Now" serves to designate what is (present-tense) actual, what segment of the temporally parameterized actual world obtains (present-tense). This A-theoretical characterization of nowness is, like Grünbaum's B-theoretical characterization, non-viciously circular. Since it uses the present-tense, it does not serve to define "now" in terms of actuality alone; but it serves to articulate nowness in terms of actuality rather than mind-dependence. The notion of present-tense actuality, like actuality itself in possible worlds semantics, seems to be a well-understood primitive notion.

Grünbaum anticipates the response that physical events possess "an unanalyzable property of presentness (i.e., nowness) over and above merely occurring at certain clock times" but finds this response "wholly unavailing."[22] It is at this point that he presents his three positive arguments in favor of a B-Theory of time. Let us then consider each one of his arguments.

THE TRIVIALITY OF A-THEORETICAL NOWNESS

Grünbaum's first argument reasserts his claim that an A-theoretical conception of nowness trivializes statements like "It is now 3 p.m.":

> It must construe the assertion 'It is 3 P.M., EST., now' as claiming non-trivially that when the clock strikes 3 P.M. on the day in question, this clock event and all the events simultaneous with it intrinsically have the unanalyzable property of nowness or presentness. But I am totally at a loss to see that anything non-trivial can possibly be

---

[21] Such an account of presentism exonerates the A-theorist from the confused charges leveled against the creationist/annihilationist by Melvin M. Schuster, "On the Denial of Past and Future Existence," *Review of Metaphysics* 21 (1967-68): 447-467. Oaklander has objected that such an account succumbs to McTaggart's Paradox or the Problem of Temporary Intrinsics (Nathan Oaklander, "Comments on William Craig's 'McTaggart's Paradox and the Problem of Temporary Intrinsics'," Central Division Meeting of the American Philosophical Association, Pittsburgh, Pennsylvania, April 23-26, 1997); cf. idem, "Bigelow, Possible Worlds, and the Passage of Time," *Analysis* 54 [1994]: 244-248. But Oaklander forgets that tensed possible worlds are not abstract objects timelessly existing, but abstract objects which exist at times and endure through time. Oaklander's objection that my explication falls prey to McTaggart's Paradox must be, therefore, that W^{t*} at t lacks the property of *being actual, whereas* W^{t*} at t^* has the property of *being actual*; therefore W^{t*} at t is not identical with W^{t*} at t^*, which is *ex hypothesi* absurd. But now the solution to the problem is obvious: the solution is just presentism once more. If t^* is present, then only entities which exist at t^* are real, including W^{t*}. Therefore, the only properties W^{t*} has are the properties W^{t*} has presently, including actuality. Thus, W^{t*}, like any entity existing at t^*, does not have incompatible properties.

[22] Grünbaum, "Status of Temporal Becoming," p. 20.

asserted by the claim that at 3 P.M. nowness (presentness) inheres in the events of 3 P.M. For all I am able to discern here is that the events of 3 P.M. are indeed those of 3 P.M. on the day in question![23]

It seems to me that our above characterization of presentness removes the teeth from this argument. Grünbaum's difficulty is analogous to those who cannot see what non-trivial claim is being made when it is asserted that α is the actual world. For every world is actual at or in itself. So to say that α is actual is to assert the triviality that α is actual in α! In a similar way, to claim that 3 p.m. is now is to assert the triviality that it is 3 p.m. at 3 p.m.. But to assert that α is actual is not the same as asserting that α is actual in α, and in the same way, to say that 3 p.m. is now is not to utter the tautology that 3 p.m. is 3 p.m. To say that it is now 3 p.m. is to pick out the one and only time of all possible times that has (present-tense) actuality. That such a claim is non-trivial is evident from the fact that it implies an ontology vastly more modest than that of the B-Theory.[24] For on a consistent A-Theory of time—what Fitzgerald calls a "reality acquisition model" of temporal becoming[25]—, only the present obtains or is actual. The past and the future can be said to exist in the sense that non-actual past and future states of affairs exist. But neither the past nor the future obtains. There is this difference between them: the past *has been* actual, whereas the future *has yet* to be actual. But both are (present-tense) non-actual. The states of the past and the future are thus like non-actual possible worlds: they exist but do not obtain. Events/things which exist at any non-present time *t* are thus the analogues of merely possible objects and cannot, accordingly, be intelligibly said to exist. The only events/things that exist are present events/things—hence, the more modest ontology of the A-Theory.[26]

---

23 Ibid.

24 For a particularly bold statement of this point see Christensen, "Source," pp. 135-137.

25 Fitzgerald, "Nowness," p. 260.

26 It seems to me inessential to the issue at hand whether we adopt an ontology of events or things. Typically, A-theorists, such as Broad and Prior, have propounded thing ontology, according to which events are changes in things, which are taken to be the fundamental existents (C. D. Broad, *An Examination of McTaggart's Philosophy*, 2 vols. [Cambridge: Cambridge University Press, 1938; rep. ed.: New York: Octagon Books, 1976], 2: 316; Arthur N. Prior, "Changes in Events and Changes in Things," in *Papers on Time and Tense* [Oxford: Clarendon Press, 1968], pp. 1-14). B-theorists, on the other hand, tend to favor the sort of event ontology propounded by Davidson or underlying substantival interpretations of relativistic spacetime, according to which the basic constituents of the material world are particular events occupying locations in spacetime (Donald Davidson, "The Individuation of Events," in *Essays in Honor of Carl Hempel*, ed. N. Rescher [Dordrecht: D. Reidel, 1969], pp. 216-234; Lawrence Sklar, *Space, Time, and Spacetime* [Berkeley: University of California Press, 1976], chap. 3). But it seems to me that Levison has correctly argued that an event ontology is quite amenable to an A-Theory of time and becoming (Levison, "Time's Flow," pp. 341-353). All one has to do is deny that events exist tenselessly. Thus for example.

(i) My falling out of a punt has receded six years into the past,

which seems to presuppose a tenseless event ontology coupled with temporal becoming, should be properly construed as

(ii) It is now six years since it was the case that my falling out of a punt is occurring.

The event of my falling out of a punt does not exist tenselessly at a time; rather it existed six years ago. This is to be understood tense logically as

(iii) It was the case six years ago that (there is an *e* such that *e* is my falling out of a punt)

Perhaps what really sticks behind Grünbaum's objection is the conviction that tense is eliminable from discourse and that tensed facts can be reduced to tenseless facts. If "It is 3 p.m. now" can be analyzed in terms of propositions like "The events of 3 p.m. are simultaneous with a certain conceptual awareness," then it might be argued that any further putative content to "now" is vacuous. But if that is Grünbaum's contention, then this is not a positive argument for the truth of the B-Theory but is a B-theoretical response to the A-theorist's positive argument on behalf of his theory that linguistic tense provides a clue to ontology. I have addressed this argument in my *Tensed Theory of Time: a Critical Examination.* As far as Grünbaum's first argument is concerned, then, we have seen no reason to regard the B-Theory as any more illuminative of tensed discourse than the A-Theory.

THE ABSENCE OF BECOMING FROM PHYSICAL TIME

Grünbaum's second argument is for him the most important: physics knows nothing of temporal becoming. He writes,

> It seems to me of decisive significance that no cognizance is taken of nowness (in the sense associated with becoming) in any of the extant theories of physics. If nowness were a fundamental property of physical events themselves, then it would be very strange indeed that it could go unrecognized in all extant physical theories *without detriment to their explanatory success.* And I hold with Reichenbach that 'If there is Becoming [independently of awareness] the physicist must know it.'[27]

Grünbaum supplements this argument with an effective refutation of Reichenbach's thesis that quantum indeterminacy serves to introduce objective becoming into physics. All that indeterminacy reveals is that relative to any of the regions of spacetime the events belonging to the absolute future of that particular region are quantum mechanically unpredictable. But all regions can be equally real or existent.

Grünbaum's second argument thus rests on two premisses:

1. The extant theories of physics take no cognizance of nowness or temporal becoming.

and

2. If temporal becoming is an objective feature of physical events, then explanatorily successful physical theories must take cognizance of it.

---

rather than as

(iv) There is an *e* such that (it was the case six years ago that *e* is my falling out of a punt).

While (iii) is true, (iv) is false. Thus, even if a thing ontology is naturally associated with the A-Theory, still, if one is persuaded by arguments such as Davidson's that events are existents, there is no obstacle to holding to the tensed existence of events. However, one may have to reject the idea that events have temporal parts; see Trenton Merricks, "On the Incompatibility of Enduring and Perduring Entities," *Mind* 104 (1995): 523-531.

[27] Grünbaum, "Status of Temporal Becoming," p. 20.

Notice that it does *not* follow from these two premisses that temporal becoming is not an objective feature of physical events; it could rather be the case that the extant theories of physics are not explanatorily successful! Grünbaum needs the additional premiss

3. The extant theories of physics are explanatorily successful.

His argument now becomes valid and the question, then, is whether these premises are true.

Against (1), it might be argued that in utilizing the temporal relations *earlier than/later than*, physics implicitly presupposes A-theoretic temporal determinations, since the former are reducible to the latter, while the reverse is not the case. Following McTaggart, Gale, for example, argues that B-theoretic relations can be reduced to what he calls "impure" A-determinations (for example, *more future than/more past than*) and therefore any B-series entails an A-series.[28] Consequently, physics does presuppose the reality of A-theoretic determinations. This sort of semantic reduction does not, however, address specifically the B-theorist's claim that if temporal becoming is objective, then physics should take *cognizance* of it. Fitzgerald replies that physical theory need not take cognizance of nowness precisely because temporal becoming characterizes all possible worlds in which time exists. Therefore no causal explanation of why the world has nowness need be given and anything that can be explained by nowness can be explained by non-temporal properties and relations.[29] Although Baker complains that Fitzgerald begs the question in assuming that there are no possible worlds in which time exists but temporal becoming does not,[30] she seems to have forgotten that Fitzgerald's position is one of *defense*; he is arguing that given A-theoretical assumptions concerning the essentiality of temporal becoming to time, there is no reason why nowness should play a role in physical theory, and therefore Grünbaum's objection finds no purchase. Rather the weakness in this defense is that it is not at all clear why the essentiality of temporal becoming to time implies that physical explanations in terms of tenseless properties and relations should obviate the need for tensed explanations. Even if we concede that all temporal worlds are characterized by nowness, it does not follow that a physically adequate account of the world may ignore that property; on the contrary, we might expect the opposite.

Whitrow and Dobbs have addressed (1) more directly, claiming that in some branches of science, such as meteorology, paleontology, tidal theory, engineering, and communications systems, nowness does play an essential role.[31] In his response to Dobbs, however, Grünbaum points out that these examples concern only the

---

28 Richard M. Gale, *The Language of Time*, International Library of Philosophy and Scientific Method (London: Routledge, Kegan & Paul, 1968), pp. 93-100, 224. For discussion see George N. Schlesinger, *Aspects of Time* (Indianapolis: Hackett, 1980), pp. 44-47 and Andros Loizou, *The Reality of Time* (Brookfield, Ver.: Gower, 1986), pp. 79-108.

29 Fitzgerald, "Nowness," p. 277.

30 Baker, "Temporal Becoming," p. 223.

31 G. J. Whitrow, *The Natural Philosophy of Time*, 2d ed. (Oxford: Clarendon Press, 1980), p. 350; H. A. C. Dobbs, "The 'Present' in Physics," *British Journal for the Philosophy of Science* 19 (1969): 317-324.

anisotropy of time, not the tensed notions of past, present, and future.[32] If time is anisotropic, then the two ordinally opposite time senses are distinguished structurally in that there are certain kinds of sequences of states of systems specified in the order of increasing time co-ordinates, such that these same kinds of sequences do not likewise exist in the order of decreasing time co-ordinates.[33] If time is anisotropic, then there is an arrow of time—or more correctly, two arrows pointing in the direction of increasing or decreasing values of the temporal co-ordinate respectively—, but this arrow is not to be confused with the flow of time or the direction of temporal becoming. According to Grünbaum, Dobbs confuses the use of past and future time in science in the sense of the anisotropy of the temporal continuum at any time with the use of past and future time in the tensed sense. Grünbaum appears to be justified in this allegation, for in a further rejoinder Dobbs appeals again to communication theory to prove the necessity of A-theoretic properties because only the past values of a certain function can be known.[34] But, as in the case with quantum indeterminacy, it is tenselessly true that at t_n the value of a certain physical quantity at t_{n+1} is unknowable from the resources available at t_n. Such considerations support temporal anisotropy, not temporal becoming.

A similar failure to discriminate between the use of "present" as a tenseless temporal indexical and as a tensed expression vitiates Smith's attempt to prove that modern cosmology takes cognizance of A-determinations like past, present, and future in its use of the "present" value of certain cosmological quantities like the Hubble time, the recessional velocity, and so forth.[35] It is tenselessly true that the values possessed by a cosmological quantity at various times t are possessed by it at those times, even if the values change over time, and it is no part of cosmological theory that a certain value of t be now. The same conflation of temporal anisotropy and temporal becoming seems to have misled those who think that the work of Ilya Prigogine in thermodynamics has reintroduced the notion of temporal becoming into physics.[36]

On the other hand, it does, indeed, seem to be the case that nowness plays a role in certain sciences like meteorology or geology. For example, in weather forecasting scientists want to know not simply the probability that, say, given certain meteorological laws and the boundary conditions obtaining off the Texas coast in late August, 1900, a hurricane *strikes* Galveston on August 27, 1900. They want to know the probability that given the relevant conditions a hurricane will, say, strike Galveston tomorrow. For if it is now August 26, 1900, then civil defense precautions will be taken and the city will be on alert, whereas if that date is long

---

32 A. Grünbaum, "Are Physical Events Themselves Transiently Past, Present and Future? A Reply to H. A. C. Dobbs," *British Journal for the Philosophy of Science* 20 (1969): 145-162.

33 Grünbaum, "Status of Temporal Becoming," p. 12.

34 H. A. C. Dobbs, "Reply to Prof. Grünbaum," *British Journal for the Philosophy of Science* 21 (1970): 275.

35 Quentin Smith, "The Mind-Independence of Temporal Becoming," *Philosophical Studies* 47 (1985): 112-115. Here Smith's failure to distinguish the uses of "now" as a tenseless temporal indexical and as a tensed expression, which I mentioned in note 11, seems to trip him up.

36 See the symposium organized by process thinkers and published as David R. Griffin, ed., *Physics and the Ultimate Significance of Time* (Albany, N. Y.: State University of New York Press, 1986).

past, such information will occasion no alarm. A similar illustration might be given for geologists with regard to forecasting events like the eruption of Mt. St. Helens or a catastrophic earthquake along the San Andreas Fault. Grünbaum might insist that such examples do not come from the realm of physics and so do not constitute counter-examples to (1). But I doubt that he would take such a recourse, for that would only serve to cast doubt on (2). For if other branches of science recognize nowness, then its absence from physics would tend to be viewed as a peculiarity of physics, rather than a proof of its objective unreality. Grünbaum could more plausibly respond that such examples do not belong to theoretical science, but only to applied science.[37] And certainly it would be expected that in the application of theoretical science to the world of human concerns nowness must enter in. But then the scientific theories themselves remain tenseless in their assertions and only become infected with tense and nowness when mind-possessing organisms employ them in their intersubjective lifeworld.

Perhaps the most effective riposte to (1) would be the demonstration that the assumption that time has a direction, which Grünbaum admits to imply temporal becoming and therefore rejects, does underlie contemporary physics. Ironically, it is B-theorists themselves who have in recent years made this point most forcefully. For example, in his *Time's Arrow and Archimedes' Point* Huw Price issues a ringing call for a wholesale reform of physics in order to make it truly tenseless by achieving what he calls an Archimedean perspective, a sort of "view from nowhen." Price argues that all of the customary, physical "arrows of time" are insufficient to give time a direction but, on the contrary, actually presuppose the directionality of time. For example, given the time reversal invariance of the laws of thermodynamics, if one were given a non-equilibrium system at a time t and asked to describe this system at t' without being told whether t' is earlier or later than t, then inability to furnish such a description would demonstrate that one's thermodynamic theory treats the two directions of time differently and so could not serve to explain this asymmetry. Absent temporal becoming, "...we have no right to assume that it is an objective matter that entropy *increases* rather than *decreases*.... What is objective is that there is an entropy gradient over time, not that the universe 'moves' on this gradient in one direction rather than the other."[38] Similarly, concerning the arrow of radiation based on the supposed impossibility of converging wave patterns, Price points out that when played in reverse a film of ordinary outgoing water waves, such as are observed when a pebble is dropped into a still pond, shows exactly the kind of apparently collaborative behavior in accordance

---

[37] See Adolf Grünbaum, "Carnap's Views on the Foundations of Geometry," in *The Philosophy of Rudolf Carnap*, ed. P. A. Schilpp, Library of Living Philosophers (LaSalle, Ill.: Open Court, 1963), p. 629, where he distinguishes between the semantic and pragmatic levels of physics. See also idem, "Operationism and Relativity," in *The Validation of Scientific Theories*, ed. with an Introduction by Philipp G. Frank (Boston: Beacon Press, 1954), p. 84, where he defines "pragmatics" as the study psychologically, sociologically, and otherwise of the activities of scientists in the pursuit of science and the investigation of the relationships between scientific symbols and the human users; semantics concerns the relationships between physical theory and the designata or denotata of that theory in the realm of nature.

[38] Huw Price, *Time's Arrow and Archimedes' Point* (New York: Oxford University Press, 1996), p. 48.

with physical laws which is supposedly impossible. "Anyone who wants to argue that it is impossible toward the future had better be prepared to tell us what distinguishes the two cases."[39] Concerning the cosmological arrow of time, it is not an objective matter of fact which end of the spacetime manifold is the Big Bang and which the Big Crunch. "Nothing in physics tells us that there is a wrong or a right way to choose the orientation of the temporal coordinates. *Nothing in physics tells us that one end of the universe is objectively the start and the other end objectively the finish.*"[40] In each case, physics does not furnish but rather presupposes a direction of time. As a B-theorist Price believes that contemporary physics needs to be thoroughly reformulated from an atemporal or Archimedean perspective. The change will be radical: "the ordinary temporal perspective is so familiar, and so deeply imbedded, that we need to be suspicious of many of the concepts used in contemporary physics. We can't simply assume that familiar concepts will carry over smoothly to an Archimedean physics."[41] Indeed, we cannot even as yet envision precisely what such an atemporal physics would look like.

> To answer this question we need to examine the ways in which the concepts ordinarily applied in physics reflect our temporal standpoint.... The conceptual apparatus of physics seems to be loaded with the asymmetric temporality of the ordinary world view. Notions such as *degree of freedom, potential,* and even *disposition* itself, for example, seem to embody the conception of an open future, for which present systems are variously prepared....the very notion of a *state* often seems to be forward-looking in this way....[42]

The reformulation of physics from such an Archimedean perspective, Price predicts, "promises profound insights into some of the deepest problems of contemporary physics."[43]

If Price's analysis is even partially correct, it explodes Grünbaum's contention that the extant theories of physics take no cognizance of temporal becoming. On the contrary, they are imbued with it. It would be unavailing for Grünbaum to answer, as he did to Withrow and Dobbs, that it is merely temporal anisotropy that is at stake here, for that is manifestly incorrect: physical cosmology, for example, treats the Big Bang as the origin, not the end, of the universe. It is time's directedness, not its anisotropy, which is at issue here. Moreover, Price's Archimedean physics would preclude even the anisotropy of time. Given time reversal invariant laws, there is just no reason for us to think of one end of the universe as expanding and the other as contracting. Since Grünbaum grants that physical theories do affirm the anisotropy of time, he is saddled with the difficult task of explaining why, absent temporal becoming, physical time is anisotropic.

Ironically, Price would probably agree with Grünbaum's (2), but would deny (3), that the extant theories of physics are explanatorily successful.[44] Their explanatory

---

39 Ibid., p. 55.

40 Ibid., p. 84. In an American Philosophical Association Central Division session devoted to his book, Price compared this situation to a carpenter who threw away half his nails because they were pointed at the wrong end!

41 Ibid., p. 234.

42 Ibid., p. 259.

43 Ibid., p. 260.

inadequacy stems, in his view, from their implicit reliance on the objective reality of temporal becoming. If it is really this assumption that does the heavy lifting in accounting for temporally asymmetric physical processes, then given the mind-dependence of becoming, the extant theories of physics must be explanatorily deficient. From the standpoint of the A-theorist, however, it is exactly this presupposition that serves to render physical theories explanatorily adequate in this regard.[45] Price's indictment of contemporary physics *presupposes* the mind-dependence of becoming, which is precisely the point to be proved. Thus, even if an Archimedean revolution were to transpire, thoroughly de-tensing physics, so that (1) did become true, the A-theorist would probably be sceptical that (3) would remain true. Price gives no reason to think that an Archimedean physics will be explanatorily adequate for the physical phenomena he discusses; and if such a physics forces us to abandon such claims as that the Big Bang occurred at the beginning of the universe, then its explanatory adequacy has been purchased only at the price of denying obvious facts about the world, in which case one has no reason to accept (2).

Finally, even given a de-tensed physics, we have reason to call (2) into question. The dubiety of (2) becomes especially evident if, as I have suggested[46], we draw a distinction between time itself and our physical measures of time. It is only because Grünbaum is presupposing the legitimacy of the positivistic reduction of time to physical time that he is able to claim that temporal becoming, being absent from physical time, is therefore non-existent. For on his view there is no time apart from physical time in which temporal becoming might lodge. But if we distinguish between metaphysical time and physical time and locate the seat of temporal becoming in the former, then why should we expect that the A-determinations of events in metaphysical time should be preserved in the abstraction of physical time? On the contrary, given the universal character of scientific theories, we should expect precisely the opposite, as Max Black explains:

> It is easy to understand why theoretical physics should express its formal results in a language that is independent of context, using formulas or sentences from which the occasion words are absent. This procedure has the great advantage of no reconstruction of the original context being required on the part of any reader.... If a scientist were to say, 'I *then* saw a green flash at the edge of the sun's disk,' anyone who was absent at the time of the original observation would need to know *who* spoke, and where and when, in order to obtain the intended information. No such supplementary information

---

44 He says, for example, that "The less charitable verdict is that the vast majority of physicists are simply mistaken, and have allowed their science to be guided by an assumption which is as groundless as the geo-centric foundations of ancient cosmology. Insofar—so very, very far—as physics relies on this assumption, then, it is likely to be grossly in error" (Ibid., p. 259). Even "the more irenic verdict" is that physics is "correct as a description of reality *from a particular standpoint*" and therefore gives us the partial truth about the world (Ibid.). At best physical theories give us the truth from the perspective of a moment in A-series time but as such do not give us the objective truth about the world.

45 See *The Tensed Theory of Time*, chap. 7.

46 See chapter 2.

> is needed in order to understand Boyle's law or any other freely repeatable scientific statement.[47]

Kroes notes that this universalizing feature of science, its abstraction from the here and now, runs at cross purposes with a scientific A-Theory of time:

> ...whereas physics generally tries to describe the universal aspects of the phenomena, the goal of an objective theory of time flow is precisely to get hold of the unique: such a theory must single out a unique moment of time as the present which separates the past from the future. Therefore, it is in principle questionable whether an objective theory of time flow is feasible. It appears that the universality of the laws of physics precludes the introduction of the notion of the present, and consequently also of the notion of the flow of time, on the basis of these laws.[48]

But if the universality of scientific theories precludes capturing nowness in physical theory, does that imply that, if temporal becoming is an objective feature of physical events, theories about those events are likely to be explanatorily inadequate, as Grünbaum alleges? No, not at all. For far from undermining the explanatory adequacy of physical theories, science's abstraction from indexical elements in the formulation of physical theories is one of its means of achieving explanatory adequacy, for the theories are then applicable at all times. That application, as we have seen, may well involve the re-introduction of A-theoretic notions. But the theories themselves, in order to achieve universality, are framed in B-theoretic terms, and this contributes to their generality and utility.

A second consideration which makes the feasibility of incorporating nowness into physical theory seem very remote, indeed, is that physical clocks, which serve to define physical time, only measure temporal intervals but are incapable of determining whether the time *t* they read is now. Again Max Black:

> The basic connection between *t* and time is not hard to discern: clocks measure only lapses of time, answer only questions about how much later one point-event (a tick, a flash, a beep) occurs than another point-event. All the other logical features that enter the constitution of the common-sense concept of time are deliberately excised. So, in order to read a clock, an observer is theoretically not required to be able to detect motion, or its sense, or to recognize the distinction between past and present. There result sets of numbers, as timeless as all sets of numbers are, in which the number corresponding to temporal position (the value of *t*) has no distinctive or privileged role. In an important sense it could be said that time and motion are deliberately omitted from such recordings and it is therefore not surprising that the applied mathematician's description of the universe is 'timeless.' Such a detemporalized, expurgated time is just what is needed in order to supply the mathematician with a compact series having the same structure as the ordered set of real numbers.... In the case of time...*t* is constructed by deliberately neglecting those features of 'becoming,' or 'passage,' that

---

47 Max Black, review of *The Natural Philosophy of Time*, *Scientific American* 206 (April 1962), p. 181. Cf. idem, "The Direction of Time," *Analysis* 19 (1958-59): 54-63. See also Alan G. Padgett, *God, Eternity and the Nature of Time* (New York: St. Martin's Press, 1992), pp. 82-95; J. R. Lucas, *Space, Time, and Causality* (Oxford: Clarendon Press, 1985), 36-38, 57, 119ff.

48 Peter Kroes, *Time: Its Structure and Role in Physical Theories*, Synthèse Library 179 (Dordrecht: D. Reidel, 1985), p. 208. See also Peter Kroes, "Physics and the Flow of Time," in *Nature, Time, and History*, ed. P. A. Kroes, Nijmegen Studies in the Philosophy of Nature and its Sciences 4/2 (Nijmegen, The Netherlands: Catholic University of Nijmegen, 1985), pp. 43-52.

> sharply distinguish time from space. Thus physical time plausibly masquerades as merely another dimension of a timeless manifold.[49]

Although Grünbaum fulminates against Black's description of the universe as changeless and timeless on a B-theoretic view,[50] the sense in which Black meant those characterizations is clear. On a B-Theory, any such changes as occur in the universe exist tenselessly at their respective spatio-temporal locations and the entire spacetime manifold exists tenselessly. The point is that physical clocks can determine only the B-theoretic determinations of events, not their A-theoretic determinations, and therefore it is hardly surprising that physical time is a B-theoretic continuum. Unless one could devise a machine which would tell of any event whether it was past, present, or future, it seems futile to expect nowness to play a role in physical time.

The role of time in physical theories can thus be very misleading philosophically. The variable *t* is only a shadow of time itself, necessary for doing science, as Cleugh emphasizes:

> The '*t*' of physics is improperly called *time*: it should be *time-interval*. It is an abstraction from lived time, and in the process all that is distinctively temporal has been eliminated. Past, present, and future have gone: in their stead remains only the logical relation of before and after, expressed in terms of numbers.... In abstracting from this, in refusing to consider diversities, which, from their point of view, are irrelevant, and in concentrating on time as serial, as a homogeneous continuum, [those who measure time] have rendered the passage of time susceptible to mathematical treatment. This is a loss, in that certain aspects are ignored, but it is also a great gain in that its clarity and simplicity make further development possible.[51]

The physical concept *t* lacks nowness, but that does not permit an inference about time itself. Moreover, since *t* represents a number, mathematical operations can be carried out on it which make no sense with respect to time itself; for example, *t* can be assigned negative or even imaginary values.[52] As Cleugh nicely puts it, "What is the wildest absurdity of dreams is merely altering the sign to the physicist."[53] If he

---

[49] Black, review of *Natural Philosophy of Time*, pp. 182-183; see also Padgett, *God, Eternity and the Nature of Time*, pp. 83-85.

[50] Grünbaum, *Problems of Space and Time*, p. 327.

[51] Mary F. Cleugh, *Time and its Importance in Modern Thought*, with a Foreword by L. Susan Stebbing (London: Methuen, 1937), pp. 46-47. Cf. the judgement of de Broglie:

> "science has always allowed, almost without discussion, the possibility of representing time as a simple variable, which can be plotted along a line like a spatial dimension, and that, by that very fact, science condemns itself to not being able to comprehend why time and space present themselves in our experience under different aspects, why, in particular, time always flows in the same direction, though any dimension of space can be pursued indifferently in two directions" (Louis de Broglie, "The Concepts of Contemporary Physics and Bergson's Ideas on Time and Motion," in *Bergson and the Evolution of Physics*, ed. P. A. Y. Gunter [Knoxville: University of Tennessee Press, 1969], p. 47).

[52] A point made by A. A. Merrill, "The *t* of Physics," *Journal of Philosophy* 19 (1922): 238-241. See also Herbert Dingle, *Science at the Crossroads* (London: Martin Brian & O'Keefe, 1972), pp. 31-32 and my remarks in "'What Place, then, for a Creator?': Hawking on God and Creation," *British Journal for the Philosophy of Science* 41 (1990): 229-234.

[53] Cleugh, *Time*, p. 46.

can find no metaphysically intelligible interpretation of such operations, the metaphysician will regard them merely as mathematical "tricks" without ontological import. So wide is the gap between the ordinary notion of time and the physicists' *t* that Black advised physicists to stop talking about "time" and to refer to their own concept simply as "*t*"![54] This is no doubt asking too much; but it is surely not too much to ask philosophers and physicists to refrain from drawing metaphysical conclusions about the unreality of objective temporal becoming on the basis of its absence from the eidolon of time which appears in physical theory.[55]

Therefore, Grünbaum's second and decisive argument rests on the reduction of time to physical time. Once that reduction is rejected, there is no reason to expect that physical time will preserve the temporal becoming characteristic of metaphysical time. Insofar as contemporary physical theories presuppose that time has a direction, they do seem to take cognizance of temporal becoming. It is not surprising that they abstract from the nowness of events, since scientific theories are constructed to hold universally and physical clocks can determine only the B-theoretic properties of events. Physical theory's omission of nowness does not render them empirically inadequate; on the contrary, it makes them universally applicable. Hence, Grünbaum's second argument is unsound.

WHY IS IT "NOW"?

Grünbaum's third argument in favor of the mind-dependence of becoming is that the B-Theory is free of an "important perplexity" which besets the view that temporal becoming is real,[56] namely, why do the events which are now happening in

---

[54] Black, review of *Natural Philosophy of Time*, p. 182. Similarly, Kroes: "...it is dubious whether coordinate time deserves to be called 'time' at all" (Kroes, "Physics and the Flow of Time," p. 49). Sklar writes,

> "If what we mean by 'time' when we talk of the time order of events of the physical world has nothing to do with the meaning of 'time' as meant when we talk about the order of time of our experiences, why call it time at all? Why not give it an absurd name, deliberately chosen to be meaningless (like 'strangeness'), and so avoid the mistake of thinking that we know what we are talking about when we talk about the time order of events in the world?" (Lawrence Sklar, "Time in Experience and in Theoretical Description of the World," in *Time's Arrows Today*, ed. Steven F. Savitt [Cambridge: Cambridge University Press, 1995], p. 226).

[55] Denbigh aptly remarks,

> "...it would be erroneous to regard the more comprehensive time-concept of conscious awareness as if it were obtained by the adding on of further features to a supposedly more basic concept originating in physics. The reverse, I believe, is closer to the truth; the t-coordinate is obtained by a process of discarding—by an elimination of these particular features of time-as-it-is-to-conscious-experience which are not required for the limited purposes of physical theory. In other words there are aspects of the more comprehensive time-concept which need not be made use of in the theories of micro-events and of reversible motions.
>
> It would be a prejudice arising from the great success of physics in its own proper field if the pared-down or minimal notion of time which it uses were to be regarded as 'more true' in some sense." (Kenneth G. Denbigh, *Three Concepts of Time* [Berlin: Springer Verlag, 1981], pp. 67-68).

[56] Grünbaum, "Status of Temporal Becoming," p. 26; cf. 21.

1965 become present in 1965 rather than at some other time? Grünbaum amplifies Smart's statement of this challenge as follows:

> If past, present, and future were real properties of events [i.e., properties possessed by physical events independently of being perceived], then it would require [non-trivial] explanation that an event which becomes present [i.e., qualifies as occurring *now*] in 1965 becomes present [now] at that date and not at some other (and this would have to be an explanation over and above the explanation of why an event of this sort occurred in 1965).[57]

Grünbaum then comments, "It would, of course, be a complete trivialization of the thesis of the mind-*in*dependence of becoming to reply that *by definition* an event occurring at a certain clock time *t* has the unanalyzable attribute of nowness at time *t*."[58] A-theorists "have not even been able to hint how they might resolve that perplexity without utterly trivializing their thesis."[59]

Let us first try to clarify exactly what the perplexity is. In the antecedent of the conditional sentence cited by Grünbaum from Smart, past, present, and future are described as real properties of events. But the pure A-theorist would disagree with this characterization because, as Broad explained, on the A-Theory there are no events which have the property of futurity.[60] Indeed, only present events (or, better, things) exist, so that the antecedent should be revised to read: "If presentness (or nowness) is a real property of events,...." How, then, is the consequent to be understood? What is it that is said to require explanation? Smart says, "that an event which becomes present in 1965 becomes present at that date and not at some other," and Grünbaum interprets this as "that an event which qualifies as occurring *now* in 1965 becomes now at that date and not at some other." Since "at that date" means "in 1965," the fact to be explained is why an event which becomes present or qualifies as occurring *now* in 1965 becomes present or now in 1965, rather than, say, 1966. But this fact appears to be a tautology—when else could an event which is now in 1965 be now than in 1965? If this is not a tautology, then the fact requiring explanation must be something like the following:

4. An event, which *occurs* in 1965, is now in 1965, rather than in 1966.

This fact is supposed to be something other than

5. An event, which *occurs* in 1965, *occurs* in 1965, rather than 1966.

The explanation of (5) would involve recounting the antecedent and contemporaneous causal conditions which exist earlier than and in 1965 and which are sufficient for the event's occurrence. But the explanation of (5) is not supposed

[57] Ibid., pp. 26-27. In Smart's original text, the final "occurred" is italicized, indicating that it is a tenseless verb (J. J. C. Smart, *Philosophy and Scientific Realism* [London: Routledge, Kegan & Paul, 1963], p. 135).

[58] Grünbaum, "Status of Temporal Becoming," p. 27.

[59] Ibid., p. 21.

[60] C. D. Broad, *Examination of McTaggart's Philosophy*, 2 vols. (Cambridge: Cambridge University Press, 1938; rep. ed.: New York: Octagon Books, 1976), 2: 316.

to suffice for (4) because nowness is something over and above tenseless occurrence at a time *t*. Accordingly, the key conditional in the argument is:

6. If nowness is a real property of events, then there must be distinct explanations of why an event, which *occurs* in 1965, not only *occurs* in 1965 instead of 1966 but also has the property of nowness in 1965 rather than in 1966.

But how does it follow from the reality of the property of nowness that the explanations must be different? For it seems to be true that

7. Necessarily, if an event occurs at t_n, then that event has nowness at t_n.

Hence, if one has an explanation for why an event has tenseless occurrence at t_n, then that explanation coupled with (7) entails that the event has nowness at t_n. Grünbaum says that such a response trivializes the mind-independence of becoming. Not at all; Grünbaum conflates the property of "nowness" with the property of "nowness at a time". His confusion is similar to that of those who conflate "actuality" with "actuality in a world." This confusion is already evident in Smart's statement, "Indeed, every event is 'now' at some time or another, and so the notion of 'now' cannot be that of an objective property in nature which singles out some events from others."[61] It is correct that "nowness at a time" cannot single out one event from another, just as "actuality in a world" cannot distinguish a world; but it does not follow that "nowness" does not distinguish an event, nor actuality a state of affairs. So while it is trivial to state that an event which occurs at t_n has nowness at t_n, that does not impugn the significance of nowness. Hence, what is trivial is to ask why an event which occurs at t_n is now at t_n. For each event is now at the time at which it occurs, just as each state of affairs is actual in the world in which it obtains.

We may conclude that the perplexity which Grünbaum alleges to attend the A-Theory is in fact no perplexity at all, but a mere triviality. Any explanation of why an event occurs at t_n will also suffice to explain why that event has nowness at t_n.

CONCLUSION

We have seen that the arguments adduced by Grünbaum on behalf of the B-Theory are not in the end very impressive. The first of these fails to understand properly the property of nowness as a sort of tensed actuality. The second and most important objection is predicated on the positivistic reduction of time to physical time and fails to show why the objective becoming which lodges in the former must appear in the latter in order for scientific theories to be explanatorily adequate. Finally, the third argument raises a question which is trivial once an explanation is given for the occurrence of an event at any time *t*.

Having examined the principal arguments for the B-Theory of time, we have seen that neither relativity theory nor the miscellaneous arguments considered in this

---

61 Smart, *Philosophy and Scientific Realism*, p. 135.

chapter successfully support that viewpoint. On the contrary, we have seen how a coherent and scientifically adequate A-theoretic conception of time, based on the distinction between metaphysical and physical time, may be constructed.

PART II

ARGUMENTS AGAINST A B-THEORY OF TIME

SECTION 1: PHILOSOPHICAL OBJECTIONS

CHAPTER 7

THE "SPATIALIZING" OF TIME

On the basis of our examination of the principal arguments supporting a B-Theory of time, we have concluded that there appear to be no compelling reasons to adopt the B-Theory of time and to reject the objectivity of tense and the reality of temporal becoming. Moreover, the B-Theory also faces a number of philosophical and theological problems which appear to be quite serious. In this section I wish to examine three philosophical problems which attend the B-Theory.

B-RELATIONS AND A-DETERMINATIONS

It has long been a not uncommon allegation among A-theorists that the B-Theory, in light of its denial of the objectivity of A-determinations, is guilty of "spatializing" time and that the theory, as so gross a distortion of reality with respect to time, therefore cannot be correct. Milic Capek, for example, asserts, "From Zeno to Russell and some contemporary misinterpretations of relativity, the fallacy of 'spatialization of time' is one of the most persistent features of our intellectual tradition."[1] In particular the representation of temporal succession as geometrical separation of two points on a line labeled "time" is misleading, according to Capek. "There is no harm in using this spatial symbolism as long as it is understood as a mere symbolism, i.e. as a static translation of genuinely successive terms into spatial imagery; but it becomes a vicious distortion of the true nature of time as soon as it is taken literally."[2]

---

1 Milic Capek, *The Concepts of Space and Time*, Boston Studies in the Philosophy of Science 22 (Dordrecht: D. Reidel, 1976), p. XXVI.

2 Ibid., p. XLIX; cf. p. XXX. Elsewhere he writes,

> "In symbolizing time by the axis *t* (of independent variables) there was at first no conscious attempt at spatializing time. The dynamic and progressive character of time was symbolized by an ideal motion of the pointlike present sliding along the time axis from the past to the future. But in contemplating a spatial diagram of temporal process it is easy and psychologically natural to forget its underlying dynamic meaning. Any spatial symbol contemplated at a given moment is *completed*, i.e., all its parts are given *at once*, simultaneously, in contrast with the temporal reality which is by its own nature *incomplete* and whose 'parts'—if we are justified in using such a thoroughly inadequate term—are by definition successive, i.e. non-simultaneous. The spatial symbolism leads us to forget the essential difference between juxtaposition and succession and to reduce the differences between the past, present, and future to simple *differences of position:* 'past' events are symbolized by the positions lying to the left of the point representing the 'present,' while 'future' events lie to the right of the same point on the same already

Of course, this allegation does not imply that the B-Theory treats time as a spatial dimension; after all, B-theorists regard time as ordered by the *earlier than/later than* relations, which are uniquely temporal relations. Nor need this objection be understood to imply that in the absence of objective A-determinations time is effectively converted by the B-Theory into a literal fourth spatial dimension. Rather the charge of the A-theorist under the metaphor of the "spatialization" of time may be taken to mean that in the absence of objective A-determinations, the temporal dimension has been robbed of that which makes it time, so that the ordering relations which obtain among putatively temporal particulars are only gratuitously labeled "earlier than/later than" and the resultant dimension, if not a mere spatial dimension, is in any case definitely not time. Thus, Peter Kroes maintains that "it is not clear at all that the occurrence of events in the tenseless sense can generate a real temporal ordering. This tenseless occurrence of events only leads to a formal ordering relation between the physical events, not to a temporal ordering."[3] Despite the protestations of B-theorists, there seems to be no reason to regard that dimension which is treated by the B-Theory, now stripped of all tense, as time—and therefore the B-Theory cannot be a correct theory of time.

In essence, then, the A-theorist's charge is that the B-Theory of time is incoherent, for it affirms the reality of time and temporal relations such as *earlier than/later than* but denies the reality of tense which is ontologically foundational to time and temporal relations. What the A-theorist wants to know is why, on the B-Theory, those relations which are said to obtain among putatively temporal particulars deserve to be called "earlier than" or "later than," why these relations so labeled should be thought to be temporal at all? After all, these relations, as usually understood by B-theorists, furnish no basis for structurally distinguishing one direction in time from the other nor for specifying a direction to time. Moreover, even a spatial continuum can be constructed which is anisotropic, based on purely spatial relations.[4] So why should we regard the B-theoretical tenseless dimension or its ordering relations as temporal?

In support of his position, the A-theorist may first note that on the A-Theory of time, the obtaining of the temporal relations *earlier than/later than* among temporal particulars can be derived from the objectivity of A-determinations and the A-series.[5] The point goes all the way back to McTaggart.[6] Although he considered

---

drawn 'temporal axis.' Thus the spatial diagram suggests the wrong idea that the successive moments already *coexist* and that their pastness and futurity is not genuine, but only 'phenomenal' or 'apparent'" (Milic Capek, *The Philosophical Impact of Contemporary Physics* [Princeton: D. Van Nostrand, 1961], p. 163).

3 Peter Kroes, *Time: Its Structure and Role in Physical Theories*, Synthèse Library 179 (Dordrecht: D. Reidel, 1985), p. 210.

4 For example, Aristotelian space was anisotropic in that the directions toward and away from the center of the universe had different causal properties.

5 Some A-theorists would deny that tenseless temporal relations obtain among temporal particulars, insisting that we should not assert, for example, that "e_1 *is* earlier than e_2" but that "e_1 was earlier than e_2" or some other appropriately tensed locution (see, for example, A. N. Prior, "Tense Logic and the Logic of Earlier and Later," in *Papers on Time and Tense* [Oxford: Clarendon Press, 1968], p. 123; Andros Loizou, *The Reality of Time* [Brookfield, Ver.: Gower, 1986], p. 95). But although such thinkers insist that the temporal relations *earlier than/later than* do not obtain tenselessly, they nonetheless concur in

both the A-series and the B-series to be essential to time, he regarded the latter as dependent upon the former, maintaining that the relation *earlier than* was definable in terms of the A-series: "The term *P* is earlier than the term *Q*, if it is ever past while *Q* is present, or present while *Q* is future."[7] A number of A-theorists have argued this same point,[8] and D. H. Mellor concurs:

> The *A* series could easily distinguish *earlier* and *later*. To be earlier is to be more past or less future, to be later is to be more future or less past. Specifically, *e* is earlier than *e** if *e* is sometime present when *e** is future and sometime past when *e** is present, whereas *e** is never present when *e* is future or past when *e* is present. *Earlier* and *later* are clearly definable by the order in which things, events and dates cease to be future and become first present and then past; and the difference between them is that everything moves from future to past, not *vice versa*.[9]

Mellor's concern in this context is not with providing some justification for the B-theorist's belief that tenseless temporal particulars are ordered by the relations *earlier than/later than*—he just assumes that they are—, but rather with providing some justification for the anisotropy of time and in particular for time's having an orientation on the B-Theory. The point he wishes to make is that on the A-Theory, the two directions of time are distinguishable and one direction is preferred; but in fact he makes here a much more fundamental point, namely, that the very temporal relations which lie at the heart of the B-Theory of time are derivable from the A-series. Mellor actually appears to present three different ways of defining "earlier than/later than" in terms of the A-Theory:

D_1: to be earlier than $=_{\text{df.}}$ to be more past or less future than

to be later than $=_{\text{df.}}$ to be more future or less past than

D_2: *e* is earlier than *e** $=_{\text{df.}}$ when *e* is present *e** is future, and when *e** is present *e* is past; and when *e* is present *e** is not past and when *e* is future *e** is not present

---

making such relations derivative from the more primitive and foundational A-determinations. It is thus somewhat misleading to refer to these temporal relations as B-relations, since, if the A-theorist is correct, they, as *temporal* relations, are parasitic upon objective tense and cannot exist independently of it, as the B-Theory postulates. I shall therefore refer to them as "so-called B-relations."

6 John McTaggart Ellis McTaggart, *The Nature of Existence*, 2 vols., ed. C. D. Broad (Cambridge: Cambridge University Press, 1927; rep. ed.: 1968), 2: 30.

7 Ibid., 2: 271.

8 C. D. Broad, *Scientific Thought* (London: Kegan, 1925), p. 303; D. Y. Deshpande, "Professor Ayer on the Past," *Mind* 65 (1956): 86; Wilfried Sellars, "Time and the World Order," in *Scientific Explanation, Space, and Time,* Minnesota Studies in the Philosophy of Science 3 (Minneapolis: University of Minnesota Press, 1962), p. 546; Richard M. Gale, *The Language of Time*, International Library of Philosophy and Scientific Method (London: Routledge, Kegan & Paul, 1968), pp. 27-28, 86-100; George Schlesinger, *Aspects of Time* (Indianapolis: Hackett, 1980), pp. 46-47; Loizou, *Reality of Time*, p. 95; F. M. Christensen, *Space-like Time* (Toronto: University of Toronto Press, 1993), pp. 123-124.

9 D. H. Mellor, *Real Time* (Cambridge: Cambridge University Press, 1981), p. 140.

e is later than *e** $=_{\text{df.}}$ when *e** is present *e* is future, and when *e* is present *e** is past; and when *e** is present *e* is not past and when *e** is future *e* is not present

D_3:*e* is earlier than *e** $=_{\text{df.}}$ *e* ceases to be future and becomes present first, and *e** ceases to be future and becomes present second; and *e* ceases to be present and becomes past first, and *e** ceases to be present and becomes past second

e is later than *e** $=_{\text{df.}}$ *e**ceases to be future and becomes present first, and *e* ceases to be future and becomes present second; and *e** ceases to be present and becomes past first, and *e* ceases to be present and becomes past second

Michael Tooley has charged that all attempts to analyze the relation of temporal priority in terms of A-theoretical concepts are circular in that they involve relational tensed concepts (like being *more past than* or *being past at t*) which can only be analyzed in terms of *earlier/later than* and in that they involve the concepts of *past* and *future*, which again can only be analyzed in terms of B-relations.[10] Let us therefore examine Mellor's three proposals more closely. Mellor's (D_1) appears to give a wonderfully simple and straightforward analysis of the *earlier than/later than* relations in terms of A-determinations. It is noteworthy that here these relations are defined in terms of absolute A-determinations, not A-determinations relative to a time. The resulting A-relations can only be ascribed to *relata* in a tensed way, for there is no sense in the tenseless proposition "*e is* more past or less future than *e**." To assert that *e* is more past or less future than *e** is to make implicit reference to the absolute present and, hence, to make a present-tense statement. For that reason, (D_1) should probably not be understood as giving a definition of *earlier than/later than*, for, unless there are no strictly tenseless truth-bearers, the ostensibly tenseless sentence "*e is* earlier than *e**" does not *mean* the same thing as the present-tensed sentence "*e* is more past or less future than *e**." For while a timeless God could know the former, He could not believe the latter. Rather (1) should be understood as giving a reductive analysis of the *earlier than/later than* relations in terms of relations involving A-determinations. If the relations *earlier than/later than* can be truly and tenselessly ascribed, it is because and only because the A-relations *more past/future than* and *less past/future than* can be truly and presently ascribed. Hence, if it is tenselessly true that "*e is* earlier than *e**," this is the case just because it is presently true that "*e* is more past or less future than *e**." Neither of these statements allows us to infer the tense of *e* or *e**: they could be both past, both future, one past and the other future, or one present and the other past or future. In

---

[10] Michael Tooley, *Time, Tense, and Causation* (Oxford: Clarendon Press, 1997), chap. 6.

the case that one of these events is present, we must think of it as possessing zero futurity or pastness and therefore less futurity or pastness than its *relatum*. Thus, for example, if e *is* earlier than e^* and it happens that e is present, then e is less future than e^*. Similarly, in the case that one of the events is past and the other future, we should think of each one has having none of the A-determination of its *relatum*. Thus, for example, if e is past and e^* is future, then e *is* earlier than e^* just because e is more past than e^*. Thus, we arrive at the logical equivalences

D_1': e *is* earlier than $e^* \equiv e$ is more past or less future than e^*

e *is* later than $e^* \equiv e$ is more future or less past than e^*

Since the right hand side of the equivalences involves disjunctions, A-determinations cannot be viewed as ontologically derivative from *earlier than/later than* relations, for these are not equivalent to any single A-determination or A-relation. But if individual tenses are real, then temporal relations also obtain. Therefore, given the truth of these equivalences, it must be the so-called B-relations which are reducible to A-determinations.

Richard Gale has objected to providing such a reductive analysis of the *earlier than/later than* relations in terms of the absolute A-relations *more/less past than* and *more/less future than* on the grounds that such analysis trivializes the thesis that so-called B-relations are reducible to A-determinations.[11] Gale's complaint is that such an analysis does not show the reducibility of the B-series to what he calls "a *pure* A-series," that is to say, a series which is determined only by the unqualified A-determinations *past, present,* and *future*, but shows only the B-series' reducibility to what Gale calls "an *impure* A-series," that is, a series which is also determined by the qualified A-determinations *more* past and *more* future. Before looking at why Gale thinks such a reduction trivial, it should be remarked at this point that the distinction he draws between a pure and an impure A-series is quite baffling.[12] For there really is no such thing—nor has anyone ever thought there was such a thing—as a so-called "pure" A-series, which would consist of just three members: the present, the unqualified past (or as Gale puts it, a "blob past"), and the unqualified future. Rather the A-series is a series of tensed positions, which include locations like "two days from now," "a little while ago," "the day before yesterday," "during the next century," and so on and so forth. There are innumerably many positions

---

11 Gale, *Language of Time*, p. 93.

12 See Loizou, *Reality of Time*, pp. 79-95. Perhaps Gale would appeal to McTaggart's statement "And this—the B series—cannot be got out of the A series alone. It is only when the A series, which gives change and direction, is combined with the C series, which gives permanence, that the B series can arise" (J. Ellis McTaggart, "The Unreality of Time," *Mind* 17 [1908]: 464). But McTaggart may be speaking loosely here, meaning that the B-series cannot be derived from A-determinations alone. The A-series only arises when A-determinations supervene on an atemporal C-series. Or he may think the A-series must be combined with a C-series in order to yield a B-theoretical ontology; but then he errs in thinking such a becomingless ontology is necessary for the existence of temporal relations. In view of his claim that the C-series cannot be derived from the A-series and the crucial role played by an ontology of permanent events in McTaggart's Paradox, this second interpretation is likely. But in neither case are we to think of the A-series as involving a "blob" past or future.

which go to make up an A-series, and the characterization of such a series as "impure" is gratuitous. The A-series just is all those temporal positions which are picked out by tensed descriptions rather than tenseless descriptions.

Why, then, is the reducibility of so-called B-relations to the A-relations specified in (D_1') trivial? Gale answers,

> The need to make use of the expression 'more past (future) than'...in our reduction of a *B*-statement to a disjunction of *A*-statements somewhat trivializes tenet (1) of the *A*-theory concerning the reducibility of *B*-relations to *A*-determinations. One feels that the only true-blue *A*-determinations are the unqualified past, present, and future: 'more future than' smells very much like a relation, since it looks like a two-place predicate. At any rate, it is a very different sort of *A*-expression than the simple 'is present (past, future)'....the need to make use of 'more past (future) than'...wins a rather hollow victory for the *A*-theory and McTaggart....[13]

This explanation only compounds one's perplexity. Gale intimates that the only genuine A-determinations are *past, present,* and *future.* But even if we regard these three as somehow foundational for the rest (indeed, I subscribe to the metaphysical primacy of the present), why think that there are no other real A-determinations? What about the past perfect, the future perfect, and all the other possible tenses?[14] What about tensed expressions like "last year," "today," and "henceforth," which are used to locate events in the A-series and to express objective tensed facts, according to the A-Theory? The B-theorist can no more admit the objectivity of these tenses than he can allow the simple tenses *past, present,* and *future.* Whether one calls all these tenses "A-determinations" or not is a mere matter of terminology; the point remains that they ascribe objective tense, according to the A-Theory, and so must be explained away by the B-theorist. Gale has misgivings about "more future than" because it appears to be a relation. This troubles Gale, I think, because he suspects that it could then be regarded as a surreptitious way of saying "later than." But such a fear is misplaced; for *more future than* is an A-relation which is ascribed *absolutely*: *e* is not more future than *e** at some time *t*; rather *e* is more future than *e**, period. Hence, its being a relation is irrelevant to the objectivity of the tense it expresses.

Tooley objects to taking the concepts *more past* and *more future* as primitive concepts and denies that they can be reductively analyzed in terms not involving *earlier than/later than* relations.[15] But it seems to me that the A-theorist is at liberty to stipulate that the above concepts are among his theoretical primitives, noting that he has good reasons to think that a tensed theory of time is correct and that no comparable analysis of tensed concepts is available in terms of B-relations alone. Tooley gives no reason for thinking *more past* is not primitive except that it is analogous to *more future*, which, he argues, cannot be a primitive concept because the concept of *future* is not primitive. But Tooley's first argument for denying that the concept of *future* is primitive is based on the empirical assumption that "a

---

13 Gale, *Language of Time*, p. 99.

14 Recall Reichenbach's analysis in Hans Reichenbach, *Elements of Symbolic Logic* (London: Collier-Macmillan, 1947), pp. 288-297.

15 Tooley, *Time, Tense, and Causation*, p. 163.

descriptive concept cannot be analytically basic for a given person unless that concept picks out a property or relation that, for the individual in question, is, or has been, either an object of direct, non-causal awareness, or an object of immediate perception"[16]—an assumption which Tooley makes no attempt to justify and which strikes me as wholly non-perspicuous. His second argument is not only based on an equally gratuitous principle (namely, that "one should attempt to show that necessary truths are analytic") but also begs the question by asserting that the proposition *any event later than a present event lies in the future* cannot be shown to be analytic if the concept of *future* is primitive, for this assertion presupposes that *later than* cannot be reductively analyzed in terms of *more future*.[17] Thus, Tooley has failed to show in a non-question begging way that *more past/more future* cannot be primitive concepts.

But even if he were right about this, the A-theorist could offer an analysis of these notions in terms of other tensed concepts. For example, *more past* could be analyzed as *longer ago* and *more future* as *further hence*, both A-theoretical notions. Or he could analyze these notions in terms of a distance scale, the present, and the orientation of time (orientation being distinct from *earlier/later than*, as is evident from the possibility of space's having an orientation). Thus, Tooley's objections are misconceived. So far as I can see, (D_1') succeeds admirably in giving a reductive analysis of *earlier than/later than* in terms of tensed facts alone.

Mellor's definition (D_2) differs from (D_1') in that it ascribes A-determinations, not absolutely, but relative to a time. According to this definition, *e* is earlier than *e** just in case there is some *t* such that *e** is future relative to *t* and *e* is present at *t*. Both the *definiendum* and the *definiens* are tenseless in their truth value. Tooley, appealing to Gale, also has objected to this sort of definition on the grounds that it is circular. For, in Gale's words,

> The predicates '___ *is* past at ___' and '___ *is* future at ___' are synonyms for '___ *is* earlier than ___' and '___ *is* later than ___' respectively, in that all four are tenseless two-place predicates which...express a timelessly true or false statement about a B-relation between two events, i.e. they make B-statements. '*Is* past (present, future) at some event or date,' unlike 'is now past (present, future),' does not indicate an A-determination; and therefore it cannot be used to determine the position of an event in the A-series.[18]

Now Gale is correct that the *definiendum* and the *definiens* both express tenselessly true or false statements. But it does not follow that no A-determination is therefore

---

16 Ibid., p. 179; cf. pp. 98-99, 379.

17 Ibid., pp. 179-180. Tooley also denies that the A-theorist can show that the proposition *any event causally dependent upon a present event lies in the future* is analytic; but I should simply say that this proposition is false. Tooley also admits that his argument that *later than* cannot be reductively analyzed in terms of *more future* is question-begging because one of his arguments on p. 163 involved the claim that the concept *future* is unanalyzable, but he insists that the other argument on p. 163 did not involve that claim, thereby avoiding a *petitio principii*. But if the reader examines the two arguments on p. 163, I am confident that he will find that they both depend crucially on the claim that the concept of *future* can only be analyzed in terms of *later than*, which on pp. 179-180 Tooley cannot justify.

18 Gale, *Language of Time*, pp. 90-91. So also Tooley, *Time, Tense, and Causation*, p. 161, *contra* Sellars and Schlesinger.

indicated in the *definiens*. This may be clearly seen by substituting for "present (past, future) at *t*" a locution which is clearly not synonymous with the tenseless expressions "simultaneous with (earlier than, later than) *t*." For example:

D_2': *e* is earlier than *e** $=_{df.}$ There is some time *t* such that at *t* the propositions *e is present* and *e** *is future* are true

e is later than *e** $=_{df.}$ There is some time *t* such that at *t* the propositions *e is present* and *e** *is past* are true

The B-theorist cannot admit that the tensed propositions in question are true at *t*, and we have elsewhere shown the failure of the New B-Theory of Language to eliminate objective tense via the device of token-reflexive truth conditions.[19] Hence, each *definiens* in (D_2') does refer to an irreducible A-determination. Or, if someone has a problem with tensed propositions, the A-theorist may propose

D_2'': *e* is earlier than *e** $=_{df.}$ There is some time *t* such that at *t* it is an objective fact that *e* has presentness and that *e** is future

e is later than *e** $=_{df.}$ There is some time *t* such that at *t* it is an objective fact that *e* has presentness and that *e** is past

According to (D_2''), at *t e* has the premier A-determination of presentness, and the *definiens* in each case refers to an objective tensed fact, which the B-theorist cannot regard as equivalent to "simultaneous with (earlier than, later than) *t*." Each *definiens* in (D_2') and (D_2'') therefore refers to an A-determination.

Gale's mistake may have been his falsely assuming that a tenselessly true statement cannot refer to an A-determination.[20] But one need not use tensed statements alone to talk about tense; for example, "The A-determination *presentness* is an absolute property, not a mere relation" is tenselessly true (or false), but refers to an objective A-determination. Gale also seemed to think that definitions like (D_2') and (D_2'') illicitly smuggle in the so-called B-relation of *simultaneous with* in their use of the expression "at *t*."[21] But I see no grounds for this suspicion; to say that at *t e* has presentness is not to assert a simultaneity relation between *t* and *e* but simply to assign a temporal coordinate to *e*.[22] The *definiens* thus should not be

---

[19] See my *The Tensed Theory of Time: a Critical Examination*, chap. 3.

[20] Cf. Tooley's odd claim that *event E lies (tenselessly) in the present at time t* is a tensed proposition (Tooley, *Time, Tense, and Causation*, p. 192).

[21] Gale, *Language of Time*, p. 97.

[22] Typically, two events are said to simultaneous with each other in that they occur at the same time. But it would be odd to say that an event and a time are simultaneous, *i.e.*, occur at the same time. Indeed, we see that being "at a time" is foundational to the notion of simultaneity, rather than the other way around.

construed in terms of the ascription of any so-called B-relations. So far as I can see, then, Mellor's second proposal also succeeds in showing how the A-Theory of time can account for the existence of the temporal relations essential to the B-Theory.

Mellor's third definition involves tenseless statements about the nature of objective temporal becoming which, if true, entail the existence of the tenseless relations *earlier than/later than.* Rather than definitions, (D_3) should probably be understood as logical equivalences, lest "*e* is earlier than *e**" be synonymous with "*e** is later than *e*."[23] Given that necessarily, every event in time is sooner or later present, it seems to me that (D_3) can be considerably simplified:

D_3': *e* is earlier than $e^* \equiv e$ becomes present first and e^* becomes present second

e is later than $e^* \equiv e^*$ becomes present first and *e* becomes present second

Neither Gale nor Tooley has any objections relevant to (D_3'). Oaklander objects that the use of "first" and "second" conceal so-called B-relations;[24] but a moment's reflection shows that this is not the case. These are ordinal numbers that are wholly atemporal and can characterize spatial or abstract objects as well as temporal particulars. Given the order of their temporal becoming, the temporal ordering of the two events in question necessarily follows.

It seems to me, therefore, that Mellor is quite right that if the A-Theory of time is true, the A-theorist can account for the existence of so-called B-relations by founding them on the reality of tensed facts; thus far, McTaggart's argument seems to be vindicated. So now the question must arise, why think that so-called B-relations will still obtain among events once the detenser has robbed time of all A-determinations? Again the point is McTaggart's: he considered that the series of events, once relieved of its A-determinations, would still exist as a series but that it would not be a B-series, that is, a temporal series. It would be an atemporal series, like the letters of the alphabet or the natural numbers, which he called a C-series. He wrote,

> The B series...cannot exist except as temporal, since earlier and later, which are distinctions of which it consists, are clearly time determinations. So it follows that there can be no B series where there is no A series, since where there is no A series there is no time.
>
> But it does not follow that, if we subtract the determinations of the A series from time, we shall have no series left at all. There is a series—a series of permanent

---

23 Perhaps we could preserve the definitions by reformulating D_3*:

e is earlier than $e^* =_{df.}$ *e* becomes present before *e** becomes present

e is later than $e^* =_{df.}$ *e* becomes present after *e** becomes present

Of course, it will be objected that "before" and "after" are just another way of saying "earlier than" and "later than." But this is not obvious. In any series with a direction, such as rank, a lower member is before a higher member in a tenseless sense. Thus, "before" and "after" express orientation and are not inherently temporal.

24 L. Nathan Oaklander, "McTaggart's Paradox and Smith's Tensed Theory of Time," *Synthèse* 107 (1996): 211.

> relations to one another of those realities which in time are events—and it is the combination of this series with the A determinations which gives time. But this other series—let us call it the C series—is not temporal, for it involves no change, but only an order. Events have an order. They are, let us say, in the order M, N, O, P. And they are therefore not in the order M, O, N, P, or O, N, M, P, or in any other possible order. But that they have this order no more implies that there is any change than the order of the letters of the alphabet, or of the Peers on the Parliament Roll, implies any change.... It is only when change and time come in that the relations of this C series become relations of earlier and later, and so it becomes a B series.[25]

The C-series of events, McTaggart acknowledged, greatly resembles the temporal series as conceived by the B-theorist (in this case, Russell).[26] The C-series will include every term which is an event in the B-series and will include all the terms in the same order as the order in which the events are ranged by the relations *earlier than/later than* in the B-series; moreover, both series exclude the objective reality of A-determinations. What then is the difference between them? Just this: the terms of the B-series are ordered by the relations *earlier than/later than* and are therefore events. The corresponding ordering relation of the C-series, however, cannot be *earlier than/later than* and thus its members cannot properly be called events. What, then, are the ordering relations among the members of the C-series which correspond to the relations *earlier than/later than* among the members of the B-series? McTaggart did not think of them as spatial relations; rather he made the ingenious suggestion,

> They are the relations 'included in' and 'inclusive of.' Of any two terms in the *B* series, one is earlier than the other, which is later than the first, and by means of these relations all the terms can be arranged in one definite order. And of any two terms in the *C* series, one is included in the other, which includes the first, and by means of these relations all the terms can be arranged in one definite order. And it seems to me possible...that it is the relations of 'included in' and 'inclusive of' which appear as the relations of 'earlier than' and 'later than'....[27]

On McTaggart's view such a C-series infused with A-determinations would yield a B-series. But without objective A-determinations, so-called B-relations could not exist. Once all A-determinations are removed from the temporal series of events, what is left over is not the B-series. Rather it is not a temporal series at all.

---

25 McTaggart, "Unreality of Time," pp. 461-462. Perhaps this is as good a place as any to say a word about Oaklander's interpretation of McTaggart as holding that time does not essentially involve B-relations (Nathan Oaklander, *Temporal Relations and Temporal Becoming* [Lanham, Maryland: University Press of America, 1984], pp. 33-37). Oaklander's cited texts only show that McTaggart took A-determinations and the A-series as more fundamental or ultimate than B-relations and the B-series but do not state that the latter are not, however, essential to time. On the contrary, McTaggart expressly says they are. His view is that if one takes a C-series and invests it with A-determinations, then the ordering relations among the terms will be the *earlier than/later than* relations and the terms will be events.

26 McTaggart, *Nature of Existence*, 2: 31.

27 Ibid., 2: 240.

THE B-THEORY AND TEMPORAL RELATIONS

It seems to me that in light of the foregoing, it is now open to the A-theorist to turn to his B-theoretical interlocutor and say, "I have an ontological foundation in my metaphysic of time for affirming the existence of the (tenseless) temporal relations *earlier than/later than*. But what entitles you, having stripped time of all its tense-determinations, to assume that what remains is really time? Why should we regard those relations obtaining between tenselessly existing events in the B-series as *earlier than/later than* rather than as some atemporal relations analogous to the *less than/larger than* relations obtaining between the members of the natural number series? Indeed, why think that any such relations exist at all, in addition to the relations which obtain between, say, the points on a spatial line? On my theory, moreover, these relations are not only entailed by the reality of tensed facts, but the existence of these relations entails the existence of tensed facts; they could not exist apart from tensed facts. So why, if there is no past, present, and future, as you claim, should we think that there are still *earlier* and *later*?"

Now what is B-theorist supposed to say at this point? One typical response is simply to deny steadfastly that there is any problem here at all. Thus, Oaklander asks,

> What distinguishes *greater than* among numbers from *later than* among events? ...the answer is not to be found in anything other than the relation itself. The temporal relation of succession is a simple and unanalyzable relation.... We can understand the difference between *later than* in time and 'later than' (or *greater than*) in a number series because we can perceive the difference between the two relations. There is no further basis for the difference....[28]

We certainly can understand the difference between the relations *greater than* and *later than*; but that leaves unaddressed the questions, "Why think that the relation analogous to *greater than* which the B-theorist claims obtains between tenselessly existing members of the B-series is in fact the *later than* relation?" and more fundamentally, "Why think that any such relation exists at all in the absence of tense and temporal becoming?" These questions are especially pressing for Oaklander, since the A-theorist will be equally dubious of his claim that on the B-Theory the "temporal relation of succession" really exists among events, which seem to be strung out like the tenselessly existing points on a geometrical line. Oaklander's sort of response to our questions seems to be little more than the assertion, "That's just the way it is." The B-series is just defined to be ordered by the *earlier than/later than* temporal relations and no further justification is required for this stipulation.[29] But such a response would be acceptable only if the A-theorist both had similarly to treat the relations as undefined and could not show that, on his theory, the existence of such relations is entailed by the reality of tense. Given the

---

[28] Oaklander, *Temporal Relations and Temporal Becoming*, p. 17.

[29] "...temporal relations are primitive and unanalyzable relations, and the difference between spatial and temporal relations is an irreducible qualitative difference" (Oaklander, "McTaggart's Paradox," p. 209). The problem is that such relations are not unanalyzable and irreducible, as the A-theorist has shown.

A-theorist's demonstration that on his theory so-called B-relations are not primitive and undefined or, more modestly, that on his theory such relations entail and are entailed by tensed facts, the B-theorist's simple stipulation of such relations as undefined and ungrounded primitives appears gratuitous and his refusal to show why such temporal relations would exist on his theory as stubborn intransigence born out of want of argument. Charging that B-theorists have shirked the task of explaining what makes the putatively *earlier than/later than* relations *temporal* relations, Mellor asserts, "Their 'block' universes have no more real time in them than McTaggart's does—the difference being that McTaggart sees this and they, by and large, do not."[30]

Where, then, shall the B-theorist turn for some justification for his contention that his tenseless series is, in fact, ordered by genuinely temporal relations like *earlier than/later than*? Some have thought to turn to science and its notion of physical time in order to provide a foundation for the *earlier than/later than* relations, as well as the anisotropy of time and time's arrow.

For example, certain philosophers and scientists find justification for regarding the relations obtaining between tenselessly existing events as peculiarly temporal in character on the basis of the difference between the time and space coordinates in the Special Theory of Relativity (SR). Capek reminds us that certain early expositors of SR carelessly treated that theory, or Minkowski's formulation of it, as literally teaching the spatialization of time.[31] Cunningham, for example, wrote, "With Minkowski space and time became particular aspects of a single four-dimensional concept.... Now if all motional phenomena are looked at from this point of view they become timeless phenomena in a four-dimensional space."[32] Similarly, Silberstein, comparing SR to the philosophy of time laid out in Wells's *The Time Machine*, observed, "There is thus far an intrinsic similarity, a kind of coordinateness between space and time, as the Time Traveller, in a wonderful anticipation of Mr. Wells, puts it: 'There is no difference between Time and Space except that our consciousness moves along it.'"[33] More careful expositors were quick to point out that SR in no way teaches that time is a fourth spatial dimension, even if under Minkowski's formulation, space and time are united in a four-dimensional spacetime continuum.[34] Einstein himself, under the prompting of a confused Émile Meyerson,[35] repudiated the "extravagances of the popularizers and

---

30 D. H. Mellor, "McTaggart, Fixity and Coming True," in *Reduction, Time and Reality,* ed. Richard Healey (Cambridge: Cambridge University Press, 1981), p. 80.

31 Milic Capek, "The Inclusion of Becoming in the Physical World," in *Concepts of Space and Time*, p. 501. I owe the following two quotations to Capek.

32 E. Cunningham, *The Principle of Relativity* (Cambridge: Cambridge University Press, 1914), p. 191.

33 L. Silberstein, *The Theory of Relativity* (London: Macmillan, 1914), p. 134.

34 Paul Langevin, for example, protested, "We do not at all mean to say that time is a fourth dimension of space; that would make no sense" (P. Langevin, "L'aspect général de la théorie de la relativité," *Bulletin scientifique des étudiants de Paris* 2 [April-May, 1922], p. 6).

35 See Émile Meyerson, *La déduction relativiste* (Payot: Paris, 1925), pp. 97-110, reprinted in *Concepts of Space and Time*, 353-362. Meyerson, and Capek after him, equated the tenseless theory of time with the literal spatialization of time, so that they mistakenly took Einstein and others' repudiation of the alleged spatialization of time by SR as an affirmation of the objective reality of temporal becoming. He

even many scientists" who construed SR to teach that time is a spatial dimension: "Time and space are fused into one and the same *continuum*, but this continuum is not isotropic. The element of spatial distance and the element of duration remain distinct in nature...."[36] The anisotropy to which Einstein refers is the fact that in the Lorentzian metric for Minkowski spacetime the temporal interval is distinguished from the spatial intervals by having the opposite sign: $ds^2 = dx^2+dy^2+dz^2-dt^2$. This difference in sign clearly distinguishes which coordinates in Minkowski spacetime are the spatial and which the temporal coordinate. Similarly, in the General Theory of Relativity spacetime possesses a metric with respect to the interval which includes three positive spatial parameters and one negative temporal parameter, so that its geometry is pseudo-Riemannian in that at every spacetime point it can be approximated by a local pseudo-Euclidean Minkowski spacetime. Accordingly, some thinkers conclude that the dimension associated with the interval having the negative sign, being thus distinguished from the other three dimensions, must be, in virtue of that difference, a temporal dimension and, hence, the relations obtaining between its occupants temporal relations such as *earlier than/later than*.

Such an inference is, however, confused. The reason that the interval having the negative sign is taken to be a temporal interval is because it has already been decided in advance that what is being fused into one and the same continuum are the three spatial and one temporal dimensions. The fact that one of the intervals in the Minkowski four-dimensional geometry has a negative sign only determines that the geometry is pseudo-Euclidean rather than Euclidean. But it does absolutely nothing to found the interpretation that what is described by this abstract 4-geometry is a spacetime rather than a four-dimensional, pseudo-Euclidean space (or anything else that can be plotted by four coordinates). The geometrical 4-space is neutral as to its interpretation; what it is taken to represent has to be brought to it by the person employing it.[37] If physical reality is conceived to be a tenselessly existing Minkowski spacetime consisting of three spatial dimensions and one temporal dimension, then the difference in sign associated with one of its dimensions could at best serve only as *evidence* that that dimension is temporal, but not as *grounds* for that dimension's being temporal.

Apart from this difference in sign between their intervals, there does not seem to be anything in SR that distinguishes the temporal from the spatial dimensions: they are structurally identical. Stephen Hawking states, "In relativity, there is no real distinction between the space and time co-ordinates, just as there is no real difference between any two space co-ordinates."[38] He points out that just as one can

---

was bewildered by such thinkers' continued advocacy of a tenseless, B-theoretical, spacetime ontology despite the fact that such thinkers had rejected the notion of time as a fourth spatial dimension.

[36] Albert Einstein, "Comment on Meyerson's 'La déduction relativiste'," *Revue philosophique de la France et de l'étranger* 105 (1928): 166, reprinted in *Concepts of Space and Time*, p. 367.

[37] Thus, Smart avers that there is no objection whatever to spatializing time in the sense of thinking of spacetime as a four-dimensional space in the geometrical sense of the word "space" (J. J. C. Smart, "Spatialising Time," *Mind* 64 [1955]: 241). What Smart fails to do, however, is provide any justification for interpreting this four-dimensional geometrical space as a spacetime.

[38] Stephen W. Hawking, *A Brief History of Time* (New York: Bantam Books, 1988), p. 24. Cf. the remark of de Beauregard with respect to Minkowski's fusion of time and space:

choose to create a new set of spatial coordinates in which the new first coordinate would be a combination of the former first and second spatial coordinates, so one can create a new time coordinate which is a combination of the former time coordinate and a former spatial coordinate. Perhaps one of the clearest illustrations of the structural identity of the space and time coordinates in relativity theory is the role reversal of space and time coordinates at the Schwarzschild radius of a black hole.[39] When the gravitational radius *r* of the collapsing object equals twice its mass, *r* assumes the role of the time coordinate while *t* becomes one of the spatial coordinates, so that the further decrease of *r* to zero represents ever later values of time. An equally startling illustration is provided by Hawking's own quantum cosmology, in which imaginary numbers are assigned to the time variable in his equations describing the universe prior to the Planck time, with the result that time is gradually converted into space and the universe becomes a four-dimensional, Euclidean space.[40] No structural changes take place in the time dimension; the conversion is accomplished simply by the mathematical trick of using imaginary rather than real numbers for the time variable.

In light of the above, I think it is clear that the B-theorist will look in vain to relativity theory for justification of his claim that the B-series is a temporal series ordered by *earlier than/later than.* SR does not teach that time is a dimension of space; but under Minkowski's formulation one dimension is temporal only because it has already been decided in advance that Minkowski 4-space is to be taken as a representation of spacetime. Apart from this stipulation, SR does not furnish any basis for taking one of the four dimensions of the pseudo-Euclidean geometry as a time dimension. In the absence of the reality of tense, then, there is just nothing in the ontological structures of the theory to constitute time. Kroes concludes,

> Through the unification of space and time into a four-dimensional spacetime manifold by means of the universal constant of homogenization *c* (=the velocity of light in vacuo), the specific character of time is completely lost. Apart from the different sign in the spacetime metric, the time coordinate is structurally indistinguishable from the space coordinates....only the introduction of the notions of past, present, and future and flow of time would turn coordinate 'time' into a real time. If these notions too have no

---

> "it must be recognized that this was really the essence of relativity theory: an equivalence of space and time, fusing geometrically to constitute the universe.... It is really impossible to push the 'spatialization of time' any further than Einstein and Minkowski did; the ancient tendency stemming from Aristotle here actually attains its final development...," (O. Costa de Beauregard, "The Principle of Relativity and the Spatialization of Time" [*Revue des questions scientifiques* 7 (1949)], rep. in *Bergson and the Evolution of Physics*, ed. P. A. Y. Gunter [Knoxville: University of Tennessee Press, 1969], pp. 246-247).

[39] See C. Misner, K. S. Thorne, and J. A. Wheeler, *Gravitation* (San Francisco: W. H. Freeman, 1973), p. 823.

[40] See J. Hartle and S. W. Hawking, "Wave Function of the Universe," *Physical Review D* 28 (1983): 2960-2975. For good discussions of the nature of this Euclidean, four-dimensional space, see Christopher Isham, "Creation of the Universe as a Quantum Process," *Physics, Philosophy, and Theology: a Common Quest for Understanding*, ed. R. Russell, W. Stoeger, and G. V. Coyne (Vatican City State: Vatican Observatory, 1988), pp. 375-408; idem, "Quantum Theories of the Creation of the Universe," unpublished paper, a preliminary version of which appears in *Interpreting the Universe as Creation*, ed. V. Brümmer (The Netherlands: Pharos, 1991).

> objective significance with regard to parameter time, then it is not clear in what sense space and time are different.[41]

Since the B-theorist denies just those notions requisite for real time, he is left without any justification in relativity theory for his claim that the B-series is a genuine temporal series.

Where else might the B-theorist turn to found the relations of *earlier than/later than?* An enormous amount of ink has been spilled in the attempt to ground the anisotropy and orientation of time in either lawlike or *de facto* physical asymmetries in nature. Fortunately, we can largely skirt that discussion here, for on the basis of what we have elsewhere seen[42], it is evident that such a project is futile, presupposing as it does a reductionistic view of time, and unsuccessful even in its own lights. Given the distinction between metaphysical and physical time, such a project cannot succeed because the physical processes at issue are at best mere sensible measures of time, not constitutive of time. They are neither necessary nor sufficient conditions for the anisotropy of time or time's arrow. We can easily conceive a possible world in which God creates a universe lacking any of the typically proposed thermodynamic, cosmological, and other "arrows" and yet God's metaphysical time is anisotropic and has a direction. And from the B-theoretic standpoint, it would be equally possible for a tenseless, becomingless metaphysical time to be wholly isotropic and without orientation even though the tenselessly existing B-series of events included states that differ systematically either *de facto* or in accordance with a natural law. (If such a scenario is impossible according to the A-Theory, it is not because the asymmetric physical processes constitute the direction or anisotropy of time, but rather because these latter properties are essential to time itself due to the reality of tense and becoming.) So once we disassociate ourselves from the positivistic reduction of time to physical time, there is no reason to see in the processes of nature anything constitutive of the asymmetry of time. Moreover attempts to ground temporal anisotropy and/or orientation in physical asymmetries are unconvincing even on the level of purely physical time.[43] Physical asymmetries are neither necessary nor sufficient for time's having an arrow, and the merely *de facto* asymmetries which exist in nature do not serve to distinguish the *earlier than* direction from the *later than* direction.

In any case, apart from being futile and unsuccessful, the attempt to found temporal anisotropy in physical processes is irrelevant to the issue at hand. For the common assumption of proponents of physically-based temporal asymmetry is that the dimension with which they are dealing is in fact a temporal dimension ordered by the relations *earlier than/later than*—and that for us is precisely the question at issue. Even if one were to agree that the fourth dimension is anisotropic and/or has an orientation, that does nothing to prove that the B-series is a *temporal* dimension

---

41 Kroes, *Time*, p. 209. Cf. idem, "Physics and the Flow of Time," in *Time, Nature, and History*, ed. P. A. Kroes, Nijmegen Studies in the Philosophy of Nature and its Sciences 4/2 (Nijmegen, The Netherlands: Catholic University of Nijmegen, 1985), pp. 43-52, where he concludes, "it is dubious whether coordinate time deserves to be called 'time' at all" (p. 49).

42 See our discussion in *Tensed Theory of Time*, chap. 7.

43 See again my *Tensed Theory of Time*, chap. 7.

or that the direction of the series from lower to higher values of the relevant coordinate is from *earlier* to *later*, since non-temporal series can equally be characterized by anisotropy and orientation. The question we face is not whether physical processes serve to render time broadly asymmetric, but whether the existence of the relations *earlier than/later than* can be founded in the B-Theory on physical asymmetries.

The effort to base the relations *earlier than/later than* on physical asymmetries is so obviously wrong-headed (even wholly apart from the distinction between metaphysical and physical time, which must render all such efforts futile) that I do not know of any philosopher who has tried to do such a thing. For suppose that there is a relation *R* which is transitive, irreflexive, and asymmetric obtaining among the members of a series. Why would the existence of natural laws which are asymmetric with respect to the two directions along the series or *de facto* asymmetries among events associated with elements in the series lend any justification to the identification of *R* as *earlier than*? When we recall that physical asymmetries are not sufficient to lend an orientation to the series, then it is difficult to see how *R* even *could* be identified with *earlier than*. For unlike relations like *to the left of*, *lower than*, or *inferior to*, the temporal relation *earlier than* seems to be a sufficient condition for time's having a direction. For if time has no preferred direction, it is purely conventional which direction in the series of tenselessly existing moments is denominated *earlier* and which *later*. It is not a fact of objective reality that *e* occurs earlier than *e**; we could reverse the situation and stipulate with equal justification that *e** shall be earlier than *e*. As Horwich admits, if time has no orientation, then "beforeness" is not wholly objective.[44] But if there is no fact of the matter concerning *e*'s temporal priority to *e**, if *e** could just as well be considered temporally prior to *e*, then *R*, if it objectively obtains, cannot be identified with *earlier than*. If *R* does not objectively obtain, then it could be identified with *earlier than*, but that is to say that the relation *earlier than* does not really obtain among ostensibly temporal particulars but is a matter of convention. What is ironic about this conclusion is that so-called B-relations thus become just as mind-dependent as A-determinations are alleged to be. But if a series lacks both objective tense determinations and objective temporal relations, it is hard to see with what justification it can be regarded as temporal. As McTaggart said, we are left with an atemporal C-series, and time becomes an illusion.

Thus, any attempt to ground the relations *earlier than/later than* in asymmetries of nature proves to be not only futile, but counter-productive for the B-theorist. Since such asymmetries cannot ground temporal orientation, and the obtaining of relations of *earlier than/later than* is sufficient for time's having an orientation, such asymmetries cannot ground those relations either. The only way to escape this conclusion would be by so emasculating the relations *earlier than/later than* that they are not sufficient conditions of time's orientation; but then they no longer appear to be temporal relations, but atemporal, akin to relations like *behind/in front of*.

---

44 Paul Horwich, *Asymmetries in Time* (Cambridge, Mass.: MIT Press, 1987), p. 43.

Are there then non-empirical, philosophical grounds for thinking that the B-series can objectively exist apart from the A-series? Mellor attempts to differentiate between *earlier* and *later* on the basis of our direct perception of one event's being later than another and the role causation plays in that perception. He contends that we do not perceive relations of temporal precedence directly; rather seeing *e* precede *e** just is seeing *e* before seeing *e** while retaining something like a memory trace of *e*. Perception of temporal priority is unique: "In no other case need perceptions resemble what they are perceptions of; they need only correlate with it. But perceiving temporal order does demand a corresponding temporal ordering of perceptions. That feature of its perception is unique to time, and clearly not accidental."[45] Mellor conceives this circumstance to hold for "not human senses only, but any sense able to perceive precedence."[46] The temporal order of my perceptions is fixed by their causal order, since I perceive *e* to precede *e** only when my seeing *e* affects my seeing *e**. Mellor then claims that any causally connected pair of events can be considered as a perception of precedence. Thus, the direction of time is always the direction of causation.

I think Mellor's account is clearly inadequate. Consider his foundational claim that perceiving temporal order demands a corresponding ordering of perceptions. If we adopt a B-Theory of time and a model of divine eternity according to which God transcends time, then He perceives the entire spacetime manifold timelessly, including the relations of temporal precedence which obtain among the various events in spacetime. He perceives which events are earlier than others, which later, and which simultaneous (whether absolutely or relative to reference frames). Of course, His perception of events is not mediated by transmission of physical influences, but He can be truly conceived to perceive in a unique way the temporal and spatial relations between things in the tenselessly existing spacetime manifold.[47] But then the fact that perceptions of temporal precedence involve a corresponding temporal precedence of perceptions is seen to be a feature accidental to creatures dependent upon physical influence chains for their perceptions of the world. There is no reason to think that temporal precedence has anything to do with the order of our perceptions of events in time.

Moreover, it is not obvious that even for physical creatures, their perceiving *e* to follow *e** entails that they perceived *e** before they perceived *e*. On the B-Theory of time, it is not obvious why causal influences cannot propagate in either or both temporal directions. Suppose that from God's transcendent, timeless perspective our perceptions of *e* precede our perceptions of *e**, but that the lines of causal influence are such that our perception of *e* involves a causal influence of *e**. In that

---

45 Mellor, *Real Time*, p. 144.

46 Ibid., p. 145.

47 Mellor himself affirms,

> "*any* kind of event could be a perception. It is not being of some special kind—e.g. electrical or chemical or organic—that makes an event a perception. Perception is simply a causal process of acquiring belief, a process from which no kind of event can be excluded *a priori*.... I am not interested only in human perception. My proposal is to apply to all perceptions of precedence, by all conceivable perceivers, among all sorts of events, things and dates, and it must be defensible as such" (Ibid., p. 153).

case, we would perceive *e** to precede *e*, but surely the objective truth of the matter is that our perception of *e* precedes our perception of *e**. In fact, it is quite conceivable on the B-Theory that causal influences are so varied that people have quite different perceptions of the temporal order of *e* and *e**. We can even imagine people whose stream of consciousness is "running in reverse" when compared to others. Rather than say, as Mellor's view implies, that in each case the person's perception of one event really precedes his perception of the other (depending on the direction of causal influence), I should insist that some people's perceptions are in a different temporal order than the order which they perceive between the events.

Even if one agrees that causal influences all run in the same temporal direction and that causal direction coincides with temporal direction, Mellor's account faces the same problem as the proponents of physical asymmetries as constitutive of time's arrow: it is wholly arbitrary which direction one denominates "earlier" and which "later." Who is to say that the direction of causation is not from later to earlier on the B-Theory? Moreover, like the proponents of *de facto* physical asymmetries, Mellor at most demonstrates an asymmetry in the series of events, not in the directions of time itself. Time itself seems to remain wholly isotropic and directionless. Finally, as an attempt to distinguish the *earlier* from *later* directions of time, Mellor's account fails to deal with the more fundamental issue of why we should think that on the tenseless theory of time, there even are *earlier* and *later* to distinguish. The proof that the series of events has an orientation identical with the direction of causation does nothing to show that this is a temporal orientation. Indeed, one cannot help but suspect that all Mellor gives us is a tenseless order of causation devoid of any real time.

CONCLUSION

In sum, I do not know of any successful attempt to prove that once tense has been taken out of time, genuine relations of *earlier than/later than* still obtain.[48] Most B-theorists are content, I am sure, simply to stipulate that this is case. But in view of the A-theorist's successful reduction of so-called B-relations to tensed facts, such a procedure on the part of the B-theorist seems dubious at best. Given that there are many other relations sharing the formal properties of the *earlier than/later than* relations, it seems incumbent upon the B-theorist to provide some justification for thinking that his denial of tense does not in truth leave us with an effete C-series. In the absence of any such justification, the B-theorist does appear to be guilty of "spatializing" time.

---

[48] Tooley does offer a causal analysis of temporal priority in *Time, Tense, and Causation*, chap. 9, but his account is based upon a dynamic conception of time according to which only the past and present are real and so is incompatible with a B-Theory of time.

CHAPTER 8

INCOHERENCE OF THE MIND-DEPENDENCE OF BECOMING

In my *Tensed Theory of Time*, I argued that the B-theorist's objections to the reality of temporal becoming were not insoluble and that, on the contrary, there were positive advantages to be had from adopting a metaphysics of temporal becoming. Moreover, we have seen in chapter 6 that the B-theoretical assertion of the mind-dependence of becoming had little by way of positive argument to commend itself. Now we may supplement those conclusions by showing that the B-theoretical denial of temporal becoming is in fact incoherent.

The B-theorist's claim that temporal becoming is purely subjective is attended by a nest of difficulties. In the first place, it is not at all clear what is meant by the "mind-dependence of becoming" or similar expressions of the subjectivity of temporal becoming on the part of B-theorists. Such expressions clearly connote that in the absence of sentient observers there would be no such thing as temporal becoming. In the absence of minds, every temporal moment and event simply exists tenselessly; there are no tensed facts; no past, present, or future; nothing comes into existence or happens except in the tenseless sense of existing at certain appointed stations as opposed to others; the spacetime manifold as a whole exists timelessly *en bloc*, time being simply a tenseless, internal dimension of the manifold; and change is merely the tenseless possession of different properties by things and/or events at their various fixed spacetime locations. This much is clear. But what is not clear is how the presence of minds serves to introduce temporal becoming into this picture.

Are we to understand that there actually exists a temporal becoming in the mental realm which is absent from the physical realm? Do mental events, unlike physical events, come to be and pass away? Does the mind-dependence of becoming mean that due to consciousness there exists what would otherwise not exist, temporal becoming, but of a peculiarly mental sort? Or are we rather to understand that temporal becoming is an illusion of consciousness, a chimera as unreal in the mental realm as in the physical? Is our experience of temporal becoming wholly non-veridical, even our experience of the passage of psychological time? Do mental events exist as tenselessly as physical events, so that becoming is absent from both aspects of reality?

As I say, B-theorists have not been very forthcoming in addressing such questions. But I shall argue that on either account, the B-Theory fails to provide a coherent account of temporal becoming.

REAL MENTAL BECOMING

Consider, first, the interpretation that temporal becoming exists in the mental realm, but is absent from physical reality. This alternative at least has the advantage of giving an accurate phenomenology of our temporal consciousness. The becoming of the contents of consciousness is a datum of experience with which any adequate theory of time must make peace. But the idea that temporal becoming characterizes mental events, but not the physical world, can easily be shown to be untenable. It leads, in the first place, to what Capek characterizes as "an absurd dualism" of the tenseless physical world and temporal consciousness, "a dualism of two altogether disparate realms whose correlation becomes completely unintelligible."[1] Imagine, for example, a possible world in which only minds exist, say, God and the angels, and that these minds experience temporal becoming of the contents of consciousness. In such a world tense and temporal becoming would be objective features of reality, since no physical realm exists and *ex hypothesi* temporal becoming characterizes the mental realm. If, then, we imagine a similar world except that some of those minds are incarnated in physical bodies, bodies which do not undergo temporal becoming but exist tenselessly, then Capek's intolerable dualism results. The absurdity of such a dualism is heightened if we reflect on the causal relationships supposedly sustained between a mind and its body. The physical states of the body at some time t tenselessly induce mental states at t, but that time may not be present for the mind and so, as past or future, such states are non-existent. Conversely, the mind may will to cause a bodily movement now, but the effect is not produced now, but tenselessly—but then the effect *exists* even when the mind is not now producing it. The problem here is essentially the same as that which theologians encounter in trying to relate a timeless God to an A-theoretical temporal universe[2], only in this case it is the universe which is supposed to exist tenselessly and God (or minds) to experience A-theoretical temporality. It seems evident that if mental temporal becoming is real, then the states of physical reality which are even loosely correlated with states of consciousness also undergo temporal becoming.

It is in this light that Frederick Ferré's troupe of objections to Grünbaum's thesis of the mind-dependence of becoming are best understood.[3] Unless Ferré is presupposing the reality of becoming in the mental realm, most of his objections are so misconceived that they are easily refuted. But if we interpret the mind-dependence of becoming to imply temporal becoming of the contents of consciousness, then Ferré's criticisms strike home.[4] Take, for example, what Ferré

---

[1] Milic Capek, *The Concepts of Space and Time*, Boston Studies in the Philosophy of Science 22 (Dordrecht: D. Reidel, 1976), p. XLVII.

[2] See my *God, Time, and Eternity* (Dordrecht: Kluwer Academic Publishers, forthcoming), chap. 3.

[3] Frederick Ferré, "Grünbaum on Temporal Becoming: A Critique," *International Philosophical Quarterly* 12 (1972): 426-445; idem, "Grünbaum vs. Dobbs: the Need for Physical Transiency," *British Journal for the Philosophy of Science* 21 (1970): 278-280.

[4] See Ronald C. Hoy, "Becoming and Persons," *Philosophical Studies* 34 (1978): 273-274, who points out that Ferré presupposes a view of persons as enduring entities and finds this incompatible with the tenseless existence of the world. On the inherently tensed nature of personhood see chapter 9.

calls the problem of temporal location. Why do I have the now-awareness of t_2, rather than of, say, t_1? The A-theorist has a ready response: because it really is t_2. But for the B-theorist all times are equally existent, so that my being aware of t_2 as present is inexplicable. Lynn Rudder Baker, who served as Grünbaum's stand-in in this debate, responded that Ferré's objection boils down to the question: why am I now at t_2 rather than at t_1?—and the answer to that question is that that now-awareness is part of the simultaneity class of events at t_2.[5] One's now-awareness associated with t_2 could not be experienced at any time other than at t_2, so of course at t_2 one has the now-awareness associated with t_2, just as at t_1 one has the now-awareness associated with t_1. This response is adequate, however, only on the assumption that all these various now-awarenesses tenselessly exist and do not come to be and pass away. But if there is temporal becoming in the mental realm, then no other now-awarenesses exist except for the present one. But then the question is, why, if all physical events exist tenselessly, am I experiencing just the now-awareness which I have? All Grünbaum can say is that such an awareness is part of the simultaneity class of events occurring at t_n. But that is true of all now-awarenesses at their respective t_i. The question is not, why do I have the now awareness of t_n at t_n, but why do I have the now awareness of t_n *simpliciter*? Why is this one now-awareness privileged? The answer seems to be: because t_n is present; but Grünbaum cannot accept such an explanation. Thus, the B-Theory on this interpretation is explanatorily deficient.

Or take Ferré's so-called problem of order in transiency. Nothing in Grünbaum's tenseless universe requires that now-awarenesses be experienced in a particular order. There is no reason why one's consciousness is not filled willy-nilly with now-awarenesses from random points on one's tenselessly existing world-line. Baker's response is that the order of one's now awarenesses is based on the order of the physical brain states correlated with them. Ferré "overlooks the fact that, on Grünbaum's view, the order of those mental events which constitute awareness of physical events depends on the order of the physical events which cause them together with the distances and velocities of the influence chains reaching the percipient."[6] This response assumes that there actually exists a tenseless series of now-awarenesses correlated with the tenseless series of brain states; but if there is only one now-awareness that exists, namely, the present one, then it is not at all obvious why the now-awarenesses associated with different brain states at random locations on one's world-line should not become successively present. None of the tenselessly existing influence chains, nor the tenselessly conceived relations of *earlier than/later than* among physical events would be violated by this bouncing about of presentness with respect to mental states. But the result would be an incredible dualism; better to say with the A-theorist that physical states associated with now-awarenesses also become and do so in the order specified by the *earlier than/later than* relations.

---

[5] Lynne Rudder Baker, "Temporal Becoming: The Argument from Physics," *Philosophical Forum* 6 (1975): 230. A Mellon post-doctoral fellow at the University of Pittsburgh during 1974-75, Baker acknowledges Grünbaum's "enormous help in working out this paper."

[6] Ibid., p. 225.

Or consider Ferré's problem of uni-directional transiency. He observes that the anisotropy of time fails to explain the transiency of temporal becoming; so why does becoming in the mental realm occur uniquely in one direction, rather than in the opposite or both? Baker answers the question in terms of entropic increase and accumulation of memories:

> The preferred direction of our experience, as well as its order, can be explained solely by reference to the physical world without appeal to an ongoing now. The anisotropic 'is later than' relation can be provided by the increasing entropy in branch systems of the universe. It is a fact that increases of entropy in branch systems generally accompany the production of traces or memories; and thus the direction of entropy increase (i.e., the direction of the 'is later than' relation for physical events) is the same as the direction of increase of stored information. And the direction of entropy increase is toward the future in 'psychological' time.... The preferred direction of our experience can thus be accounted for without appeal to a mind-independent 'now.'[7]

It seems to me that the efficacy of this response is moot even if we deny that mental states become,[8] but it is clearly inadequate on the assumption that temporal becoming does characterize the contents of consciousness. For there is no reason that events in the direction of decreasing entropy should not become successively present, even if that should involve a progressive loss of information content on the part of the sentient subject. Recall that such *de facto* physical asymmetries are not sufficient for time's having a direction. How much less do they determine the direction in which consciousness experiences some tenselessly existing physical state as occurring now! Even if the tenseless, ordinal ordering of events were determined by thermodynamic considerations, that order could be experienced by consciousness in either direction.

This leads to another of Ferré's concerns: the problem of inter-subjectivity. How is it that we all experience the tenseless series of physical events in the same order and direction, if they do not in fact become present in just that way? Why do we share the same now? Baker replies that an event at *t* is the terminus of two causal chains, one issuing in my awareness of it and the other in someone else's awareness of it; since the causes are simultaneous, so are our respective awarenesses. This answer is completely satisfactory on the assumption that mental events do not become. But if there is becoming in the mental realm independently of the tenseless existence of the physical world, there is no reason why there should be a single time order among the community of minds. Each person could have a different A-series of mental states with differing directions. They might on occasion share the same now, only to discover that they have differently ordered pasts—or, their paths through successive now's might never intersect (even if they both on occasion shared the same tenseless spacetime coordinates, since that moment might not ever be experienced by both as present) and so they remain mutually unknown to one

---

7 Ibid., pp. 225-226.

8 The correlation between higher entropy states and greater memory will not yield a direction to psychological time, since such a correlation is directionally neutral. See also comments by James McGilvray, "A Defense of Physical Becoming," *Erkenntnis* 14 (1979): 281.

another. This is a bizarre picture, but it comes from trying to wed a tenseless physical world to a mental realm in which temporal becoming occurs.

I have probably belabored this point unnecessarily, for it is rather obvious that tense and tenselessness cannot be combined coherently in the way described.[9] It would be as if a modal realist were to espouse the mind-dependence of being actual and thereby mean that although the physical universe is actual only relative to a world, our conscious life has actuality *simpliciter*. This is clearly incoherent, and the first interpretation of the mind-dependence of becoming is equally so.

DENIAL OF MENTAL BECOMING

But do B-theorists mean to affirm the second interpretation, that there is no becoming of mental states? This would seem to be Smart's position when he writes,

> Now most people do seem to be subject to the metaphysical illusion of an absurdity (flow of time, advance through time).... However, the description of this absurdity need not itself be done absurdly: the illusion of the septic sort of change may involve change, i.e., the sort of change that can be described four-dimensionally in terms of non-similarities of time-slices of the brain of the metaphysician who has the illusion of time flow.[10]

The difficulty with this position is that it flies in the face of our inner experience of temporality. Eddington, who struggled to understand the relationship between thermodynamic irreversibility, the direction of time, and temporal becoming, finally came down squarely on the side of temporal becoming as the true source of the arrow of time on the basis of the phenomenology of time consciousness: "We have direct insight into 'becoming' which sweeps aside all symbolic knowledge as on an inferior plane. If I grasp the notion of existence because I myself exist, I grasp the notion of becoming because I myself become. It is the innermost Ego of all that *is* and *becomes*."[11] In Eddington's view, then, temporal becoming is as obvious as one's own existence; indeed, one perceives one's existence as becoming, not as static being.

Now according to the second interpretation of the mind-dependence of becoming, this fundamental intuition is wholly non-veridical. But then what of our experience of temporal becoming? Grünbaum seems to deny that we do, in fact,

---

[9] But see Michael Lockwood, *Mind, Brain, and the Quantum* (Oxford: Basil Blackwell, 1989), pp. 263-282, who seems to affirm just such a view. For though adopting a static spacetime approach, he posits a "phenomenal frame" which is transported along the world-line of the brain. Lockwood emphasizes that he is talking about a single phenomenal perspective, rather than a spatio-temporally extended series of phenomenal perspectives, whose contents are a function of its location on the world-line. Thus, from our point of view there would be temporal progression, even though there is no overall, perspective-independent, temporal progression.

[10] J. J. C. Smart, "Spacetime and Individuals," in *Logic and Art*, ed. Richard Rudner and Israel Scheffler (Indianapolis: Bobbs-Merrill, 1972), p.11. Smart is a materialist, but he allows that the dualist may instead speak of the metaphysician's soul.

[11] Sir Arthur Eddington, *The Nature of the Physical World,* with an Introductory Note by Sir Edmund Whittaker, Everyman's Library (London: J. M. Dent & Sons, 1964), pp. 102-103.

experience temporal becoming, allowing us at most our experience of diverse "now's" and the (tenseless) order of events, but not of temporal becoming:

> Hence it is a matter of fact that the 'Now' shifts in conscious awareness to the extent that there is a diversity of the Now-contents, and it is likewise a fact that the Now-contents are temporally ordered. But since these diverse Now-contents are ordered with respect to the relation 'earlier than' no less than with respect to its converse 'later than,' it is a mere tautology to say that the Now shifts from earlier to later....a non-directional or directionally neutral claim of the transiency of the Now and of the temporal order of the various 'Now'-contents does codify factual truths pertaining to psychological (common sense) time.[12]

Here Grünbaum appears to assert that although it is a fact that psychological time does involve an experience of "nowness" and temporal order, it does not involve any experience of becoming or the direction of time. Such an assertion is so obviously false a description of psychological time that it is hard to believe that this is Grünbaum's meaning. In speaking of the psychological phenomenon of melody awareness, Grünbaum recognizes that we do not have merely the succession of states of awareness, but also an instantaneous awareness of succession, which, he says, "is an essential ingredient of the meaning of 'now': the now-content, when viewed as such in awareness, includes an awareness of the order of succession of events in which the occurrence of that awareness constitutes a *distinguished element.*"[13] If succession were understood in the A-theoretical sense of becoming, this would be an admirable analysis of now-awareness. But Grünbaum only speaks of the tenseless order of events and does not mean to affirm any experience of becoming. He claims that the transiency of the now arises from the diversity of the now-contents of awareness and the differences among them concerning what is remembered of earlier events.[14] By "arises" he must mean that the transiency of the now just consists in those two facts, for he does not allow any further experience of temporal becoming beyond this. Not only is this a phenomenologically incomplete analysis of time consciousness or psychological time, but it is far from clear how one could *not* have an experience of temporal becoming in the sense of moments elapsing or events becoming present and then passing away, if there are for a conscious subject a diversity of now-awarenesses and the consciousness of temporal order in which one awareness (the last) is distinguished. For such an awareness of changes in present-tense consciousness engenders the experience of temporal becoming. Moreover, as events or moments become successively present, then past, a direction of time inevitably arises in experience. It thus seems futile to try to avert the inevitable clash with an adequate phenomenology of temporal consciousness by

---

[12] Adolf Grünbaum, *Philosophical Problems of Space and Time*, 2d ed. Boston Studies in the Philosophy of Science 12 (Dordrecht: D. Reidel, 1973), pp. 315-316. Peter Geach's argument against the illusory character of temporal becoming based on the incompatibility of present contents of consciousness at best proves a diversity of now-awarenesses (Peter Geach, "Some Problems about Time," in *Logic Matters* [Berkeley: University of California Press, 1972], p. 306), which fact is admitted by Grünbaum, and so is not decisive as an argument for the objective reality of becoming.

[13] Ibid., p. 325.

[14] Ibid.

claiming that our experience only involves a diversity of now-contents and an awareness of the order of those contents.[15]

The question, then, is what sense it makes to say that we experience temporal becoming in consciousness but that mental events do not become. The very awareness of a "now" is an awareness of tense, and if we are successively aware of different experiences as present, how is it that there is no mental becoming? This question arises due to the so-called presentness of experience.[16] Grünbaum himself affirms the presentness of experience when he says such things as: "presentness or nowness of an event requires conceptual awareness of the presentational immediacy of either the experience of the event or...of the *experience* of another event simultaneous with it."[17] The notion of "presentational immediacy of experience" coincides with Mellor's "presence of experience," but whereas Mellor strove to give some tenseless account of this crucial notion, Grünbaum is strangely silent. All we get is Baker's dismissal of the query, "Why are events ever characterized as occurring now?" as a pseudo-question, equally inappropriate to mind-dependence or mind-independence theorists.[18] But since the presentness of experience, affirmed by Grünbaum, lies at the heart of our experience of temporal becoming, it is crucial that the B-theorist offer some account of this notion if he is to explain away temporal becoming, as Mellor recognized and struggled vainly to do. The A-theorist answers Baker's question straightforwardly in terms of the objectivity of A-determinations and a presentist ontology. Grünbaum's failure to account for our experience of "nowness" or better, presentness, signals a singular weakness in the position that temporal becoming is as absent from the mental as from the physical realm. Grünbaum has not explained away the phenomenological data that mental events become.

Moreover, as McGilvray points out, Grünbaum's own account of the mind-dependence of becoming implies the reality of temporal becoming in the mental realm.[19] It will be recalled from chapter 6 that Grünbaum explicates the notion of an event's occurring now in terms of its simultaneity with a sentient subject's experiencing the event while being aware that he is experiencing the event. This characterization is, as Grünbaum came to realize, circular, since the verbs contained in this explication are present-tensed. But he dismissed the circularity as non-vicious because he claimed that he was not trying to provide a tenseless analysis of "nowness," but merely to articulate the mind-dependence of "nowness." We let that explanation pass in chapter 6, but now is the time to re-examine it. If the characterization is essentially tensed, how can it possibly articulate successfully the mind-dependence of becoming? On the contrary, by presupposing the objectivity of

15 As Lockwood recognizes, "...one could think of these multiple awarenesses simply as existing in parallel, distributed as a continuous, perhaps overlapping, series along one's world-line. What one needs, in addition, is a conception of oneself – more than that, a *sense* of oneself—as progressing through the series" (Lockwood, *Mind, Brain, and the Quantum*, p. 227).

16 See my *Tensed Theory of Time*, chap. 5.

17 Adolf Grünbaum, "The Meaning of Time," in *Basic Issues in the Philosophy of Time*, ed. E. Freeman and W. Sellars (La Salle, Ill: Open Court, 1971), p. 206.

18 Baker, "Temporal Becoming," pp. 223, 230.

19 McGilvray, "Defense of Physical Becoming," pp. 291-294.

tense, does it not rather underline the fact that presentness is *not* mind-dependent? In order to articulate what it is for some event *e* to be present, Grünbaum has to presuppose that a sentient subject is experiencing (present-tense) *e*. If this is an objective matter of fact, then the presentness of *e* cannot be merely illusory. McGilvray formulates a tenseless rendition of Grünbaum's conditions for the truth of "*e* is now for *M*":

a. *E* be an experience of *e*
b. *E* be had by *M*.
c. *M* be conceptually aware that *E* be had by *M* (*C* = this event of conceptual awareness)
d. the content of *C* include an awareness that *E* be coincident with *C*.

But (a-d) as they stand are not sufficient conditions for "*e* is now for *M*," since (a-d), if ever true, are always true. What, then, would make these conditions sufficient for the truth of "*e* is now for *M*"? McGilvray answers,

> There is a simple answer, and it is that *C* and perhaps *E* become. This is a legitimate answer, consistent with Grünbaum's account of the 'mind-dependence' of becoming, and it has the advantage of clarifying his cheerful use of present progressives throughout the argument. Here is where one must look for the sense of tensed as opposed to tenseless occurrence. Moreover, it is difficult to imagine any other way in which to get a sufficient condition for the nowness of *C* and *E*; it is apparent that conditions (a-d), if true at some time, are always true.[20]

Thus, Grünbaum seems forced by his own account of what it is for an event to occur now to regard becoming as real in the mental realm, even if unreal in the physical realm. But then his account of the mind-dependence of becoming lies open to all the objections discussed above.

Moreover, it seems to me that from the standpoint of the theist, Grünbaum's account of the mind-dependence of becoming is objectionably anthropocentric. Let us assume for the sake of argument that the B-Theory of time is correct and that God is temporal (if the temporality of God on the B-Theory seems objectionable, we can substitute incorporeal angels). Since God is sentient and *ex hypothesi* temporal, He has a diversity of now-awarenesses of varying content correlated with the B-series of moments of time. But now recall Ferré's problem of unidirectional transiency, or the direction of time. Grünbaum attempts to explain the direction of psychological time in terms of the direction of entropy increase and memory. Wholly apart from the fact that such an account yields only anisotropy, not orientation, even for psychological time, this explanation is inadequate because it fails to explain God's experience of unidirectional transiency. Since God is incorporeal, increased entropy states are just irrelevant to His mental processes and events. Thus, there is no reason why God could not be remembering higher entropy states in the universe and anticipating lower entropy states to come, experiencing the flow of time in a direction opposite to the thermodynamic arrow. It cannot be rejoined that God,

---

20 Ibid., p. 293.

therefore, has no experience of temporal becoming, for, as Mellor has emphasized, without such phenomenal becoming God would be grossly impaired in His ability to act in a timely fashion, but would be frozen into immobility. Besides, as one who experiences a succession of tensed contents of consciousness, it is unavoidable that God should also experience temporal becoming. Yet God's experiential "arrow of time" remains wholly unexplained by Grünbaum's physicalistic account. Or consider the problem of inter-subjectivity. In the divine Trinity all three persons share similar now-contents at each successive moment of time. But clearly this coordination of now-awareness cannot be based on the similar velocities and distances traveled by causal chains from the events to the percipients' brains, since such spatial notions are inapplicable to God. Similar problems attend Grünbaum's explanation of order in transiency. Perhaps an amended, non-physicalistic account could solve these last two difficulties, but it is hard to see how the B-theorist could hope to account for God's experience of the direction of time. Grünbaum's account is therefore metaphysically inadequate as an explanation of the supposed illusion of temporal becoming.

Finally, it seems to me that the position that the temporal becoming of mental as well as physical events is purely illusory is self-referentially incoherent. The point here, stated simply, is that the very illusion of becoming involves becoming.[21] The German metaphysician Herrmann Lotze, contemplating whether the Principle of Identity entails the unreality of becoming, made this point when he remarked, "This consideration might lead us to repeat old attempts at a denial of all Becoming, or—since it cannot be denied—to undertake the self-contradictory task of explaining at least the becoming of the appearance of an unreal becoming."[22] The position of the B-theorist is reminiscent of Buddhist, Parmenidean, and Idealist denials of the reality of temporal becoming and their consigning becoming to the realm of appearance and illusion. The principal difference is that the B-theorist insists that time is real, even though temporal becoming is not—to that extent these temporal illusionists were more consistent than the B-theorist, as I have argued. But never mind; the point remains that for the B-theorist as well as for atemporalist schools of philosophy, temporal becoming is real neither in the physical world nor in the life of the mind but is somehow wholly illusory. But then the B-Theory labors under the same incoherence as all temporal illusionist theories, namely, that an illusion or appearance of becoming involves becoming, so that becoming cannot be mere illusion or appearance. A Buddhist philosopher can consistently deny the reality of space and spatial objects and hold these to be mere illusions of the mind, for an illusion of spatiality does not entail spatiality, one's thoughts and perceptions not being spatial in nature. But an illusion of temporal becoming, by contrast, is itself

---

[21] For brief statements of this point, see A. N. Prior, "Some Free Thinking about Time," in *Logic and Reality*, ed. B. J. Copeland (Oxford: Clarendon Press, 1996), p. 47; Milic Capek, *The New Aspects of Time*, Boston Studies in the Philosophy of Science 125 (Dordrecht: Kluwer Academic Publishers, 1991), p. 341; Mauro Dorato, *Time and Reality: Spacetime Physics and the Objectivity of Temporal Becoming*, Collana di Studi Epistemologici II (Bologna: CLUEB, 1995), p. 25.

[22] Hermann Lotze, *Metaphysic*, 2 vols., 2d ed., ed. Bernard Bosanquet (Oxford: Clarendon Press, 1887), 1: 105. On the problem of identity and time, see chapter 27.

an experience in which becoming plays a part. We can conceive of a person accepting tenselessly and with no experience of becoming the putatively false proposition *Temporal becoming is real*, and such a one would be genuinely deceived about the reality of becoming. But such a person would not experience the illusion or appearance of becoming. We do. And therein lies the inescapable difficulty for the B-theorist.[23]

To make the same point in a different way, imagine a state of consciousness existing at a particular position in McTaggart's atemporal C-series. How could the illusion of temporal becoming—as opposed to merely false belief—possibly arise in such a static consciousness? How could a C-series person experience the appearance of temporal becoming? The B-theorist may protest that the C-series is atemporal, whereas the B-series is temporal. But on the B-theoretical account, the only difference between them is that the irreflexive, transitive, and asymmetric relations which order the terms of the two series are in the one case *included in/inclusive of* and in the other *earlier than/later than*; moreover, on the B-Theory the latter relations are so emasculated as to make the difference between earlier and later merely conventional, and time isotropic and directionless. So whence the illusion of becoming?

To say that there is an illusion of becoming just is to say that things become in consciousness, though not in reality. Becoming in consciousness is thus undeniable and, hence, the denial of becoming self-referentially absurd. John Laird writes:

> Take the supposed illusion of change. This must mean that something, X, appears to change when in fact it does not change at all. That may be true about X; but how could the illusion occur unless there were change somewhere? If there is no change in X, there must be a change in the deluded mind that contemplates X. The illusion of change is actually a changing illusion. Thus the illusion of change implies the reality of *some* change. Change, therefore, is invincible in its stubbornness; for no one can deny the appearance of change.[24]

Laird is speaking here of tensed change, or becoming, not of tenseless change; so it is no use opposing to his argument the familiar B-theoretical riposte to McTaggart that in the B-series there exists tenseless change. His argument is two-edged: (i) that the undeniable experience of becoming cannot be wholly non-veridical, since, *qua* experience, the illusion of becoming involves the becoming of experience, and (ii) if there were no becoming whatsoever, then the basis of the illusion of becoming with which we live becomes inexplicable.

---

23 B-theorists have been largely silent in response to this objection. I have seen no more than Savitt's remark: "So does not the mere illusion of time imply that there is succession, i.e., time? This response relies on the unstated premiss that illusions have precisely the sort of temporal characteristics that Gödel's argument is designed to show they do not have" (Steven F. Savitt, "The Replacement of Time," *Australasian Journal of Philosophy* 72 [1994]: 471). The problem is that temporal illusions *do* have the property of becoming and this phenomenological feature of time consciousness is far more obvious than the premisses of Gödel's argument for the B-Theory.

24 John Laird, *Theism and Cosmology* (London: George Allen & Unwin, 1940), pp. 145-146.

CONCLUSION

In summary, then, the B-theoretical denial of the reality of temporal becoming cannot be sustained in the light of our experience. If the B-theorist concedes that becoming is real in the realm of consciousness, then he is saddled with a metaphysical dichotomy between the external, physical world and the inner life of the mind which is intolerable. Such a bifurcation of reality leads to intractable problems concerning the temporal location of the now, the order of transiency, the direction of becoming, and the inter-subjectivity of the "now." On the other hand, if the B-theorist regards even mental becoming as unreal, he flies in the face of our most deeply-seated experiences of temporal becoming and the presentness of experience. Moreover, it seems that he actually needs to affirm temporal becoming in consciousness in order to explain the mind-dependence of physical becoming. His account of the illusion of temporal becoming is in any case metaphysically inadequate, since it could not account for God's own experience of temporal becoming. Finally, the B-theorist's position is self-referentially incoherent, since the posited illusory experience of becoming entails becoming. It therefore seems to me that once again the B-Theory succumbs under the weight of its own internal weaknesses.

CHAPTER 9

THE PROBLEM OF TEMPORARY INTRINSICS

ENDURANCE VS. PERDURANCE

Adoption of an A- or B-Theory of time seems to involve commitment to radically different understandings of what it is to be an object or thing. Borrowing terminology suggested by Mark Johnston,[1] we may say that an *endurantist* holds that objects exist wholly at any time at which they exist, that is to say, an object which exists at any time *t* does not have at *t* parts which do not exist at *t*. More positively, an endurantist holds that an object which exists at one time is identical to that object existing at another time.[2] For the endurantist, persistence through time consists in an object's existing at several successive times, being wholly present and remaining self-identical at each of those times. By contrast, what we may call a *perdurantist* holds that an object does not exist wholly at any one time (unless its total temporal history lasts only for an instant) but is identical with the whole temporally extended entity; for a spatio-temporal object this would be what is marked off as its world line on a spacetime diagram. Sections of the line correspond to temporal parts of the object, of which the object itself is composed.

---

1 See Mark Johnston "Is There a Problem about Persistence?" *Aristotelian Society Supplementary Volume* 61 (1987): 112-113. Johnston's terminology was brought into the philosophical mainstream by David Lewis, *On the Plurality of Worlds* (Oxford: Basil Blackwell, 1986), p. 202.

2 For these two features of enduring objects, I am indebted to the first draft of Trenton Merricks' "Endurance and Indiscernibility," published in revised form in *Journal of Philosophy* 91 (1994): 165-184. In his more recent "Persistence, Parts, and Presentism," *Noûs* 33 (1999): 421-438 Merricks gives a presentist characterization of endurance and perdurance:

> E. For any presently existing object *O*, *O* endures iff *O* persists and all of *O*'s parts *simpliciter* exist at the present time.
>
> P. For any presently existing object *O*, *O* perdures iff *O* persists and some of *O*'s parts *simpliciter* do not exist at the present time.

I think it is evident that this account, apart from the requirement that *O* persist, does not differ essentially from the earlier account, since the earlier account merely generalizes to all times *t* and does not speak of *O*'s parts at *t*, but of *O*'s having at *t* all its parts *simpliciter*. The more general account is preferable, since (E) does not exclude the existence of some not presently existing object *O′* which does not have all its parts at a single moment of time, that is to say, there could be objects at non-present times which have temporal parts, which Merricks wants to deny. Hence, we should revise (E) to read:

> E′. For any persisting object *O* and time *t*, *O* endures iff *O* exists at *t* and all of *O*'s parts *simpliciter* exist at *t*.

Perdurance would be the negation of (E′). Merricks's account is significant because it serves to expose the inadequacy of attempts by Sider, Markosian, and others to define perdurance by means of appeal to a notion of "temporal part" defined in terms of size. Their account not only fails to capture the intuitive idea of perdurance but fails to solve the problem of temporary intrinsics, as Merricks shows. I am indebted to Prof. Merricks for his comments on this chapter.

For obvious reasons, perdurantism is often called four-dimensionalism, and perduring objects four-dimensional objects. This nomenclature is, however, a misnomer, for it gratuitously assumes that all temporal objects are also spatial objects.[3] But abstract objects such as propositions which change their truth values, God, and angels may be conceived as non-spatial temporal objects, in which case we should speak of one-dimensionalism. The salient point of perdurantism is that an object is to be regarded as the entire temporally extended whole and the temporal slice of that object at any moment as a temporal part of the object. Persistence through time consists in just being temporally extended.

Now obvious affinities exist between endurantism and the A-Theory on the one hand and between perdurantism and the B-Theory on the other. To a man, A-theorists have come down on the side of endurantism. In their thinking, an object which is present at a time is not merely partially present then; it is the whole object which is then present, and this whole object continues to be present at each successive time until it ceases to exist.[4] Given the reality of temporal becoming, all that exists is what presently exists, so that all the parts that an object has are those that it presently has—like merely possible objects, non-actual, non-present temporal parts do not exist. Thus, A-theorist Peter Geach blasts what he calls the four-dimensional view as abolishing real change by reducing change to a mere variation of properties between different temporal parts, thereby abolishing time itself.[5] On this view, he charges, reality is a changeless arrangement of four-dimensional solids, and the time axis is only inappropriately so called—in actual fact time is an illusion.

B-theorists, for their part, have typically endorsed a perdurantist view of objects. Because temporal becoming is, in effect, an illusion of consciousness, because there is no property of presentness which is successively instanced by events or things, because the distinctions between past, present, and future have no validity outside of consciousness, all that objectively exists (apart from timeless entities) are objects having tenseless temporal extension just as, if they are also spatial objects, they have spatial extension. D. H. Mellor would be one of the very few B-theorists to dissent from this picture; he holds things to be continuants which exist wholly at every time at which they tenselessly exist.[6] The vast majority of B-theorists would agree instead with J. J. C. Smart when he advocates taking physical objects to be four-dimensional in extent rather than merely three-dimensional.[7]

But is this natural and nearly ubiquitous association between the A-Theory and endurantism and between the B-Theory and perdurantism obligatory? Smith has

---

3 Moreover, a realist interpretation of certain theories of particle physics would require many more than four dimensions, so that we should be required to speak, for example, of eleven-dimensional objects.

4 Some A-theorists may be willing to countenance events or processes as composed of temporal parts, but they deny that such talk is appropriate or even makes sense with respect to things. B-theorist D. H. Mellor (see below) makes a similar distinction.

5 Peter Geach, "Some Problems about Time," in *Logic Matters* (Berkeley: University of California Press, 1972), pp. 304-305.

6 D. H. Mellor, *Real Time* (Cambridge: Cambridge University Press, 1981), p. 105; so also J. Butterfield, "Spatial and Temporal Parts," *Philosophical Quarterly* 35 (1985): 32-44.

7 J. J. C. Smart, "Spatialising Time," *Mind* 64 (1955): 239-241; idem, "Spacetime and Individuals," in *Logic and Art*, ed. Richard Rudner and Israel Scheffler (Indianapolis: Bobbs-Merrill, 1972), pp. 3-20.

recently argued that it is not; indeed he asserts that this traditional association is "irrational" and expresses "merely prejudices that have not been sufficiently thought through."[8] Smith's assertions are based on his argument for the compatibility of endurantism (particularly with respect to persons) and the B-Theory of time. He argues that if it is true in tensed terms that a person is an enduring object, then it is also true in tenseless terms that a person is an enduring object. After providing tenseless analogues of tensed sentences referring to enduring persons, Smith concludes,

> Thus if the theory that persons are continuants is true, it does not entail that the tensed theory of time is true. The theory that persons are continuants can be expressed in a tensed idiom or in a tenseless idiom, and consequently does not provide evidence for or against either one of these theories of time.[9]

Wholly apart from the success of his tenseless analogues, however, Smith's argument is manifestly inadequate.[10] The fact that one can tenselessly refer to enduring objects or that it can be tenselessly stated that enduring objects exist does nothing to prove that endurantism is compatible with a B-Theory of time. One might as well argue that one's ability to refer tenselessly to tensed facts, as, for example, in the sentence "At t_1 there *are* tensed facts," proves that tensed facts are compatible with a tenseless theory of time! Despite one's ability to refer tenselessly to enduring objects, like persons, it may well be the case that things can endure only if tense is an objective feature of reality.

D. H. Mellor disagrees. As a B-theorist and an endurantist, Mellor distinguishes between things and events precisely on the grounds that while events have temporal parts, things do not.[11] Change, which he defines tenselessly as something's having incompatible properties at different dates, characterizes things, but not events. For apparent changes in a temporally extended event are really just differences between its temporal parts, so that the event as a whole does not change, whereas a thing, being entirely existent at the times at which it exists, loses or acquires properties as a whole. Mellor provides arguments for the endurantist viewpoint, arguments with which the A-theorist would largely agree. He later goes on to argue against fellow B-theorists who would, as he puts it, "reduce things to events" by taking things to be temporally extended objects having temporal parts.[12] Philosophically, he contends that although the temporally successive "stages" of a thing are causally connected, they should not for that fact be construed as events nor can a thing be non-circularly

---

8 Quentin Smith, "Personal Identity and Time," *Philosophia* 22 (1993): 156, 166.

9 Ibid., p. 165. Smith's argument seems to be directed against the claim that an endurantist theory of personal identity cannot be stated in tenseless sentences. This, however, is a straw man, since no one to my knowledge has made such a claim, including Delmas Lewis, "Persons, Morality, and Tenselessness," *Philosophy and Phenomenological Research* 47 (1986): 305-309, whom Smith misrepresents by quoting out of context.

10 A minor inadequacy of Smith's argument is that it fails to demonstrate the compatibility of the A-Theory with perdurance. Perhaps we should all be endurantists. Earlier, Smith did present independent arguments aimed at rebutting objections to A-theoretical perdurance, including the problem of temporary intrinsics, but we shall see that his proposed solution to this problem is implausible.

11 Mellor, *Real Time*, pp. 103-107.

12 Ibid., pp. 104, 127-132.

defined as a collection of temporal stages or parts. Scientifically, Mellor argues that relativity theory, even when interpreted by the B-theorist along the lines of a four-dimensional "block universe," does not imply that physical objects are four-dimensional entities composed of temporal parts. Whether Mellor's arguments on this score are entirely convincing may be left for the reader to investigate;[13] at least they are a salutary correction to the naive and facile appeals to relativity theory sometimes made on behalf of perdurance.[14] Even if Mellor's arguments are successful, he has not thereby shown the B-Theory to be compatible with endurantism, for he has left out of account what I conceive to be the major philosophical objection to a B-theoretical endurantism, an objection which springs from the very nature of change as he defines it and has come to be known as the problem of temporary intrinsics.[15]

THE PROBLEM OF IDENTITY AND INTRINSIC CHANGE

Set forth by B-theorist David Lewis in defense of perdurantism, the problem of temporary intrinsics not only discloses the logical connection between the B-Theory and perdurance, but, I shall contend *contra* Lewis, is also most plausibly resolved by

---

13 On occasion, his arguments strike one as convincing only if one is presupposing an A-Theory of time, but not at all obvious on the presupposition of a B-theoretical block universe. For example, he remonstrates against those who infer that because the world-line of an object on a spacetime diagram can be divided into parts, therefore there exist corresponding temporal parts in the thing represented. He asserts, "When spatial dimensions are used to represent pressures or temperatures, for example, no one infers that these quantities must have parts corresponding to the spatial parts of lines representing them; nor should we infer it with time" (Ibid., p. 129). The point appears to have been borrowed from Geach, "Problems about Time," p. 303: "We might as well be asked to believe that the use of temperature charts requires the physical existence of 'world lines' in a 'temperature-time' continuum." But Geach was using the analogy *against* the inference of a block universe from a spacetime diagram; by contrast Mellor already admits the existence of the block universe depicted on the diagram. Given the existence of such a block universe, it is hard to see why there would be no object corresponding to the world line on the diagram. Granted that one of the dimensions is temporal and the others spatial, we may still ask why the temporal dimension is privileged, why an object's extension along that dimension alone does not count as constitutive of the object, why the object is confined to its three spatial dimensions rather than includes its temporal dimension as well. Mellor rejects "the itch to spatialise" time, advising, "If world-lines were represented by animated cartoons in which time itself was used to represent time, the itch would disappear" (Ibid., p. 130). This advice is persuasive, however, precisely because of the temporal becoming involved in animation; but in a block universe no such becoming exists. In a block universe an object's extension in time is every bit as rigid and real as its spatial extension. It seems quite arbitrary to restrict the object to only some of its dimensions. But if an object has temporal extension as well as spatial extension, it would seem to have temporal parts as well as spatial parts. Indeed, the problem of temporary intrinsics requires this, as we shall see.

14 See, *e.g.*, Johnston, "Problem about Persistence," pp. 114-115, 129, who erroneously assumes that relativity theory refutes presentism, or the A-Theory. Those who appeal to relativity theory to justify perdurance must first show that relativity requires a B-Theory, which I have denied, and then show that the block universe requires spatio-temporal objects, which Mellor denies.

15 See further William R. Carter and H. Scott Hestevold, "On Passage and Persistence," *American Philosophical Quarterly* 31 (1994): 269-283. Their argument that static time implies perdurance on pp. 276-277 is based on the problem of temporary intrinsics. See also Merricks, "Persistence, Parts, and Presentism."

an A-theoretical, endurantist solution.[16] I hope to show that the most plausible solution to the problem of temporary intrinsics is endurantism and that endurance entails a tensed theory of time.

As a springboard for his discussion, Lewis takes up the question of whether something can exist in more than one possible world. He argues that an object cannot literally be part of two possible worlds if its accidental, intrinsic (as opposed to extrinsic, or relational) properties differ in those two worlds. A person's size and shape, for example, are not essential properties and therefore differ from world to world; moreover, these are not relational properties which may differ for an identical object merely in virtue of its changing surroundings. Lewis's examples of size and shape are not felicitous, since on the received interpretation of SR, at least, these are not intrinsic properties: the selfsame object will vary in size and shape wholly as a function of its relation to reference frames moving at various velocities with respect to it. Still, Lewis's meaning is clear, and his example of having a certain number of fingers on one's left hand is a good candidate for an accidental, intrinsic property. Thus, if an object is conceived to be a part of two possible worlds, if the worlds overlap, so to speak, at that object, then one faces the insoluble problem of accidental intrinsics: the object in one world is the same object as in the other world and so cannot differ in its intrinsic properties. In order to make room for accidental, intrinsic properties, one must adjust one's conception of how an object exists in multiple worlds: either by Lewis's preferred Counterpart Theory, according to which an object and its counterparts can differ in their accidental, intrinsic properties; or Serious Actualism, which posits non-actual *ersatz*-worlds which, in Lewis's thinking, misrepresent objects as having intrinsic properties they do not actually have; or else some theory which represents things as transworld objects having differing parts in different worlds.

Now parallel to the problem of accidental intrinsics is the problem of temporary intrinsics, which Lewis identifies as the traditional problem of change, and the proffered solutions to the two problems are also for the most part parallel. The problem is, how can an object which persists through time be self-identical if it possesses different intrinsic properties at those times? Lewis writes,

> The principal and decisive objection against endurance, as an account of the persistence of ordinary things such as people or puddles, is the problem of temporary intrinsics. Persisting things change their intrinsic properties. For instance shape: when I sit, I have a bent shape; when I stand, I have a straightened shape. Both shapes are temporary intrinsic properties; I have them only some of the time. How is such change possible?[17]

As Merricks's formulation of this objection makes clear, the problem of temporary intrinsics arises from the application of the principle of the Indiscernibility of Identicals to identity over time:

1. *O* at *t* is identical with *O* at *t**. [assume for *reductio*]

---

[16] Lewis, *Plurality of Worlds*, pp. 199-204. In what follows I am indebted to Merricks, "Endurance and Indiscernibility."

[17] Lewis, *Plurality of Worlds*, pp. 203-204.

2. *O* at *t* is bent. [premiss]

3. *O* at *t** is not bent. [premiss]

4. If *O* at *t* is identical with *O* at *t**, then *O* at *t* is *F* iff *O* at *t** is *F*. [Indiscernibility of Identicals]

5. Therefore, *O* at *t* is bent and is not bent. [RAA, (1), (2), (3), (4)]

Just as the problem of accidental intrinsics revealed that no object can be part of two possible worlds in which it differs in its intrinsic properties, so the problem of temporary intrinsics purports to show that no object can exist at two times at which it differs in its intrinsic properties.

There are, says Lewis, only three solutions which he knows of for this problem:

> First solution: contrary to what we might think, shapes are not genuine intrinsic properties. They are disguised relations, which an enduring thing may bear to times. One and the same enduring thing may bear the bent-shape relation to some times, and the straight-shape relation to others. In itself, considered apart from its relations to other things, it has no shape at all. And likewise for all other seeming temporary intrinsics; all of them must be reinterpreted as relations that something with an absolutely unchanging intrinsic nature bears to different times. The solution to the problem of temporary intrinsics is that there aren't any temporary intrinsics. This is simply incredible, if we are speaking of the presence of ordinary things.... If we know what shape is, we know that it is a property, not a relation.
>
> Second solution: the only intrinsic properties of a thing are those it has at the present moment. Other times are like false stories; they are abstract representations, composed out of the materials of the present, which represent or misrepresent the way things are. When something has different intrinsic properties according to one of these ersatz other times, that does not mean that it, or anything else, just *has* them—no more so than when a man is crooked according to the *Times* or honest according to the *News*. This is a solution that rejects endurance; because it rejects persistence altogether. And it is even less credible than the first solution. In saying that there are no other times, as opposed to false representations thereof, it goes against what we all believe. No man, unless it be at the moment of his execution, believes that he has no future; still less does anyone believe that he has no past.
>
> Third solution: the different shapes, and the different temporary intrinsics generally, belong to different things. Endurance is to be rejected in favor of perdurance. We perdure; we are made up of temporal parts, and our temporary intrinsics are properties of these parts, wherein they differ one from another. There is no problem at all about how different things can differ in their intrinsic properties.[18]

The first solution denies that intrinsic properties exist and in that respect is analogous to the solution rejected by Lewis to the problem of accidental intrinsics according to which supposedly accidental, intrinsic properties turn out upon analysis to be relational properties. The second solution is the A-theoretical or presentist solution, analogous to the solution of Serious Actualism to the problem of accidental intrinsics. And the third solution represents the B-theoretical or tenseless solution, analogous to the solution to the problem of accidental intrinsics which posits the existence of transworld individuals composed of parts in each world, or somewhat more remotely, to the Counterpart Theory, which ignores the transworld individual

---

[18] Ibid., p. 204.

in favor of the various parts. I shall argue that while Lewis is correct in rejecting the first solution, he errs in rejecting the A-theoretical solution, which is immune to the shortcomings of the B-theoretical solution and which is therefore the preferable solution.

INADEQUACY OF SOLUTION I: DENIAL OF TEMPORARY INTRINSICS

In discussing the problem of accidental intrinsics, Lewis rejected the view that all the accidental properties of an object which is part of two worlds are really relational in nature. How could the selfsame object shared by two worlds W and W^* have the property of *being five-fingered* in W and the property of *being six-fingered* in W^*? Having five or six fingers on a hand just is not a relational property, but something that one has intrinsically. Similarly, if the identical object O is shared by two moments of time t and t^* and yet at t has the property of having five fingers and at t^* the property of having four fingers, then in order to avoid a contradiction the relationalist has to say that no change of O's intrinsic properties has taken place from t to t^*. Rather O has remained intrinsically the same; only O's external surroundings have changed. O's digitality is a matter of the relation *having n fingers* obtaining between O and a time. In relation to t O has five fingers, but in relation to t^* O has four fingers, without any intrinsic change on O's part. Indeed, O is an absolutely immutable object, both in its intrinsic nature and in its external relations. If even its relational properties were to change, O would not be self-identical. Since the relations in question are to times, however, this is not problematic for the relationalist. O always has five fingers relative to t and four relative to t^*. Even O's temporary extrinsic properties, like *being envied by P*, can be eternalized by making them relative to times. O undergoes no change between t and t^*.

The relationalist would escape the problem of temporary intrinsics by interpreting (1)-(4) as follows:

1.′ O exists relative to t and t^*.

2.′ O is bent relative to t.

3.′ O is not bent relative to t^*.

4.′ If O exists relative to t and t^*, then at t O is F relative to t iff at t^* O is F relative to t.

It then does not follow that relative to t O is bent and not bent, for (1) and (2) are interpreted as ascribing only relational properties to O, and according to (4′) a thing's relational properties do not change.

But, as Lewis asserts, such a solution is quite bizarre. How can the selfsame, intrinsically immutable object have a different number of fingers simply in virtue of being related to different times? As Merricks points out, even if such relational properties as *having five fingers at t* exist, there also seem to be non-relational properties like *having five fingers* as well. In order for a relational difference between *O* and *t* and *t** respectively to exist, intrinsic change seems necessary as well.

I am not aware of anyone who has explicitly propounded a relationalist solution to the problem of temporary intrinsics, but a number of philosophers have embraced another version of solution I. In response to Ronald Hoy's raising what was in effect the problem of temporary intrinsics,[19] both Oaklander and Smith propose to treat temporary, intrinsic properties as analogous to world-indexed properties like *being bent-in-W*, that is to say, as time-indexed properties, like *being bent-at-t*.[20] The hyphens are meant to signal that what is at issue here is not a relational property specifying a relation between an object *O* and a time *t*, but rather a monadic property which is ascribed to *O simpliciter*. This solution counts as a version of solution I because it denies that there are temporary, intrinsic properties. Just as the relationalist denied that temporary intrinsics are really intrinsic, so the time indexer denies that they are really temporary. The indexer would escape the problem of temporary intrinsics as follows:

1.″ *O* at *t* is identical with *O* at *t**.

2.″ *O* is bent-at-*t*.

3.″ *O* is not-bent-at-*t*.*

4.″ If *O* at *t* is identical with *O* at *t**, then *O* at *t* is *F*-at-*t* iff *O* at *t** is *F*-at-*t*.

No contradiction follows, since *O* is always *F*-at-*t* regardless of *O*'s temporal location.

Merricks notes two difficulties with the time-indexed solution. First, it only works if all ostensibly temporary, intrinsic properties are in fact time-indexed properties. But as in the relational case, it seems plausible to think that if there are properties like *being bent-at-t*, then there are also properties like just *being bent*. It is surely just the sober truth that, if $t<t^*$, then at *t O* was bent and that by *t** *O* was not bent. Second, if one denies that there is such a property as simply *being bent*, then one is at a loss to explain the more complex property of *being bent-at-t*. The latter is most perspicuously to be explained in terms of possessing the property of *being bent* at a certain time *t*, but the time-indexer, denying non-indexed properties

---

[19] Ronald C. Hoy, "Becoming and Persons," *Philosophical Studies* 34 (1978): 275-278.

[20] L. Nathan Oaklander, "Delmas Lewis and Responsibility: a Critique," *Philosophy Research Archives* 13 (1987-88): 183; Smith, "Personal Identity and Time," pp. 162-163; cf. Peter Van Inwagen, "Four-Dimensional Objects," *Noûs* 24 (1990): 247, 249. Oaklander in fact does come dangerously close to endorsing the relational version, for he writes, "If, for example, at t_1 P is thinking of Plato and at t_2, P is not thinking of Plato (but thinking of Descartes), then P has (tenselessly) the properties of thinking of Plato at t_1 and of thinking of Descartes at t_2"—a strictly relational statement, although context makes clear that Oaklander intends to express the time-indexed version.

of this sort, is forced to take *being bent-at-t* as a primitive notion.[21] These difficulties suffice to render this solution implausible.

A final observation on both these versions of solution I should be made: both of them are incompatible with the A-Theory of time, at least with any A-Theory which construes A-determinations as temporary, intrinsic properties possessed by things or events.[22] For on the relational version nothing can be said to be present *simpliciter*; something is only present in relation to a time. But then there is no privileged present; everything is present relative to the time at which it exists. What we have, in effect, is a B-Theory according to which every time (along with its contents) is present at itself, but no time is simply present. Similarly, on the time-indexed version, nothing has the property of simply *being present*; things or events are at best *present-at-t*. But once again, everything is *present-at-its-respective-time*. This represents a B-Theory of time in which presentness is reduced to a tenselessly held property characterizing no unique moment. Were it not for the implausibilities of these versions, the B-theorist might find them quite congenial.

There is another proffered solution to our problem which Lewis, at least, claims is yet another version of solution I, namely, adverbialism. Championed by Johnston and Haslanger, the adverbial solution draws attention to the *way* in which a property is exemplified, namely, temporally rather than timelessly. The time at which an object possesses a property is neither a *relatum* nor a constituent of the property, but characterizes the possessing of the property by the object, just as temporal adverbs modify verbs. Johnston explains, "Temporal qualification has to do with the ways individuals *have* properties.... Temporal qualifiers are often adverbs. Sam is presently fat. But he is *t*ly thin. And we do not get committed to simple contradictions so long as we do not in general allow temporal adverb dropping."[23] Johnston here appropriates A. N. Prior's tense logical insight into the adverbial nature of tense in order to argue that properties are not had *simpliciter*. The having is temporally qualified so that incompatible intrinsic properties of an object are not possessed in the same way, thereby averting a contradiction. Haslanger says, "The intuitive idea behind the so-called 'adverbial' option is that objects have properties at times, and that time should modify this 'having' rather than the subject or the property."[24] The A-theorist will no doubt applaud this fundamental intuition at the base of adverbialism, and, indeed, it may be thought at first blush that adverbialism really is a version of solution II. Unfortunately, a further look dispels this impression.[25] When Johnston says, "Instantiating a property, it turns out, is

---

21 In a personal communication Merricks also notes that the time-indexer has difficulty explaining the evident similarity of the properties *bent-at-t* and *bent-at-t**, since these are diverse, unanalyzable monadic properties.

22 See my *Tensed Theory of Time*, chap. 6. If one holds that presentness alone is a property, then the A-Theory renders relationalism otiose.

23 Johnston, "Problem about Persistence," p. 128.

24 Sally Haslanger, "Endurance and Temporary Intrinsics," *Analysis* 49 (1989): 120.

25 Indeed, in his *Real Time II* (London: Routledge, 1999), Mellor himself puts forward an account of intrinsic change which is basically adverbial. He thinks of truthmakers as *a*-is-*F* facts which are located at times and intrinsic change as the absence of the fact *a*-is-*F* at some time and its presence at some other time. In response to Hinchliff's objection that Mellor must have relativized truthmakers or otherwise he

instantiating at some time the property,"[26] the temporal qualifiers are not A-determinations, such as Prior had envisioned, but tenseless dates. —Similarly for Haslanger when she states, "Lewis suggests that a property is intrinsic iff the object has the property in virtue of the way *it* is independently of anything else. An endurance theorist will demand a temporally sensitive construal of this condition, e.g. that a property is (at a time) intrinsic iff the object has (at that time) the property in virtue of the way *it* is (at that time) independently of anything else."[27] That the times referred to here are tenseless is evident from her semantical treatment of tense and time in language.[28] We can, she says, regard propositional content as devoid of temporal elements, this content then being evaluated with respect to times, or else we can include the temporal elements in the propositional content: "Treating time as

---

runs into the problem of temporary intrinsics (Mark Hinchliff, "McTaggart, Change, and Real Tense," in *Essays on Time and Related Topics*, ed. L. Nathan Oaklander, Philosophy of Time Society Proceedings 1995-1999), Mellor denies "that *a*'s having this property (when it does) is relative to anything, since none of my *a*-is-*F* facts depends for its existence on anything other than spacetime" (D. H. Mellor, "Reply to Hinchliff," in *Essays on Time and Related Topics*). But then it seems to me that Mellor simply begs the question in assuming that it is, indeed, *a* which has *F* at *t* and lacks *F* at *t'*. Without an adverbial move, the Identity of Indiscernibles precludes *a*'s existing *simpliciter* at both locations with different properties.

[26] Johnston, "Problem about Persistence," p. 129.

[27] Haslanger, "Endurance and Temporary Intrinsics," pp. 123-124.

[28] The same is true of her re-formulated "factualist" account, presented in response to Mark Hinchliff, "The Puzzle of Change," paper presented at the Pacific Division meeting of the American Philosophical Association, April 2, 1994. The factualist alternative sounds at first like a genuinely tensed account, since she writes, "To say that the candle is bent is to say that the candle's being bent presently obtains; to say that the candle was straight is to say that the candle's being straight obtained in the past" (p. 4). This sounds very much like what the defender of a tensed theory of time wants to say: that the states of affairs comprising the actual world do not simply obtain tenselessly, but that they are actualized *seriatim*. This seems to imply a presentist ontology. However, she goes on to say, "then we can make sense of the idea that the present isn't all there is: to say that there are other ways things are besides how they presently are, is to say that types of states of affairs different from the ones that presently obtain, obtain at other times (in the past or future)" (p. 5). Here we still have the affirmation of tensed time, but now coupled with a tenseless ontology, a sort of hybrid of McTaggart's A- and B-series. But then it seems to me that Haslanger is landed square in McTaggart's Paradox: *all* the various types of states of affairs obtain presently at their respective times, so that in order to distinguish the presently obtaining state of affairs among all the others, she must say that it presently obtains at the present moment, and we are off on an infinite regress. The only alternative would seem to be to say that the property of presentness *simpliciter* "moves" along the series of tenselessly obtaining states of affairs, each of which has presentness at its respective time. But then I simply have no idea of what this property of "presentness" is anymore. Worse still, perhaps, for the factualist account, this simple property of presentness would not obtain relative to a time—a state of affairs would just have it, and then we seem to be back to Hinchliff's account. So it is difficult to see how the factualist account is to be combined with an affirmation of real tenses. Indeed, Haslanger speaks of tenses as though they have only a subjective reality: "The token states of affairs are eternal, but we make reference to them using tensed discourse. So the candle is (presently) bent because the candle's being bent (presently) obtains, and the candle was straight because the candle's being straight obtained in the past. Here we're referring to two different token states of affairs and locating them with respect to us in time" (p. 5). Although tensed terminology continues to be employed here, her gloss seems to make clear that tense is a token-reflexive feature of our language only. The state of affairs consisting of a tyrannosaurus devouring a trachodon never presently obtained; it only obtained in the past. Therefore, the factualist account, despite its misleading use of tensed language, is in the end as tenseless as the adverbialist account. When Haslanger speaks of presently obtaining, or obtaining in the past or future, what she really means to assert is the obtaining of a state of affairs simultaneously with the truth of some sentence token, or earlier or later than it.

part of the circumstances of evaluation...offers a temporalist view of propositions such that propositions can change truth value from time to time; treating time as an additional constituent of the propositions (perhaps not explicit in the sentence expressing the proposition but determined via the context of utterance) offers eternalist propositions."[29] The fact that she takes time-laden propositions to be eternalist makes it clear that the times she has in mind are mere tenseless dates (or clock times); the same would accordingly hold for the times of the circumstances of evaluation in the alternative case. The adverbial approach, as defended by its leading exponents, then, represents a B-theoretical *Ansatz*, which proposes to dissolve the problem of temporary intrinsics as follows:

1.‴ *O* exists *t*ly and *O* exists *t**ly.

2.‴ *O* is *t*ly bent.

3.‴ *O* is *t**ly not bent.

4.‴ If *O* exists *t*ly and *O* exists *t**ly, then when *O* exists *t*ly, *O is t*ly *F* iff when *O* exists *t**ly, *O* is *t*ly *F*.

The reformulated Principle of the Indiscernibility of Identicals does not mandate that if *O* exists *t*ly and *t**ly, then at both moments *O* is simply *F*, for it must be *F* in some way. What way? The same way, of course; it must be, for example, *t*ly *F* at both times. In a sense, then, intrinsic change does not occur in *O* between *t* and *t**, since *O* only has intrinsic properties in timely ways and those timely possessed properties never change.

Now Lewis charges that adverbialism is just a variant of relationalism, the relationality being put, not in the shapes themselves, but in the having of them. He explains,

> There is a three-place relation of instantiation, this relation holds between me and bentness and some times, and it holds between me and straightness and some other times. I ask: what does standing in some relation to straightness have to do with just plain being straight? And the variant still claims that to be shaped is to stand in relation to other things, *inter alia* to times. I say it still amounts to a denial that things have temporary intrinsics.[30]

In Lewis's opinion, then, to say that *O* is *t*ly *F* is just a disguised way of saying that *O*, *F*, and *t* are so related that *O* is *F* at t.

E. J. Lowe disputes this verdict. Underlining the adverbial character of the solution, Lowe states that relativizing the having of properties to times is not a three-place relation; rather it is to say that the *holding* of the two-place relation of property exemplification like *O*'s being *F* is related to a time. Relative to *t O*'s being bent obtains, while relative to *t** *O*'s being bent does not obtain. The worry about *O*'s just being bent or just being straight unrelated to a time is therefore bogus. Lowe asks,

---

29 Haslanger, "Endurance and Temporary Intrinsics,", p. 121.

30 David Lewis, "Rearrangement of Particles: Reply to Lowe," *Analysis* 48 (1988): 66.

> But what does *Lewis* understand by 'just plain being straight'? And why does he attach such importance to the notion?—why should there be anything to do with 'just plain being straight'? I can only assume that this phrase is supposed to denote some atemporal mode of property exemplification (such as is enjoyed, perhaps, by abstract geometrical figures). But in that case, why can't we just happily deny that changeable, persisting physical objects can ever correctly be described as 'just plain being straight'?[31]

In response to Lowe's questions, it seems to me that by "just plain being straight" Lewis understands that *O*'s being straight is not, as Lowe avers, a matter of a two-place relation obtaining between an intrinsically shapeless object and the property of straightness. *O* is straight and does not take on a shape in virtue of standing in some relation to an abstract entity. In construing *O*'s being straight as a relation, Lowe seems to have already succumbed to relationalism. Moreover, Lewis's statement that "to be shaped is to stand in relation to...times" does accurately characterize Lowe's position that the holding of the above two-place relation stands itself in relation to times. For apart from this relation, on Lowe's view, physical objects have no shape. The reason Lewis attaches such importance to just plain being straight is that relationalism solves the problem of temporary intrinsics only at the high price of implausibly denying the intrinsicality of properties like *being straight.*

Uncomfortable with Lowe's construal of "*O* is *F*" as a two-place relation obtaining between *O* and *F*-ness, Haslanger maintains that *O*'s being *F* need not be construed relationally, but is a basic notion of an object's instantiating its properties.[32] The timeless proposition expressed by "*O* is *F*" is true or false in relation to a time. She notes Lewis's objection that even in such a case there is a three-place relation between *O, F,* and *t*, such that at *t* the proposition expressed by "*O* is *F*" holds. Haslanger concedes that this is correct, but finds it unproblematic because the relation between *O, F*, and *t*, such that the proposition expressed by "*O* is *F*" is true at *t*, is a relation which is defined partly in terms of the more basic, non-relational notion of *O*'s being *F*. My point remains, however, that *O*'s being *F* still leaves *O* without any shape at all until the relation to *t* is taken into account. Paradoxically, the *O* that has the property of being bent really is not bent except in relation to certain times.[33]

It seems to me that any plausibility which does attend the adverbialist view derives solely from the tacit assimilation of adverbialism to the tensed view of time which inspired it and from which it was abstracted. The A-theorist need not

---

31 E. J. Lowe, "The Problems of Intrinsic Change: Rejoinder to Lewis," *Analysis* 48 (1988): 73.

32 Haslanger, "Endurance and Temporary Intrinsics," pp. 121-122.

33 This same problem remains unresolved in Haslanger's factualist formulation of her position. According to that view, the state of affairs *O's being bent* does not imply that *O* is bent until that abstract state of affairs is related to a time at which said state obtains. As Haslanger explains, "to be a property in the relevant sense is to be a constituent in a token state of affairs....to say that an object has a property is to say that a state of affairs type obtains" (Haslanger, "Comments," p. 9). If only the present state of affairs obtained, such an account might go through. But since her factualist account is tenseless, she must therefore deny that saying *O* is bent is to predicate unqualifiedly a property of *O*. Hence, it seems to me that Hinchliff is justified in regarding her position as another version of relationalism (Hinchliff, "Puzzle of Change").

construe "just plain being straight" as an atemporal mode of predication, as Lowe surmises, since "being straight" may be taken as a tensed locution. Thus, changeable, persisting physical objects may be correctly described as "just plain being straight." Adverbialism must construe such a predication as atemporal only because it has forgotten (or omitted) tense. It is true, as Lowe states, that no physical object just *is* (tenselessly) bent, in abstraction from any time at which it is bent; but that does not imply that *O*'s shape derives from the relation obtaining between *O*'s *being* (tenselessly) *bent* and some time. Rather the time of *O*'s being bent may be inherent in the tense of *O*'s exemplifying bentness. Thus, when Haslanger states, "The endurance theorist denies that the description which characterizes the object 'timelessly' is the description which captures all of the intrinsic properties of the object,"[34] the A-theorist will say, "Amen!" But he will not concur that the correct description concerns relations obtaining between the object, a property, and a tenseless date.

Moreover, a deeper problem now surfaces. Haslanger proposes to define the relation between *O, F*, and *t*, such that the proposition expressed by "*O* is *F*" is true at *t*, in terms of the more basic notion of *O*'s being *F*. But as Merricks points out, this is just what adverbialism does not permit one to do. He asks,

> What are we to make of esoteric concepts like *having a property at a time*, or *having a property tly?* The obvious answer: analyze them in terms of the more basic notions that we already understand—for instance, having a property at a time might be analyzed simply as having a property when the relevant time is present. But the adverbialist is forced to *start with* concepts like *tly bent*, and analyze our everyday concepts in terms of her adverbial ones.... Adverbialists must take oddities like *tly bent* as primitive.[35]

Once again, any plausibility which surrounds adverbialism is an afterglow of adverbial tenses, which have been expelled from the theory. What is left is, indeed, odd. What do we mean by affirming that Lincoln has *1860*ly the property *being President*, or that I have *6:00*ly the property *being unshaven*? Given adverbial tenses, these complex notions can be analyzed in terms of having properties *simpliciter*. The A-theorist can affirm with Merricks, "if I have a property now, it follows that I simply have it."[36] But tenseless adverbialism has to explain having a property *simpliciter* in terms of having a property at a time. We doubtless understand having a property at a time because we already understand the notion of having a property. Merricks thus seems justified in his complaint that "we don't have a good grasp on what the adverbialist must mean by an object *being p tly* (a technical notion she has introduced) because what we take that to amount to...cannot be analyzed into, or explained in terms of, other concepts we clearly understand."[37]

---

[34] Haslanger, "Endurance and Temporary Intrinsics," p. 124.

[35] Merricks, "Endurance and Indiscernibility," p. 169-170.

[36] Ibid., p. 170. Similarly, we understand having a property at a place in terms of the more primitive notion of simply having a property. Hence, Lowe's spatial analogy to adverbialism fails (Lowe, "Problems of Intrinsic Change," p. 75).

[37] Merricks, "Endurance and Indiscernibility," preprint. Michael Rea misses the point when he objects that from the fact that *being F* is an ingredient of *being* t*ly F* it does not follow that it is possible for something to be simply *F* (*running Φly*, he says, has *running* as an ingredient, but it does not follow that

The various versions of solution I, therefore, are implausible and should be rejected if a more plausible solution can be found. Adverbialism, which took tense seriously enough to get an argument, but not seriously enough to make it go through, already forms a bridge to solution II and points to a better way.

SUCCESS OF SOLUTION II: THE A-THEORETICAL SOLUTION

The adverbialist correctly saw that the having of a property needs to be temporally modified; but he erred in turning to tenseless times as the qualifiers rather than to tense. The insight of the A-theoretical solution is that the having of a property *simpliciter* is a tensed having and therefore inherently temporal. Not that the A-theoretical solution is simply to embrace adverbialism with a proviso that tensed expressions rather than dates be substituted for *t* in "having *t*ly *p*"—for on a pure A-Theory neither past nor future events/things exist and so can no more have properties than merely possible objects can have properties.[38] Thus, it would be

---

running can exist *simpliciter*, without being slow, fast, *etc.*) (Michael Rea, "Temporal Parts Unmotivated," *Noûs* [forthcoming]). The issue is not one of ontology, but semantics. The question is what notion is semantically primitive. *Running Φly*, for example, is not semantically primitive because it can be analyzed in terms of the more primitive notion *running*. But on adverbialism there is no such notion as having a property *simpliciter*. The apparently complex notion of having a property *t*ly must be taken as semantically primitive, which to all appearances it is not.

Balashov points out a further difficulty attending both indexicalism and adverbialism within the context of SR. Consider, first, a Flatland analogy of various two-dimensional figures which persist through time. The perdurantist would regard these figures as various two-dimensional slices of a three-dimensional box. But the adverbialist Flatlander denies that there is any such object as the box; there is just a multitude of two-dimensional figures which have perspective-indexed shapes, appearing to one person as triangular, to another as square, and so forth. When appropriately arranged in three-dimensional space (whose existence the adverbialist Flatlander accepts, denying only that objects are three-dimensional), all these various, two-dimensional configurations line up neatly and fill up a compact, three-dimensional, box-shaped region. But how, asks Balashov, can the adverbialist explain such a harmonious unity among the different two-dimensional shapes? Since the box is not given as an object to be dissected, but was constructed, as it were, "from bottom up," as an arrangement of a multitude of perspective-indexed, two-dimensional shapes stacked up in a certain way in three-dimensional space, the harmonious unity "must be a complete mystery"(Yuri Balashov, "Relativistic Objects," *Noûs* [forthcoming]). Analogously, on three-dimensional adverbialism, why is it that some collections of different, reference frame relative three-dimensional shapes can be arranged in four-dimensions into nice volumes? The adverbialist "will have a hard time explaining how 'separate and loose' 3D shapes come together in a remarkable unity, by lending themselves to an arrangement in a compact and smooth 4D volume....the three-dimensionalist must regard it as a brute fact, indeed, as a complete mystery" (Ibid.). Notice, however, that Balashov gratuitously presupposes that "We have already been forced to abandon the idea of absolute simultaneity" (Ibid.) But if the adverbialist adopts a neo-Lorentzian interpretation of relativity theory—which Balashov elsewhere recognizes as viable—, then he must explain only the stack composed of a *single* series of two-dimensional shapes, *viz.*, the object's absolute shape. It could be bound together causally perhaps. The presentist has it even better, however; for he will face no mystery at all, since there exists neither the four-dimensional spactime volume nor the varying three-dimensional shapes. There is just the single, absolute shape which endures, however distorted our measures of it may be due to absolute motion.

38 This answers the query of Lawrence Sklar, "'What makes you think there is any choice to be made other than a choice of how to represent one and the same situation in a chosen language?'...doesn't it seem pretty compelling to say in this case that the decision on how to talk about the situation is nothing but that, a decision on how to talk?" (Lawrence Sklar, "Comments on Mark Hinchliff: 'The Puzzle of

erroneous to say that Napoleon has pastly or two centuries ago the property of being alive. Rather as a presentism, the A-theoretical solution holds that "having a property" is paradigmatically a present-tense notion: to have *p* is to presently have *p*. Thus, the A-theorist meets Lewis's *desideratum* of giving primacy to a notion like "just plain being straight," for "being straight" is a tensed, not a tenseless, locution. With respect to Napoleon, then, one must say of him that he had the property of being alive, even though he does so no longer. On a presentist ontology, the problem of temporary intrinsics cannot even arise because there are (present tense) no times which overlap in sharing an object *O*. For *O* exists (present tense) only at one time, the present time, and so does not have (present tense) incompatible properties, as it would if it existed tenselessly at two different times in the B-series with different properties at each time. All the properties *O* has are the ones it presently has, and so no contradiction can arise. For even if *O* undergoes intrinsic change between *t* and *t**, it nonetheless does not have (present tense) incompatible properties: *O* has only the properties it has presently and these are mutually compatible. To insist that *O* has incompatible properties at *t* and *t** makes no sense on presentism, for this can only mean (since *O* does not have tenselessly any properties) that when *t* was present *O*'s properties were mutually incompatible and when *t** was present *O*'s properties were mutually incompatible, both of which are *ex hypothesi* false. Hence the A-theorist will propose the following interpretation of (1) to (4):

1.'''' *O* existed at *t* and exists at *t**.

2.'''' *O* was bent at *t*.

3.'''' *O* is not bent at *t**.

4.'''' If *O* existed at *t* and exists at *t**, *O* which existed at *t* is *F* iff *O* which exists at *t** is *F*.

By tensing all the verbs in (1'''') to (4''''), the A-theorist avoids any contradiction.[39] (The particular tenses assigned are, of course, arbitrary.) For although *O* possessed different intrinsic properties at *t* than it does at *t**, nevertheless the *O* which existed at *t* is (present tense) *F* just as the *O* which exists at *t** is *F*; they are the selfsame *O*.

This seems like a wonderfully perspicuous and common sensical solution; why, therefore does Lewis reject it? Remarkably, he says that this solution actually rejects endurance because it rejects persistence. This allegation is more than mildly surprising, since the A-Theory of time is typically contrasted with the B-theory in that the latter affirms at best temporal extension with respect to things' persisting in

---

Change'," comment at the Pacific Division meeting of the American Philosophical Association, April 2, 1994). The difference between tensed and tenseless solutions to the problem is not just a matter of semantics, but of metaphysics. Altogether different ontologies of time are presupposed.

[39] I take "*O* was bent at *t*" to mean P*t* (*O* is bent), where "P" is the tense logical operator "It was the case." Thus, one avoids Merricks' objection to rewriting (2) as "At *t*, *O* was bent" that "*O* was bent" is true now, whereas, at *t*, "*O* is bent" was true. When one says, "At *t*, *O* was bent," one is not affirming that the past tense statement "*O* was bent" was true at *t*. One could rephrase this premiss "At *t* *O* existed and was bent." One should not be too wooden in expressing correctly the sense of "*O* at *t*."

time, whereas the A-Theory involves the notion of things' enduring from one moment to the next, an idea which is expressed metaphorically as the passage of time. But Lewis says that on A-theoretical endurantism, things do not really endure because they do not persist. He defined persistence by saying "something *persists*, iff, somehow or other, it exists at various times."[40] Since according to presentism, only the present time exists, things do not exist at various times, if "exists" is present-tensed; they only exist at one time, the present. Therefore, they do not persist! But even Lewis's definition allows that persisting things exist at various times "somehow or other," and the A-theorist can surely give some good sense in which things exist at various times, namely, they existed at past times, they exist at the present time, and they will exist at future times. Or again, *O* exists at various times in the sense that when *t* is present it is then true that "*O* exists" and when *t** is present it is then true that "*O* exists." In any case, if we insist on the present tense of "exists" in the definition of persistence, neither is it true on a B-Theory that things exist (presently) at various times. That would be to assert the contradiction that other times earlier or later than the moment picked out by the token-reflexive expression "present" are simultaneous with that moment. The B-theorist must take the "exists" tenselessly. But the A-theorist can affirm this same tenseless truth. Here, I think, is the rub as far as Lewis is concerned. When the B-theorist says that things exist at various times, he means this quite literally. But when the presentist says a thing exists (tenselessly) at some other time *t*, this affirmation is to be parsed in a way analogous to the affirmation of the Serious Actualist that a thing exists in some other possible world *W*. The Serious Actualist does not posit the concrete existence of another world in which the object exists; rather he means that there is a state of affairs *W* which is such that if *W* were actual, then it would be true that the thing exists. Similarly, the presentist holds that there is a state of affairs W^t which is such that if W^t were present (which is a sort of tensed actuality), then it would be true that the thing exists.[41] Lewis finds these *ersatz*-times as repugnant as *ersatz*-worlds.[42] He wants chronal realism as well as modal realism. "In saying that there

---

40 Lewis, *Plurality of Worlds*, p. 202.

41 Sklar objects,

> "When we are dealing with possible worlds semantics for modality, we have two ways of going with 'actual.' We can take it to be token-reflexive or we can take it as being a property of one and only one possible world. But if we think of time instants on the analogy of possible worlds, 'now' doesn't work like 'actual.' It can be thought of as token-reflexive, to be sure, but the alternative is not 'now' as picking out a single instant of time. Rather it must, at each time, pick out that moment of time as the only 'real' moment of time" (Sklar, "Comments").

If we stick to tensed discourse, however, it is true that 'now' picks out (present tense) a single moment of time. It is only through the imperialism of tenseless discourse that difficulties (most notably McTaggart's Paradox) arise. The moment which 'now' picks out is not the same as the moment which 'now' will pick out.

42 He adds,

> "What passes for persistence and change, on this solution, does not really involve other times. Rather there are 'abstract' ersatz times, to go with the one 'concrete' genuine time. These represent, or misrepresent, the present. If I am bent now, and straight later, there is an abstract misrepresentation of the present according to which I am straight. 'Persistence' and 'change,' so understood, do not deserve their names. This solution

are no other times, as opposed to false representations thereof, [solution II] goes against what we all believe."[43] One can only smile at the audacity of this pronouncement. It is universally acknowledged that the common sense view of time is the A-Theory, and it takes a good deal of schooling to convince oneself of the chronal realism of the B-Theory. Only philosophers and physicists—and only some of them—believe, for example, that future times literally exist; most of us believe in the ontological primacy of the present. That does not imply that we believe we have no future and no past, as Lewis alleges. I have a past in that I existed at and lived through times which once were present, and I have a future in that I shall exist and live through times which will be present. If most of us find modal realism fantastic, I see no reason why we should feel impelled to embrace chronal realism. These two do seem quite parallel, and Lewis is at least consistent. The philosophers who really find themselves in tension on this score are those B-theorists who accept chronal realism, but recoil at modal realism, despite the obvious parallels between the two.

In any case, the perdurantist can hardly be said to affirm unequivocally that things exist at various times and therefore to persist either. For he denies that things exist at various times in a straightforward sense; they do so only derivatively by having temporal parts at various times. If the perdurantist can be said to affirm "somehow or other" that things exist at various times, courtesy must be given to the endurantist to affirm the same.

Objections to endurance raised by other perdurantists are no more impressive. Zemach, for example, in arguing for a doctrine of temporal parts, presents three objections against presentness (or as he puts it, "now") being a property ascribed to times: (1) If presentness is a property of moments of time, then so are pastness and futurity. But this leads to the postulate of a super-time in which these properties change, which is "pure fantasy and crackpot physics."[44] But, as I have elsewhere argued,[45] construing presentness as a monadic property of events/things/times does not imply a similar construal of pastness and futurity, which are better construed as relational predicates which ultimately reduce to past- or future-tense statements. By affirming the metaphysical primacy of the present, the A-theorist averts what is, in essence, McTaggart's Paradox. (2) The change of A-determinations by a time requires a physical cause. But since physics knows of nothing which changes when tense changes, we should suspect the reality of the change. Again, we have already seen the narrow and unjustified scientism on which such an argument rests.[46] Zemach must regard it as impossible that God, being non-physical, could exist alone in A-series time, which is a presumptuous assumption to make *a priori*. (3) According to SR, if any point event is objectively now, then it is meaningless to say that any other point event is objectively now. We should then also have to say that a point event is objectively here. If by "objectively now," Zemach means "possessing

---

amounts to a denial of persistence and change, and therefore is untenable" (Lewis, "Rearrangement of Particles," p. 66).

43 Lewis, *Plurality of Worlds*, p. 204.

44 Eddy M. Zemach, "Time and Self," *Analysis* 39 (1979): 143.

45 See my *Tensed Theory of Time: a Critical Examination*, chap. 6.

46 See chap. 6, pp. 134-142.

the property of presentness irrespective of the consciousness of some sentient being," then it is false that SR regards such a notion as meaningless. On the contrary, the burden of SR is to provide a means expressly to determine which events relative to a reference frame are simultaneous with a given point event, so that if one is willing to ascribe to that event the property of presentness, one will have a means of determining all other point events which are present relative to that frame. On the other hand, if by "objectively now," Zemach means "absolutely present," that is, "present irrespective of (non-privileged) reference frames," then his is simply a naive relativity which knows nothing of the various empirically equivalent interpretations of the mathematical core of SR which are compatible with absolute presentness, as we have seen at length.[47] The same goes for Johnston, who asserts that it is difficult to maintain that only what is present constitutes concrete reality, since what is now is a frame relative matter, whereas what is concretely real cannot be relative to reference frames.[48] As for Zemach's *tu quoque* argument about spatial tenses, I have elsewhere exposed its inadequacies.[49] It is interesting in this connection that Zemach holds to the equivalence of "It is 2:00 now" with "I am located at 2:00,"[50] but he fails to observe that "am" is in this case present-tensed, so that the assertion does not merely attribute some B-series position to its author.

Sklar is disturbed by the "wildly counter-intuitive, paradoxical nature" of the presentist metaphysic which underlies endurantism: "If any extended time is composed of future (which is non-being) and past (which is non-being), how can there be anything real at all? Given that the present is just the dividing line between past and future, aren't we consigning all reality to nothingness by taking a presentist line?"[51] But I have elsewhere argued that the presentist need not—indeed, should not—construe the present as just a single instant constituting the boundary point between the past and future.[52] Rather all expressions like "the present" are ambiguous and incomplete; we should speak properly only of some present (non-degenerate) interval, such as the present minute, the present hour, and so forth. There is no such thing as simply *the present*. Because the present interval, whatever it is, is potentially infinitely divisible, one never arrives at a present moment which is instantaneous or unreal.

The A-Theory of time thus embraces a metaphysic on which the problem of temporary intrinsics cannot even arise. Moreover, it does not have the drawbacks

---

47 See chaps. 2-4 and chap. 6. —Not to mention the fact that if one is willing to concede that a point event is absolutely present, then one has already abandoned Einsteinian relativity and, hence, implicitly admitted the existence of other simultaneous, and hence, absolutely present events. What Zemach really wants to say, I think, is that if an event is "objectively now" in the first sense, then it cannot be "objectively now" in the second sense, which allegation I refute in the text.

48 Johnston, "Problem about Persistence," pp. 114-115. Johnston's argument ought to lead him to question SR's presupposition of the non-existence of absolute space, which leads to the relativity of simultaneity and of concrete reality.

49 See my *Tensed Theory of Time*, chap. 4.

50 Zemach, "Time and Self," p. 145.

51 Sklar, "Comments."

52 See my *Tensed Theory of Time*, chap. 7.

alleged against it. If, then, the B-Theory cannot provide an equally plausible solution of the problem, this constitutes a sound argument for the A-Theory of time.

FAILURE OF SOLUTION III: THE B-THEORETICAL SOLUTION

The B-theorist denies that the only properties an object has are the properties it presently has. For since there are no tensed facts, the properties an object has "presently" are simply those which it has tenselessly at the B-theoretical moment at which the present-tensed sentence token ascribing said properties is true. But then it is obvious that in this tenseless sense of having, an object which persists longer than the interval of the present has properties which it does not presently have. That lands us almost immediately in the problem of temporary intrinsics. For if intrinsic change is real, then an object, in virtue of existing at more than one time, has incompatible properties. The selfsame object *O* is shared by two moments of times, they "overlap" at *O*'s existence, and therefore, given the Indiscernibility of Identicals, *O* must at both times have all the same intrinsic properties: *O* cannot intrinsically change.

A B-theoretical endurantism, such as that espoused by Mellor, is therefore self-contradictory. Given the reality of intrinsic change, the problem of temporary intrinsics goes through with a vengeance. Endurantism requires that *O* exists at t and t^*, and the B-Theory requires that the mode of *O*'s existence is the same tenseless existing. The Indiscernibility of Identicals requires that if *O* which exists at t is identical with *O* which exists at t^*, then *O* must have all the same properties at t and t^*, and the B-Theory again requires this having to be a tenseless mode of possession in both cases. So if *O* is (tenselessly) bent at t and is (tenselessly) not bent at t^*, contradiction ensues. *O* which exists at t is said to be identical to and hence indiscernible from *O* which exists at t^*, and yet *O* which exists at t is bent and *O* which exists at t^* is not bent, so that there is a discernible difference in *O* at t and at t^*. Therefore, *O* which exists at t is not the same object as *O* which exists at t^*. What we have here are two different objects, and *O* did not endure from t to t^*.

Mellor affirms the reality of intrinsic change, argues for endurantism, and refutes attempts to prove that things are temporally extended wholes. But he says nothing to explain how the selfsame object could exist tenselessly at two times and yet there be a discernible difference between the object at one time and the identical object at another time. About the only recourse left is to deny the principle of the Indiscernibility of Identicals, but this would be a move of desperation, since the alleged exceptions to the principle are generally agreed to turn out upon analysis not to be violations at all, but to concern statements involving, for example, an "is" of constitution, rather than an "is" of identity.[53] It is not merely by accident that virtually all B-theorists are perdurantists.

---

[53] For a helpful discussion of alleged exceptions to the principle, see David Wiggins, *Identity and Spatio-Temporal Continuity* (Oxford: Basil Blackwell, 1971). Concerning the justification of the principle, Wiggins comments,

> "If Leibniz' Law were not controverted, I should remark that it is as obvious as the Law of Non-Contradiction, and simply ask: 'How if a is b could there be something true of

Indeed, the above reasoning constitutes an argument for the existence of temporal parts. *O* which exists at *t* turns out not to be the same object as *O* which exists at *t**. Rather they are distinct temporal parts of a temporally extended meta-object, if you will, the reality which is represented by a world-line on a spacetime diagram. Intrinsic change of this temporally extended object consists only in its temporal parts' having different intrinsic properties among themselves.

It is not uncontroversial, however, whether the doctrine of temporal parts can even be intelligibly formulated. There is a real danger of circularity here: a perduring object is defined as an aggregate of temporal parts, and a temporal part is defined as a segment or subdivision of a perduring object.[54] The problem becomes particularly acute with respect to personal identity over time. We normally think of persons as enduring things, not as extended meta-objects. There is a temptation to say that persons are a collection of person-stages or temporal parts, and to characterize the latter in terms of persons. Thus, Lewis, after defining a person as an "aggregate of person-stages" which are appropriately interrelated, explains,

> A person stage is a physical object, just as a person is. (If persons had a ghostly part as well, so would person stages.) It does many of the same things that a person does: it talks and walks and thinks, it has beliefs and desires, it has a size and shape and

---

the object a which is untrue of the object b? After all, they are the same object.' Since it is controverted, I shall add these points:

(i) The principle marks off what is peculiar to identity and differentiates it in a way in which transitivity, symmetry and reflexivity...do not.

(ii) The counter-examples to Leibniz' Law are scarcely more impressive than the counter-examples to the Law of Non-Contradiction....

(iii) If Leibniz' Law is dropped then we need some formal principle or other...to justify the valid instances (if these are not all the instances) of the intersubstitution of identicals.... It is extraordinarily difficult to find or formulate a weaker principle, or amend Leibniz' Law suitably" (Ibid., pp. 4-5).

One putative counter-example is of especial interest to us. On behalf of the relativization of identity, it is said that the identity between the little boy John and the Lord Mayor John cannot be absolutely affirmed, since the Lord Mayor is not a boy. Rather one must say something like "The boy John is the same person as the Lord Mayor John." Wiggins rejoins that all this counter-example shows is "the necessity for care about tenses" in the interpretation of Leibniz's Law (Ibid., p. 6). We should say "John will be the Lord Mayor" or "The Lord Mayor was the school boy." Wiggins remarks,

> "We only thought we had a case of R [the relativization of identity thesis] because we confused the timeless and the tensed way of speaking within one utterance. If 'g (Sir John Doe)' is a tensed statement it should be read as saying that Sir John *was* a boy and it is true. If it is a tenseless statement then it says of Sir John Doe that at some time or other he (timelessly) is a boy. This again is true. If we take tenses seriously it is impossible to say 'Sir John Doe *is* the same boy as John Doe' since it is false that Sir John Doe is now a boy. But it is true and perfectly unproblematic that Sir John Doe *was* the same boy as John Doe" (Ibid., pp. 6-7).

Interpreting the principle of the Indiscernibility of Identicals as a tensed statement provides the key for unlocking the problem of temporary intrinsics.

[54] See E. J. Lowe, "Substance Identity and Time," *Aristotelian Society Supplementary Volume* 62 (1988): 61-78; David S. Oderberg, *The Metaphysics of Identity over Time* (New York: St. Martin's Press, 1993), pp. 88-91, 109-113; cf. George Graham, "Persons and Times," *Southern Journal of Philosophy* 15 (1977): 309-315.

> location. It even has a temporal duration. But only a brief one for it does not last long.... It begins to exist abruptly, and it abruptly ceases to exist soon after.[55]

This characterization is unhelpful precisely because a person, on Lewis's account, does *not* talk and walk and think, nor does it have beliefs and desires. The four-dimensional object which is a person does not go anywhere or talk to anybody. Rather its temporal parts do. But then Lewis's characterization amounts to saying that a person-stage does many of the same things that a person-stage does, which is uninformative.

A similar problem seems to attend Mark Heller's attempt to give a lucid and straightforward account of temporal parts.[56] Whereas Lewis began with the temporal parts and defined an object in terms of them, Heller begins with an object and defines its temporal parts in terms of it. According to Heller, a physical object is the material content of a region of spacetime. This characterization, however, is immediately inadequate for an understanding of temporal parts, since it will not apply to non-physical objects like God or persons, which are temporally, but not spatially, extended. But let that pass; the characterization is implausible even as a description of a physical object, for it allows that the material content of the region of spacetime occupied by some of my left arm and shirt sleeve, some of my desktop, and a quantity of adjacent air constitutes a physical object, which is an unacceptable reification. To avoid such pseudo-objects' getting into his ontology, Heller has to revert to talk of three-dimensional objects, talk which he wanted explicitly to avoid, to whose boundaries is added a fourth dimension so as to outline a four-dimensional object which fills all its boundaries.[57] In a later piece he makes it clear that on his view the identity conditions of a four-dimensional object are fixed by its spatio-temporal boundaries.[58] But those boundaries, we have seen, must seemingly themselves be established on the basis of an object's three-dimensional boundaries. This procedure is obviously problematic. According to Heller, three-dimensional objects exist at instants. Now clearly in order to fix the boundaries of a four-dimensional object, one will have to take into account not merely the three-dimensional boundaries of an object at its initial instant of its existence, but rather—because objects change in shape, size, and orientation—its three-dimensional boundaries at *every* instant of its existence. If temporal parts are then defined in terms of three-dimensional (spatial) cross-sections of the four-dimensional whole, it seems that a circle has already been closed.

Such a conclusion might seem hasty, however, since on Heller's view three-dimensional cross-sections are not temporal parts. For he stipulates that a temporal part of a four-dimensional object is a spatio-temporal part that is the same size as the

---

55 David Lewis, "Survival and Identity," in *Philosophical Papers,* vol. 1 (Oxford: Oxford University Press, 1983), pp. 59, 76.

56 Mark Heller, "Temporal Parts of Four Dimensional Objects," *Philosophical Studies* 46 (1984): 325.

57 This seems to bear out Loizou's contention that so-called four-dimensional objects are parasitic upon three-dimensional objects which endure through time (Loizou, *Reality of Time*, pp. 13-15).

58 Mark Heller, "Varieties of Four Dimensionalism," *Australasian Journal of Philosophy* 71 (1993): 59.

object for as long as that part exists. This definition needs qualification,[59] but the salient point is that temporal parts have temporal extension and are themselves four-dimensional objects composed of temporal parts, which is not the case for three-dimensional objects. Thus, on Heller's view, no instantaneous temporal parts exist. But a little reflection shows this, too, to be problematic. If a four-dimensional object's boundaries are fixed by its instantaneous three-dimensional spatial cross-sections and yet those cross-sections are not temporal parts of the object, then what determines which three-dimensional objects should be taken together to fix the boundaries of the four-dimensional object? On the endurantist view, the various three-dimensional objects are taken as the identical object enduring through time, so there is no difficulty in determining its world-line. But on perdurantism, there are in the place of the one enduring object literally infinitely many, distinct, three-dimensional objects. Even if no intrinsic change occurs between two instants, no object endures from one instant to the next; we have two objects at those instants. But now it becomes impossible to justify why this series of distinct objects should be collected together rather than some other series of distinct objects in order to fix the boundaries of a four-dimensional object. This raises again the danger of the arbitrary reification of pseudo-objects in spacetime. On the other hand, if, as is usually accepted, instantaneous cross-sections are legitimate temporal parts, then we seem to be caught in a circle, fixing the identity conditions of a four-dimensional object via its three-dimensional temporal parts and identifying a three-dimensional object as a temporal part because it is a subdivision of a four-dimensional object.

In order to avoid circularity, it seems that the perdurantist is driven to embrace *conventionalism* with respect to physical objects, that is, the doctrine that no mind-independent physical objects exist. As Michael Jubien affirms, there really are no (physical) things – there is just "stuff" spread around spacetime, and we carve the world into things conventionally.[60] On this view there are neither three-dimensional nor four-dimensional physical objects, so that no problem of inter-definability arises, and the strange pseudo-objects we can create by carving up the material content of spacetime arbitrarily do not objectively exist, so that no question of objectionable reification arises either. In his book *The Ontology of Physical Objects*, Heller, too, advocates conventionalism with respect to physical objects, even going so far as to affirm that which collection of temporal parts goes to constitute a certain person is conventional.[61]

Conventionalism is an extraordinarily high price to pay for the adoption of a temporal parts ontology. Heller himself feels exposed to "those jokes people cannot help making when they are confronted with someone who claims that there are no

---

[59] It is not clear how relativity theory plays into Heller's view, since the size of an object, on the received interpretation, is relative to a reference frame. Moreover, one would have to take into account other factors as well, such as shape and orientation, to get our recognizable 3-D objects from slicing up spacetime regions. For a critique of this definition of "temporal part," see Merricks, "Persistence, Parts, and Presentism."

[60] Michael Jubien, *Ontology, Modality, and the Fallacy of Reference*, Cambridge Studies in Philosophy (Cambridge: Cambridge University Press, 1993), p.1.

[61] Mark Heller, *The Ontology of Physical Objects: Four-Dimensional Hunks of Matter* (Cambridge: Cambridge University Press, 1990), p. 23.

people."[62] Worse, conventionalism seems to face a problem of self-referential incoherence. For Heller affirms that (i) conventions require convention makers, (ii) convention makers, including people, are themselves conventional, and (iii) conventional objects do not really exist at all.[63] If people are conventional objects and such objects do not exist, then, given that conventions require convention makers, it follows that there are no conventions! Although Heller tries to elude this paradox by admitting that many of his own statements are strictly false, but insisting that they are nonetheless appropriate,[64] it is not clear to me, at least, how this move serves to qualify (i)-(iii) in any way sufficient to avert the paradox. Thus, conventionalism seems a counsel of desperation.

In general, then, the perdurantist seems to face a problem parallel to the difficulty with adverbialism which was pointed out by Merricks: he seems, on pain of circularity, to take the notion of temporal part as primitive and to construct our concept of an object, and implicitly a person, on that basis, even though we have no clear understanding of this primitive notion.[65]

We have no clear grasp of what it is for a thing to be either a temporal *part* or a *temporal* part. With respect to temporal *parts*, it is controversial what exactly deserves to be called a part, insofar as a part is taken to be an object, as temporal parts are.[66] Consider a spatial object like a table. I think we should all agree that the legs, leaves, and top of the table are parts of the table. But does just any cross-section of the table deserve to be called a part? Consider, for example, a cross-section that bisects the tabletop. Obviously the matter of this slice belongs to the table; but is this cross-section properly called a "part" of the table? Or consider a cross-section running through the table's four legs. In this case we have four spatially separated wooden discs. Is this a "part" (or perhaps "parts"?[67]) of the table? It seems clear that though these are pieces of the table, they are not a part or parts of the table in the way that the legs and leaves are. Now with respect to a four-dimensional object—let us consider here a Euclidean four-dimensional spatial object, so as not to draw in the problem of time—why should we consider three-dimensional cross-sections of this object to be parts of the object? Why would the object rather not be composed of four-dimensional parts? And would not a four-dimensional table be composed of four-dimensional legs and a four-dimensional top as its parts, rather than just any four-dimensional slices, not to speak of three-

---

62 Ibid., p. 111.

63 Ibid., p. 66.

64 Ibid., p. 141.

65 Alternatively, he can begin with the primitive idea of a temporally extended object and construct temporal parts from that. But the notion of an object which has temporal extension is also strange, and I have not yet encountered an adequate account of how temporal parts are constructed from such an object. See remarks by Oderberg, *Metaphysics of Identity*, pp. 94, 131.

66 For a good discussion see Peter Van Inwagen, "The Doctrine of Arbitrary Undetached Parts," *Pacific Philosophical Quarterly* 62 (1981): 123-137.

67 This raises the issue of whether a spatially scattered object can exist. In an interesting discussion by Richard Cartwright, "Scattered Objects," in *Analysis and Metaphysics*, ed. Keith Lehrer, Philosophical Studies Series in Philosophy 4 (Dordrecht: D. Reidel, 1975), pp. 153-171, many of the issues in the *Auseinandersetzung* between Van Inwagen and Heller (see below) are pre-figured.

dimensional parts? A three-dimensional orthogonal cross-section of the legs of the four-dimensional table would appear to us as four spatially separated wooden spheres. These three-dimensional hyper-surfaces belong to the table all right, but that does not suffice to show that they are parts of the table. In the same way, why would three-dimensional hyper-surfaces of a four-dimensional object in which one dimension is time be parts of that object? Why should our three-dimensional table be thought to be a part of a four-dimensional object? Perhaps the perdurantist would not insist on temporal parts' being parts in the sense at issue here but would be content to construe what are normally considered to be three-dimensional objects as mere spatial hyper-surfaces of four-dimensional spatio-temporal objects. But if so, he owes us a clearer explication of what he means by temporal "part." And he would still face the problem of how to specify the identity conditions of a four-dimensional object before it is sliced into its three-dimensional foliations.

Similar difficulties attend the notion of *temporal* parts. Most persons who believe in the reality of spatial objects agree that such objects may have (spatial) parts which are objects in their own right. My lungs and liver and heart are parts of my body, even if a region comprised of some of my left arm and side and cutting through my torso to a point on my right side is not, properly speaking, a bodily part. But does it make sense to talk of a spatial object's having *temporal* parts as well? The problem here lies not in our inability to visualize a four-dimensional object, but in the obscure idea of parts that include both spatial and temporal dimensions. Consider the case of a three-dimensional table being lowered through two-dimensional Flatland. An inhabitant of Flatland will observe four spatially separated planar discs suddenly appear in his world. If the table is then withdrawn, the discs will after a time suddenly blink out of existence. Being unaware of the third spatial dimension, the perdurantist Flatlander will say that he has observed the temporal parts of four three-dimensional objects. Let us ignore his excusable error about the number of the objects and the number of spatial dimensions. Let us concede that the discs he observed were legitimate spatial parts of the table. In addition to these parts, did he also observe temporal parts of the table? What would these be? For the Flatlander they would be cylindrical three-dimensional objects, where two dimensions are spatial and one is temporal. It is at once evident that the term "temporal part" is a misnomer so far as physical objects are concerned. A purely temporal part of an object would be a one-dimensional line of points along the temporal dimension of the hypothetical object, that is to say, a one-dimensional series of instants. Only an immaterial being could occupy such a series of instants. What the perdurantist is really talking about with respect to physical objects, as Heller saw, is spatio-temporal parts, which are or occupy spacetime points. But has the Flatlander really experienced any such thing? We can readily see how the Flatlander has observed two-dimensional spatial cross-sections of a three-dimensional spatial object, our table, but those are not so-called temporal parts. The cylindrical temporal part is not the table leg, but two-dimensional cross-sections of the leg at different times. Does this strange, mixed object exist, or is it another reification, a pseudo-object? Similarly, is the three-dimensional table really a part of a four-dimensional object which extends across time?

The perdurantist can most plausibly make sense of his contention that the objects we observe are (spatio-)temporal parts of a four-dimensional object by adopting Minkowski's proposal to think no longer of space and time as distinct entities, but as merged into a single entity spacetime.[68] In this way, (spatio-)temporal parts are not a strange admixture of spatial parts and a temporal part, but are spacetime parts, parts which occupy regions of spacetime. The spatialization of time which already occurs in the B-Theory permits the perdurantist to think of an object as not merely spatial, but spatio-temporal.

But then we are brought round again to the problem of the identity conditions of objects composed of these (spatio-)temporal parts. It has been claimed that one possible advantage of perdurantism is that it avoids problems of determinate identity in cases of fission, as, for example, when an amoeba divides into two single-celled organisms.[69] The perdurantist can avoid the dilemma of having two objects each with an equally justified claim to be identical with the pre-fission object. Rather the two post-fission amoebae, as well as the pre-fission amoeba, are all (spatio-) temporal parts (and, hence, distinct objects) which do not stand in relations of identity with one another. But then it seems to become arbitrary how the perdurantist fixes the identity conditions of a four-dimensional object which undergoes fission.[70] Consider the following four-dimensional depiction of a dividing amoeba (Figure 9.1):

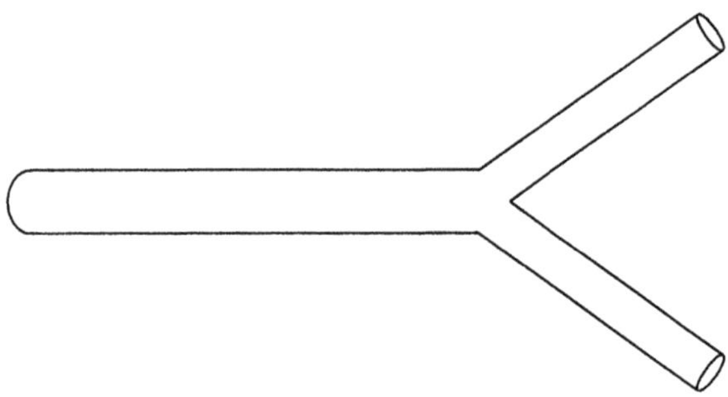

Figure 9.1. A four-dimensional depiction of an amoeba undergoing fission.

Is this one forked four-dimensional object? Or is it two distinct four-dimensional objects which somehow simultaneously occupy precisely the same initial space? Or is it a four-dimensional object which ends at fission into two, new, contiguous four-dimensional objects? Or is one leg part of the original four-

---

68 Oderberg rightly points out that the *concept* of spacetime does not serve to justify a temporal parts ontology (*Metaphysics of Identity*, pp. 105-113), but our concern here is with the coherence of such an ontology. If one is a spacetime realist, then, barring conventionalism, things must have spatio-temporal parts.

69 For discussion see Trenton Merricks, "Fission and Personal Identity over Time," *Philosophical Studies* 88-2 (1997): 163-186; Oderberg, *Metaphysics of Identity*, pp. 185-195.

70 See Judith Jarvis Thomson, "Time, Space, and Objects," *Mind* 74 (1965): 6-13.

dimensional object and the other a distinct object? What objective truth is there to answers to such questions? In fact, the problem of fission seems even worse for the perdurantist than for the endurantist, since *everything* in the universe is spatio-temporally contiguous and fission seems to require division into two non-contiguous objects. So is not everything from the Big Bang singularity on just one, huge four-dimensional object? (Shades of Spinoza!) Perdurantism seems to offer no advantages over endurantism in understanding cases of fission or resolving traditional problems of composition.

I do not know if these difficulties pose insuperable obstacles for the perdurantist; but he does need to address them more fully. Perhaps he can maintain, as Heller suggests, that the identity conditions for a four-dimensional object can be based on some individuating principle of unity, rather than on its spatio-temporal boundaries—though Heller thinks any such principles necessarily tend to be somewhat "arbitrary and artificial."[71] Perhaps he can appeal to individual essences or haecceities belonging to four-dimensional individuals. What we need from him is a fuller account.

According to perdurantism, temporally extended objects exist tenselessly at their locations determined by their temporal end points in the B-series, and they change intrinsically only in the sense that different (spatio-)temporal parts of them possess different properties in comparison with one another. These parts are themselves temporally extended objects (unless they are instantaneous slices of the original object) composed of (spatio-)temporal parts, which are in turn themselves temporally extended objects, *ad infinitum*. Since no object ever exists at any temporal location other than the interval marked by its temporal endpoints, no contradiction arises from intrinsic change, since such change concerns, not the object as a whole, but its (spatio-)temporal parts. Thus, the problem of temporary intrinsics is solved as follows:

1.'''' *O*-at-*t* is not identical with *O*-at-*t**, but these are (spatio-)temporal parts of *O*.

2.'''' *O*-at-*t* is bent.

3.'''' *O*-at-*t** is not bent.

4.'''' If O-at-*t* is identical with *O*-at-*t**, then *O*-at-*t* is *F* iff O-at-*t** is *F*.

By construing "*O*-at-*t*" and "*O*-at-*t**" to refer to two distinct objects, namely, temporal parts of the object *O*, which extends at least from *t* to *t**, the perdurantist avoids any contradiction, since, as Lewis says, there is no problem in asserting that different objects differ in their intrinsic properties.

There are, however, good reasons to reject perdurantism as implausible. Perdurantist views on four-dimensional objects, the nature of change, and personal identity have been widely criticized even by philosophers who carry no brief for the A-Theory of time.

---

71 Heller, "Varieties of Four-Dimensionalism," p. 59.

1. *Perdurantism's account of change is metaphysically counter-intuitive.* It will be recalled that Lewis rejected every variant of solution I because all of them held that "The solution to the problem of temporary intrinsics is that there aren't any temporary intrinsics," which Lewis rejected as "simply incredible."[72] But it seems clear that solution III itself falls victim to this criticism. Although Lewis says, "we are made up of temporal parts, and our temporary intrinsics are properties of these parts, wherein they differ from one another,"[73] it is evident that strictly speaking a temporally extended object has no temporary intrinsic properties at all. If *O* is an object temporally extended from *t* to *t** and *O* has a certain shape, then that shape never changes. Even if *O* is bent at *t* and not bent at *t*, O* is (tenselessly) just crooked at one end and straight at the other and that fact never changes. This is not a question-begging complaint based on a McTaggartian conception of change as inherently tensed.[74] Rather the point reflects the perdurantist's denial that *O* exists at *t* and endures until *t**. As Mellor explains, tenseless change is something's having incompatible properties at different dates. But what has the intrinsic property of being bent is not the same thing as what has the property of being straight. As an extended spatio-temporal object, *O* does not even have a three-dimensional shape; it has a four-dimensional shape, and that never changes. At best *O* can be said to change in its relational properties. As Lewis explains, solution III does not reject shape-at-a-time relations altogether.[75] *O* can be said to be bent at *t* and to be straight at *t** in virtue of the fact that it has temporal parts which exist at *t* and *t** and are bent or straight respectively. Thus, in relation to *t O* is bent and in relation to *t* O* is not bent. But in terms of its intrinsic properties *O* has a changeless, four-dimensional shape which is bent at one end and straight at the other. No temporary intrinsics here! What, then, of *O*'s extended (spatio-)temporal parts? Since they, like *O*, are themselves distinct, spatio-temporal objects, neither can they have temporary intrinsic properties. Frozen in their spacetime positions, these discrete objects never alter in their intrinsic properties. What of *O*'s instantaneous (spatio-)temporal parts? The same is true of them: not one lasts more than the instant at which it tenselessly exists; therefore, none ever changes in its intrinsic properties. Lewis is therefore misleading, if not mistaken, in claiming that temporary intrinsics are properties of an object's temporal parts. That is not true; there are no temporary intrinsic properties on perdurantism.

---

72 Lewis, "Plurality of Worlds," p. 204.

73 Ibid.

74 As alleged by Mark Heller, "Things Change," *Philosophy and Phenomenological Research* 52 (1992): 695-704. As Lawrence Brian Lombard, "The Doctrine of Temporal Parts and the 'No-Change' Objection," *Philosophy and Phenomenological Research* 54 (1994): 371 points out, for the perdurantist who says a four-dimensional object can be said to change in virtue of its having different properties at different times, temporal parts have ceased to have any explanatory role at all in change. Moreover, Heller's attempt to justify the apellation "change" for temporal, but not spatial, differences in properties on the basis that time alone has a direction sits very ill with a B-Theory of time, for on that theory even if time is anisotropic (itself a moot point), it is hard to see that it has an orientation, especially if there are no continuants.

75 Lewis, "Rearrangement of Particles," p. 66.

Is this an incredible view? I think it is. As Haslanger observes, Lewis's view is counter-intuitive because "there is nothing such that it persists through a change in *its* intrinsic properties."[76] Not only is this contention extremely bizarre in and of itself, but it essentially leaves the problem of temporary intrinsics unsolved, since that problem just was to explain how persisting things change their intrinsic properties. If this is an important and pressing philosophical problem, then perdurantism, in denying that intrinsic change occurs, amounts to denying that the problem is real, and the plausibility of its denial will be inversely related to the starkness of the problem itself.

Moreover, according to perdurantism, when an object appears to change its intrinsic properties over time, what we really have is a temporal series of definite and discrete objects having respectively different intrinsic properties at the time that each exists. Thomson reacted,

> this seems to me a crazy metaphysic. It seems to me that its full craziness comes out only when we take the spatial analogy seriously. The metaphysic yields that if I have had exactly one bit of chalk in my hand for the last hour, then there is something in my hand which is white, roughly cylindrical in shape, and dusty, something which also has a weight, something which is chalk, which was not in my hand three minutes ago, and indeed, such that no part of it was in my hand three minutes ago. As I hold the bit of chalk in my hand, new stuff, new chalk keeps constantly coming into existence *ex nihilo*. That strikes me as obviously false.[77]

Heller attempts to alleviate the counter-intuitiveness so powerfully evoked by Thomson, first, by reducing the sense of craziness attending the perdurantist's view and, second, by turning the tables on the endurantist. With respect to the first, Heller focuses on the claim that new chalk keeps coming into being *ex nihilo*. Why should it be thought that the temporal parts of the perduring chalk come into existence *ex nihilo?*—not because nothing existed before each temporal part, since each temporal part is preceded by others, nor because everything that exists before each temporal part continues to exist, since there are temporal parts that go out of existence at just the moment that the part in question comes into existence. According to Heller, it is the passage of time that is responsible for the temporal part's coming into existence and explains why the matter's being there at that time constitutes a new object's coming into existence. "At t_0 the temporal part in question did not exist. At t_1 it did exist. The change which brought this object into existence was just the change in time from t_0 to t_1."[78]

This response needs some untangling. First, Heller says nothing to alleviate the counter-intuitiveness of the central claim that the piece of chalk one holds in his hand at any moment is not the same object that existed at the previous moment, regardless of its mode of origin. Second, the talk of "coming into existence" and "passage of time" ill-suits a discussion of the perdurantist view, since a pure A-theorist, as a presentist, does not hold to perdurance. Perhaps a hybrid A-B-theorist might conceivably want to hold to the reality of time's flow and to perdurantism, but

---

76 Haslanger, "Endurance and Temporary Intrinsics," p. 119.

77 Judith Jarvis Thomson, "Parthood and Identity over Time," *Journal of Philosophy* 80 (1983): 213.

78 Heller, "Temporal Parts," p. 332.

then he would be caught in the clutches of the self-contradiction that a temporal part has no temporary intrinsic properties (as we have seen) and yet that it does have the temporary intrinsic property of presentness (not to speak of pastness and futurity). So we should interpret such tensed talk in this context tenselessly. Third, when Thomson says that according to the perdurantist the object in her hand at any moment has come into being *ex nihilo*, she means minimally that it has (tenselessly) a beginning of existence at that moment without any material cause.[79] For the matter of the prior chalky spatio-temporal object does not endure through the time at which the next object begins to exist. If a (spatio-)temporal part begins to exist at the instant t_1, then the prior part exists at every instant in the open interval leading up to t_1, but does not exist at t_1. *Nothing* endures from $t < t_1$ to t_1.

Heller's response seems to take the matter's existence as logically prior to the impositions of temporal divisions within it, so that a new part and, hence new matter, originate simply in virtue of temporal divisions' being made. We can, I think, make sense of the four-dimensional object's existing as a whole logically prior to its division into parts. When divisions are made in the next logical moment, there would then exist a series of parts. (Notice that in such a case one cannot consider the spatio-temporal whole to be the construction of its parts.) To take a spatial analogy, one could begin logically with the piece of chalk as a whole and then make demarcations along its length to create parts. One could then say that a certain part has a beginning simply in virtue of the fact that at location l_0 one part exists and at l_1 a different part exists, and one could further point out that the matter in one part is different matter than that in another part. Just as there is nothing mysterious about this, it might be said, neither is there anything mysterious about (spatio-)temporal parts' beginning to exist and having different matter than that in an earlier part simply in virtue of divisions' having been made in the four-dimensional object.

But notice one thing we should *not* want to say in the spatial case: that the parts created as a result of our demarcations are distinct objects, that the single piece of chalk one holds in his hand is in fact a collection of such objects. Similarly, in the temporal case, it seems fantastic to think that the (spatio-)temporal parts which we conceive the chalk to possess are discrete objects, rather than divisions of the history of one enduring object. Yet the perdurantist is committed to the metaphysically counter-intuitive view that the chalk I hold in my hand at this second is a wholly different object from the piece of chalk I held in my hand a second ago and that the matter of that former piece of chalk did not endure through the present second but has been succeeded by a portion of wholly distinct matter.

---

[79] A further interesting question is whether a temporal part can have any efficient cause in the preceding temporal part. For the two parts do not at all overlap, not even at a common instant. So how could a prior temporal part cause its successor to begin to exist, even *ex nihilo*, if it goes out of existence before the successor even begins to exist? At most it seems that the temporal part could have an efficient cause in God as its timeless creator and conserver, which seems to lead to just the sort of occasionalism Jonathan Edwards defended (see intriguing exposition by Roderick Chisholm, *Person and Object*, Muirhead Library of Philosophy [London: George Allen & Unwin, 1976], pp. 138-139).

These considerations form a natural transition to Heller's second point, in which he indicts endurantism as counter-intuitive. For now we discover that Heller *does* believe, in the spatial case, that a piece of demarcated chalk is composed of a number of distinct objects. Little wonder, then, that the notion of (spatio-)temporal parts as non-identical objects presents him with no difficulty! Heller takes as his point of departure Peter Van Inwagen's critique of the Doctrine of Arbitrary Undetached Parts,[80] which Van Inwagen formulates as follows:

> DAUP: For every material object *M*, if *R* is the region of space occupied by *M* at time *t*, and if sub-*R* is *any* occupiable sub-region of *R* *whatever*, there exists a material object that occupies the region sub-*R* at *t*.

Van Inwagen emphasizes that he does not deny that undetached parts exist, but he does deny that any arbitrarily designated piece of undetached matter constitutes an object, or a part, belonging to some larger object. DAUP is false, he contends, because it entails that it is impossible for any material object to lose any of its parts; if a part is irreplaceably removed from it, the object ceases to exist. He argues as follows: assume that there are two objects *O* and *P*, such that *P* is a part of *O* and *O* could survive the loss of *P*. Let *O*- be the object filling the sub-region occupied by *O* minus *P*. Clearly *O* and *O*- are not identical, since they occupy different regions of space and have different parts. Suppose, then, *P* to be annihilated. Now *O* and *O*- would seem to be identical, even though they were once non-identical. We thus arrive at a violation of the principle of the Transitivity of Identity, as is illustrated by Figure 9.2.

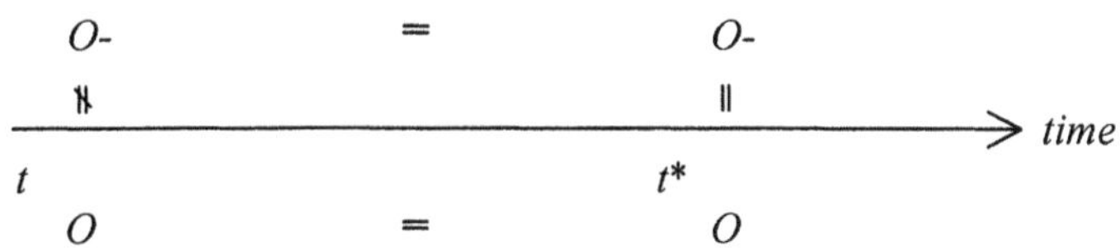

Figure 9.2. Violation of Transitivity of Identity by DAUP. At *t* $O \neq O$-, since they occupy different regions. Since *O* can, *ex hypothesi*, survive the loss of *P*, *O* at $t = O$ at t^*. Similarly, *O*- *at* $t = O$- at t^*, since *O*- remains intrinsically unchanged. But since $O\text{-} = O\text{-}P$, at t^* $O = O$-.

The proponent of DAUP will most plausibly deny that *O* can survive the loss of *P*, that is, he will deny that *O* at $t = O$ at t^*. Since Van Inwagen believes that things can survive the loss of a part, he rejects DAUP.

Since DAUP entails that it is impossible for an object to lose any of its parts, DAUP entails that no object endures through intrinsic change of parts, that is, that endurantism is false. If he holds to a doctrine of persistence, then, the proponent of DAUP endorses perdurantism. So it is not surprising that Heller rejects Van

[80] Van Inwagen, "Arbitrary Undetached Parts," pp. 123-126.

Inwagen's position as counter-intuitive. Heller finds it "unpleasant" to affirm that there is no physical object in the space that is now exactly occupied by all of me other than my left hand.[81] He therefore rejects endurantism in favor of the view that at t^* $O- \neq O$, not because at t^* we have two distinct objects O and $O-$ occupying the same region, but because O no longer exists at t^*. Perdurantism rescues DAUP by maintaining that O at t is not the same object as O at t^*; rather both are temporal parts of a persisting, four-dimensional object.

Heller's brand of perdurantism also entails DAUP. For Heller's definition of a physical object is a spatio-temporal hunk of matter, the material content of a region of spacetime. Such a definition reifies, as we have seen, all sorts of pseudo-objects, not only objects like all of me other than my left hand, but also objects like an undulating slice of me which haphazardly cuts across organs and bones. So there must exist arbitrary, undetached parts. Far from rendering perdurantism more intuitive than endurantism, however, the warranting of such objects makes perdurantism all the more incredible. It seems to me, therefore, that Heller's counter-argument completely backfires. In any case, I think it can be safely said that the endurantist's denial of DAUP does not count *against* the plausibility of his metaphysic. Hence, Heller fails to mitigate the craziness which Thomson justifiably sees in perdurantism.

It seems to me, therefore, that in denying the reality of intrinsic change and endorsing the objectification of (spatio-)temporal parts over against temporal continuants, perdurantism is metaphysically counter-intuitive.

2. *Perdurantism flies in the face of the phenomenology of personal consciousness.* Certainly one of the most bizarre features of perdurantism is the account it gives of personhood and personal identity. On the perdurantist view, persons are not what we normally take them (or ourselves) to be: self-conscious individuals who act and react with other things in space and time. Such individuals are just temporal parts or stages of persons. Persons are temporally extended objects which have no consciousness and, hence, no intelligence, volition, or emotion, no interactions, no agency, no moral responsibility, no aesthetic intuition, virtually none of the distinctive properties which we normally associate with personhood. For this reason, it is improper, as Geach points out, to say, for example, that McTaggart believed in Hegel's dialectic in 1901 and no longer did so in 1921.[82] For McTaggart, according to perdurantism, does not believe anything or hold any philosophical positions, since he is a four-dimensional object which has no self-consciousness. Rather the temporal slices of McTaggart supposedly believed these things.

But surely such a doctrine is absurd. I cannot imagine any sane individual who would say, if asked, that he or she is not a person. We know ourselves, in the phenomenon of self-consciousness we apprehend ourselves as persons, and we experience interaction with other persons as well. Van Inwagen's remark is apt: "If a philosophical argument leads us to deny something that every human being in

[81] Heller, "Temporal Parts," p. 323.

[82] Geach, "Problems about Time," p. 310.

history has believed, then it is a pretty good bet that something is wrong with the argument."[83] In the case of perdurantism, we do not even have an apparently compelling argument for an obviously false conclusion (as in Zeno's Paradoxes, say); rather perdurantism is just an alternative solution to a problem (temporary intrinsics) which is dissolved by another solution (presentism). Why would anyone want to embrace so outlandish a solution as one that denies that I, as a self-conscious individual, am a person?

The perdurantist could perhaps revise his view such that we are in fact persons and the temporally extended object is a meta-person or hyper-person or what have you. But such a revision would lead to an absurd doctrine of personal identity. Since the proper temporal parts of a meta-person are non-identical objects, a meta-person is therefore composed of a series of distinct persons! But am I seriously to believe that I am not the same person who was here a second ago and wrote the previous sentence, that my memories are really the recall of some other person's experiences, that my consciousness of enduring from one moment to the next is an illusion?

In fact, talking about temporal parts of persons does little to alleviate the problem of personal identity occasioned by perdurantism. For if my self-conscious self is a temporal part of a person, I am still a distinct object from the person-part who thought of himself as "I" a moment ago. He and I are not identical, he did not continue to exist, indeed, the "I" who finishes this sentence is not the same entity which began it. Lewis says that it is not disconcerting to realize that my present person stage will not survive the battle; what I want to know is, will *I* survive?[84] But what Lewis fails to see here is that an adequate phenomenology of personal consciousness requires that "I" refer, not to the tenseless person, who has no self-consciousness in any case, but to the present individual. On his ontology that refers to a person stage, and each of these is numerically distinct. Therefore, I did not exist in the past, nor shall I exist in the future—an interesting conclusion, since, it will be recalled, Lewis indicted endurantism precisely for denying that I have a past or a future. Perdurantism thus fails to make good sense of personal identity over time.

Finally, there is again the problem of intrinsic change. Perdurantism denies that any object survives intrinsic change. But I have every reason to believe that there is at least one object which endures through its intrinsic changes across time; namely, I myself. Despite the intrinsic changes in me, I existed a second ago and I still exist now. By taking tense seriously, endurantism explains how diachronic personal identity despite intrinsic change is possible.[85] By contrast, perdurantism has to take

---

83 Van Inwagen, "Arbitrary Undetached Parts," p. 127.

84 Lewis, "Survival and Identity," p. 58.

85 Chisholm poses and answers the problem of intrinsic personal change:

> "('But a thing *x* is not identical with a thing *y* unless they have all their properties in common. And if the present you has one set of properties and the past you another, how can they be the same thing?') The answer is, of course, that there aren't two you's, a present one having one set of properties and a past one having another. It is rather that you *are* now such that you have these properties and lack those, whereas formerly you *were* such that you had those properties and lacked these. The 'former you' *has* the same properties that the 'present you' now has and the 'present you' *had* the same

the absurd recourse that the object now referred to by "I" did not exist a second ago.[86]

It seems to me, therefore, that perdurantism just makes nonsense of the phenomenology of personal consciousness with respect to the nature of persons, personal identity, and intrinsic change—or perhaps we should say that the phenomenology of personal consciousness makes nonsense of perdurantism.

3. *Perdurantism is incompatible with moral facts of personal responsibility, praise, and blame.* One of the implications of the perdurantist conception of persons is that persons are not moral agents and, hence, not bearers of moral duty, virtue, or responsibility. Pointing out that "This tenseless person is not itself conscious, although many parts of it are," Del (Kiernan-)Lewis asks, "What would it mean to assign responsibility to such an individual?...the tenseless view implies that persons cannot perform any right or wrong actions and so implies that any judgment to the effect that someone acted rightly or wrongly is necessarily absurd."[87] I think that Kiernan-Lewis's conclusion is undeniable; and since our moral judgements are plausibly not absurd, it follows that the perdurantist conception of persons is false.

Of course, the perdurantist can claim that temporal parts of persons are moral agents responsible for their decisions. He will probably not want to say that temporal parts ought, therefore, to be redefined as persons, for the reason explained earlier: we should then have a series of diverse persons comprising a temporally extended hyper-individual, and it would be immoral to punish one person for the crimes performed by somebody else and also misplaced to bestow praise on someone because of meritorious deeds performed by a wholly different person. But, once again, even taking temporal parts of persons to be responsible individuals creates the same difficulty. It was partly for this very reason that Mellor affirmed endurantism:

> the first prerequisite for moral and legal responsibility is identity through time. Nothing and no one can be held responsible for an earlier action unless he, she or it is identical with whoever or whatever did that earlier action. That is the point of an alibi, to show the defendant to be not the same individual...as the earlier doer of the deed complained of. Now whatever identity through time may call for elsewhere, here it evidently requires the self-same entity to be wholly present both when the deed was done and later when being held accountable for it—a condition satisfiable by things, but not by events.... In particular, whatever temporal parts of them caused the deed in question will never be the same as the parts which might be later brought to court for it.[88]

Similarly, Oaklander concedes to Kiernan-Lewis the truth of endurantism as necessitated by moral facts and is forced, as we have seen, to preserve the B-Theory

---

properties that the 'former you' then had" (Chisholm, *Person and Object*, p. 92).
The italics are meant to emphasize the tenses of the verbs involved. The only properties you have are those you presently have.

86 As Hoy remarks with respect to Chisholm's statement in the previous note, "In this context, 'the former you' should properly refer to an entity tenselessly existing earlier than the entity referred to by 'the present you.' Moreover, the former you has (tenselessly) properties that are (tenselessly) different from those possessed (tenselessly) by the present you" (Hoy, "Becoming and Persons," p. 277).

87 Lewis, "Persons, Morality, and Tenselessness," pp. 307, 306.

88 Mellor, *Real Time*, p. 106. Mellor also allows corporate things like nations and corporations to be morally responsible.

of time by denying the existence of temporary intrinsic properties in favor of tenseless, temporally-indexed properties.[89]

It seems to me, therefore, that the perdurantist can consistently maintain his view only at the expense of denying that persons are responsible moral agents and that individuals can be praised or blamed for actions performed at earlier times. This seems to be a compelling reason to reject solution III, especially when we recall that there is no good reason to adopt it. The theist in particular must find this objection compelling, since he affirms not only that human persons are responsible moral agents but also that God is just in holding them responsible and is Himself a virtuous agent Who is to be praised for His gracious deeds.

4. *Perdurantism entails implausible modal commitments.* Van Inwagen has developed a powerful argument against perdurantism which is analogous to his objection to the Doctrine of Arbitrary Undetached Parts.[90] Take Descartes as an example of a persisting object, one that perdured from 1596 to 1650. Descartes's temporal extent is not essential to him: he might have died younger or lived longer. This obviously does not mean that Descartes might have had the same temporal parts he in fact had except that those parts would have been greater or lesser in extent. The extent of a temporal part is essential to it; an hour-length temporal part could not be an hour and a half in some other possible world and still be the same temporal part. Rather it means that Descartes might have been the sum of a different set of temporal parts, a set formed, say, by adding more parts to or by subtracting some parts from the set of temporal parts he in fact had. Let *D* be Descartes, *D-* be that temporal part of Descartes stretching from 1596-1649, and *L* be the last year of his life. Descartes could have lacked *L*. There are therefore possible worlds in which *D-* exists and *L* does not. Now *D* and *D-* are not on the perdurantist view identical. But if Descartes had died one year earlier than he did, then *D* and *D-* would have been identical. But if *D* and *D-* could therefore have been identical, then there are two things that could have been one thing. This not only violates the modal principle about identity that $x=y \supset \Box(x=y)$, but also violates the principle of the Transitivity of Identity. Therefore, Descartes is not a perduring object composed of temporal parts.

The same conclusion can be more directly reached. Consider *D*, which is the largest temporal part of Descartes. *D* has its temporal extent essentially. But that entails that Descartes could not have lived longer or shorter than he actually did—which is patently false.

Van Inwagen does provide a way of escape for the perdurantist, one that is apt to delight David Lewis but prove to be a source of dismay to most philosophers: one

[89] Oaklander, "Lewis and Responsibility," pp. 181-183. Oaklander thinks that endurantism is consistent with the postulate of temporal parts, that in virtue of psychological connection or physical continuity, two temporal parts may be said to be the same person. But this is just a self-contradiction. As Van Inwagen observes, no causal process can turn one hunk of matter into a diverse hunk of matter (Peter Van Inwagen, *Metaphysics*, Dimensions of Philosophy Series [Oxford: Oxford University Press, 1993], p. 172).

[90] See Van Inwagen, "Arbitrary Undetached Parts," pp. 133-137; idem, "Four-Dimensional Objects," pp. 252-254.

may adopt Counterpart Theory. Counterpart Theory would permit *D* and *D-* in the actual world to have a common counterpart in another world *W*. Since that common counterpart (equal in extent to *D-*) is not, strictly speaking, identical with either *D* or *D-*, the Transitivity of Identity is not violated nor any other modal principle. In fact no appeal to a common counterpart is necessary: the object in *W* which is an intrinsic duplicate of *D-* will not be a counterpart to *both D* and *D-*. The Counterpart Theorist's solution to Van Inwagen's second argument is more complicated. He must suppose that there are two counterpart relations with respect to *D*: a personal counterpart relation and a temporal part counterpart relation. On this view a temporal part counterpart of *D* must have the same temporal extent as *D*. So *D-* in *W* is not a temporal part counterpart of *D*. But *D-* in *W* could still be a personal counterpart of *D*. Thus, *D* could not have had a different temporal extent *qua* temporal part, but he could have *qua* person.

Such an escape route, however, is highly implausible, since, as Plantinga has shown, Counterpart Theory, despite its claim to provide a plausible account of trans-world identity, entails the outlandish consequence that I have all my properties essentially, since it is not I who exist with different properties in other worlds, but counterparts of me.[91]

In his *The Ontology of Physical Objects*, Heller in response to Van Inwagen bites the bullet: Since four-dimensional hunks of matter have, on his view, their spatio-temporal boundaries essentially, Descartes could not have lived for one year less than he did.[92] Although this seems wildly implausible, Heller argues that the "standard ontology" of enduring, three-dimensional objects is demonstrably untenable and therefore one has no choice but to adopt a four-dimensional ontology with its attendant consequences. Heller's case against the ontology of enduring objects is based on traditional sorites paradoxes concerning an object's diachronic identity as tiny parts of it are successively removed until it ceases to exist as that object.

The central difficulty with this line of argument, however, is that four-dimensionalism *per se* does nothing at all to solve the sorites paradoxes; it is only by conjoining additional metaphysical hypotheses to four-dimensionalism that the perdurantist escapes the paradoxes, and these same hypotheses, which do all the explanatory work, are equally available to the endurantist. This is immediately evident by considering the case of a spatial four-dimensional object, from which chips are successively removed – the addition of an extra spatial dimension makes no difference to the problem. It might be thought that in the case of a spatio-temporal object, the paradoxes cannot arise because successive removal of pieces is impossible for an object which is extended along the time dimension. But what such a consideration fails to appreciate is that the sorites paradox in question is essentially modal in nature and only accidentally temporal. It asks, "How many chips (and which chips) could be removed from an object and yet that object exist? If such-

[91] Alvin Plantinga, *The Nature of Necessity*, Clarendon Library of Logic and Philosophy (Oxford: Clarendon Press, 1974), pp. 115-119.
[92] Heller, *Ontology of Physical Objects*, p. 28.

and-such chips were removed from this object, would it be identical with the whole object?" Such a modal problem applies to spatio-temporal objects as well as spatial ones. Because the problem is modal, a sorites paradox could be pressed even against a spatial object existing at a last moment of time. Heller's escape from the paradox is to adopt either mereological essentialism or conventionalism. Insofar as four-dimensional objects objectively exist, they are said to have their spatio-temporal boundaries essentially, which is just mereological essentialism writ large. A similar solution is available to the endurantist. A mereological essentialist who is an anthropological dualist-interactionist can consistently maintain that while Descartes's body has its parts essentially and therefore does not survive through intrinsic change, the person Descartes does endure through time and could have undergone death a year sooner than he did. Heller cannot make a similar affirmation, for if Descartes is an immaterial, perduring object, he still has temporal parts and has them essentially, so that premature death is impossible. On the other hand, if one takes the conventionalist route, then this same route is available to the three-dimensionalist. He will maintain that "stuff" is spread out in space and endures through time and that we conventionally carve up three-dimensional stuff into spatial objects. Since for the conventionalist, there really are no objects – whether three- or four-dimensional—, the sorites paradoxes do not arise. Thus, Heller's four-dimensionalism is an extraneous feature to his solution to the sorites paradoxes and so cannot be justified by them.

In any case, less radical solutions to the sorites paradoxes are available which are endurantist in character.[93] For example, Van Inwagen himself holds that the only physical objects which exist are either living organisms or certain fundamental particles which have no parts and that inert, macroscopic, physical realities like chairs and tables and the moon are but collections of physical objects.[94] Van Inwagen's ontology entails the strange conclusion that tables and chairs do not exist; but Heller himself affirms this conclusion, and he goes further to deny the existence of mice and men, which makes Van Inwagen's ontology look conservative by comparison.

Finally, one might note that the sorites argument is itself suspect, being susceptible to a sort of anti-sorites sorites paradox.[95]

Jubien, in response to Van Inwagen, finds himself driven, if possible, to adopt even more radical hypotheses. Taking the conventionalist line, Jubien would escape Van Inwagen's argument by denying that there is any such thing referred to by the name "Descartes".[96] Van Inwagen, he asserts, commits what Jubien calls the Fallacy of Reference, which is just to assume that a proper name like Descartes has

---

93 See Oderberg, *Metaphysics of Identity*, pp. 166-173.

94 Peter Van Inwagen, *Material Beings* (Ithica, New York: Cornell University Press, 1990), pp. 5, 91-105, 228-229.

95 See Heller, *Ontology of Physical Objects*, pp. 113, 147; Oderberg, *Metaphysics of Identity*, p. 173. Basically the argument is that if chips of size 1.0 will generate a paradox, whereas chips of some large size *n* will not, then there is no clear point of demarcation between 1.0 and *n* where the paradox ceases to arise. Heller's response to this paradox seems no more convincing than a similar response to the original.

96 Jubien, *Ontology, Modality, and the Fallacy of Reference*, pp. 35-36.

a referent. On Jubien's view to say that something is Descartes is not to make an identity statement but is to assert that something has the property *being Descartes*. This thing, which is just the material "stuff" occupying a region of spacetime, is temporally inductile and incompressible, but what accommodates the fact that Descartes could have lived longer or shorter is the fact that many *other* mereological sums of stuff having greater or smaller temporal extent could have had the property *being Descartes*. Moreover, Van Inwagen's claim that the perdurantist must adopt Counterpart Theory also commits the Fallacy of Reference, for it assumes that the person Descartes is identical with the entity *X*, so that questions about Descartes's modal properties are questions about *X*'s modal properties. Rather we should deny, advises Jubien, that "*X* is Descartes" is an identity statement. We should interpret questions about what properties Descartes may have as questions about what properties may be had by *any* thing which has the property of *being Descartes*. Thus, on Jubien's view there is no entity denoted by "Descartes" at all; there is merely some entity *X* with the property *being Descartes*, and there may be some other entity *Y* which might have had that property instead.

So strange an understanding of statements of reference and identity will hardly commend itself to most philosophers as more plausible than Van Inwagen's premisses. It requires us to say that Descartes as such does not exist, that there is no object which is Descartes. Even the four-dimensional chunk of stuff which has the property *being Descartes* is not Descartes. But this seems crazy; how could something have the property of *being Descartes* without being Descartes? It requires us to hold that identity statements are really disguised ascriptions of predicates. But identity statements are necessary truths, whereas on Jubien's analysis the corresponding predicate ascriptions are contingent. Thus the modal principle of identity is still violated. It requires us to say that someone else might have had the property of *being Descartes*. But is it plausible to think that I could have been Descartes? The fact that Van Inwagen's detractors are driven to such desperate expedients is, I think, merely testimony to the power of his argument.

In his later article Heller attempts to escape Van Inwagen's argument by denying that on perdurantism a thing's temporal boundaries are essential to it. Heller argues that a perdurantist can avoid Counterpart Theory if he conceives an object's identity conditions to be given not by its boundaries, but by some individuating principle of unity.[97] Such a perdurantist will deny that if something has temporal parts then it has those very parts in every world in which it exists. This denial breaks, Heller thinks, the link between perdurantism and Counterpart Theory. Heller's point is obscure, however, for Van Inwagen's argument does not depend on the premiss that something has the very same temporal parts in every world in which it exists. Rather it depends on the premiss that the *temporal extents* of a thing's (spatio-) temporal parts are the same in every world in which it is composed of those parts. The temporal parts themselves could vary in other properties from world to world, and Van Inwagen's argument actually relies on the intuition that Descartes could have lacked some of his temporal parts. Thus, it is a veritable premiss in the

---

[97] Heller, "Varieties of Four Dimensionalism," p. 50.

argument linking perdurance to Counterpart Theory that if anything has temporal parts, then it does not have those very parts at every world in which it exists.

In any case Heller goes on to argue that the perdurantist described above can consistently hold that an object's parts can have different spatio-temporal boundaries while remaining self-identical:

> The title...'Philip's drunken part,' if it is understood as the name for a temporal part of Philip, works like the title 'Philip's heart'. What makes an object Philip's heart is not its boundaries. It is that the object satisfies the conditions for being Philip's heart (whatever those conditions are). Similarly, what makes an object Philip's drunken part is not its boundaries. It is that it satisfies the relevant conditions (whatever those conditions are—drunkenness is presumably a necessary condition). If a different collection of atoms had satisfied the heart-conditions, a different collection would have been the heart. Hearts have their locations, precise shapes, and parts only accidentally. Similarly, if Philip had drunk less or more at different times, his drunken part would have had different boundaries. There is no more reason to suppose that his drunken part has its location, shape, or parts essentially than there is to suppose the same about his heart.
>
> If Philip's drunken part, Philip's heart, or Philip himself have their spatiotemporal structures accidentally, then, unless we accept counterpart theory, there are worlds at which those very objects exist with different structures.[98]

This argument is presumably meant to show that on perdurantism the spatio-temporal shape or structure of a self-identical object can differ by its temporal parts' having different temporal extents. But does it succeed? That depends on how we take the key sentence

6. If Philip had drunk less or more at different times, his drunken part would have had different (temporal) boundaries.

(6) entails that it is possible that Philip's drunken part have a greater temporal extent. Now we can take this possibility *de dicto* or *de re.* Taken *de dicto* the affirmation that Philip's drunken part could have had a greater extent is undoubtedly true. Intuitively, we take this affirmation to be true, not because that temporal part of Philip which is drunk in the actual world could somehow have dilated so as to have a longer temporal extent, but because other temporal parts of Philip, which are not drunk in the actual world, could have been drunk. Thus, "Philip's drunken part" could have referred to a temporal part which is greater in extent that the part actually referred to. On the other hand, taken *de re* the affirmation of Philip's drunken part that it itself could have been longer seems intuitively false. How could an hour-length temporal part, for example, expand to last longer than an hour? To say that temporal parts could have had different temporal extents is like saying that an hour could last longer or less than an hour.

Heller's interpretation of (6), on the other hand, has a counter-intuitive air. Because he takes *drunkenness* to be part of the identity conditions for "Philip's drunken part," (6) is true when taken *de dicto* or *de re.* Taken *de re*, the affirmation that Philip's drunken part could have been longer is true, not because the hours could have dilated to last longer, but because the extent of the part in question is

---

98 Ibid., pp. 50-51.

accidental to it; that very temporal part could have lasted several hours longer than it did, since its identity is determined, not by its length, but by its drunkenness. Now this, as I say, strikes me as counter-intuitive. Consider a similar claim like

7. If Philip had drunk less, his drunken part would not have been drunk.

According to (7) it is possible that Philip's drunken part not be drunk. Taken *de dicto*, this is false, being a contradiction. But taken *de re*, it seems intuitively true; that temporal part of Philip which is drunk in the actual world could have been sober and would have been, had Philip been more circumspect. But Heller must regard (7) as false, since it is not possible, *de dicto* or *de re*, for Philip's drunken part not to be drunk. For drunkenness is an essential property, one of the identity conditions, of Philip's drunken part. Hence, Heller must say that if Philip had drunk less, then that temporal part of him would not have existed. But surely this is very strange. Would we not rather intuitively say that his temporal part would have had the property of sobriety rather than that it would not have existed? Compare an analogous claim with regard to spatial parts:

8. If Philip had worn a hat, his sunburned face would not have been sunburned.

Would we not naturally take (8) *de re* to affirm that Philip's sunburned part might lack the property of being sunburned? It would be counter-intuitive to regard (8) as false *de re* because if Philip had worn a hat, he would have lacked one of his spatial parts, namely, his sunburned face. The latter view would require us to say that Philip's face cannot be either pale or sunburned; he literally has another face if fails to wear his hat. Analogously, if we think of Philip's drunken (or sunburned) temporal part, it seems bizarre to think that had Philip had one less drink or worn his hat that certain subsequent temporal parts—which might look exactly the same as in the actual world with the exception of the different relevant property—would not have existed, rather than to think that those temporal parts would have existed with different accidental properties. Drunkenness, then, is an accidental property of a temporal part, not one of its identity conditions, and hence a claim like (6), though true *de dicto*, must be regarded as false *de re*.

Therefore, I find Heller's refutation unconvincing. Van Inwagen's argument is a persuasive demonstration of the startling fact that perdurantism has the implication that I have my temporal extent essentially and therefore cannot live any longer or shorter than I shall. This conclusion is so wild that perdurantism's hopes of being the rationally preferred solution to the problem of temporary intrinsics seem bleak, indeed.

In the final analysis, then, perdurantism turns out to be manifold craziness. It offers a metaphysically counter-intuitive account of intrinsic change (namely, no such change exists), it gives the lie to our self-experience of personhood and diachronic identity, it invalidates moral responsibility, and it commits us to modal theses which are incredible. Perdurantism is so bizarre that even if there were no competing successful solution to the problem of temporary intrinsics in sight, we should still be advised to reject it in favor of agnosticism. But given that on a metaphysic of A-theoretical endurantism, the problem does not even arise, then the choice of endurantism must be preferred. Since the B-Theory cannot plausibly affirm endurantism, it follows from the truth of endurantism that the A-Theory of time is correct.

SUMMARY OF PHILOSOPHICAL OBJECTIONS

In summary, we have seen that the B-Theory of time is burdened with truly serious philosophical problems. The B-theorist is at a loss to explain why that dimension which in his theory is denominated "time" is truly temporal at all. By abstracting tense from time, the B-theorist has removed all rationale for thinking that the relations which order events are *earlier than/later than*, rather than some non-temporal, structurally similar relations. To appeal to these relations as primitive temporal relations must appear gratuitous in light of the A-theorist's demonstration that tenseless temporal relations are derivable from the A-series. Having "spatialized" time, the B-Theory winds up not being a theory about time at all.

Moreover, the B-Theory cannot explain away the reality of temporal becoming. To admit becoming exclusively in the realm of consciousness leads to an untenable dualism of static and dynamic interactive realities. To deny the reality of becoming altogether contradicts the undeniable reality of the presentness of experience, obscures the thesis of the mind-dependence of becoming, and is self-referentially incoherent.

Finally, the B-Theory of time cannot explain how even tenseless intrinsic change can occur and entails a theory of persistence through time which is so counter-intuitive as to be incredible. Therefore, philosophically we have abundant reason for rejecting the B-Theory of time.

SECTION 2: THEOLOGICAL OBJECTIONS

CHAPTER 10

CREATIO EX NIHILO

Secular philosophers may find it odd or at least rather quaint to lodge theological objections against a particular theory of time. But Christian philosophers, such as the author, take such objections with utmost seriousness. A view which is philosophically coherent but theologically untenable cannot be true. It is entirely appropriate, therefore, to reflect on the theological implications of one's theory of time with a view to assessing its adequacy.[1]

It seems to me that the B-Theory of time does compromise theological truth in a significant way with respect to the doctrine of creation, namely, it emasculates the biblical doctrine of *creatio ex nihilo*.

The biblical conception of God's relation to the world is one of Creator to creature. Dualistic conceptions of God confronted with a co-eternal, uncreated material which He fashions into a cosmos are alien to the biblical writers, who think of God as all-powerful and the source of all reality external to Himself. He speaks, and the universe springs into being, created out of nothing by His incomparable power. "Before" the beginning, so to speak, only God existed, and creation, we learn from the New Testament, results from His Word, which is the pre-incarnate Christ. Thus, it is not only inadequate to conceive of creation as a mere fashioning, but also even as a conserving of the universe in being. The biblical doctrine of creation is inextricably bound up with temporal considerations and asserts that the universe began to exist at a point in the finite past at which it sprang into being out of nothing by God's almighty Word.[2]

Unfortunately many contemporary theologians evince an unseemly timorousness concerning the biblical affirmation of *creatio ex nihilo*. Claiming that "creation is concerned with ontological origin, not temporal beginning,"[3] John Polkinghorne

---

1 Oddly, few have done this. But see Richard Gale, "A Reply to Paul Helm," *Religious Studies* 29 (1993): 258, who blasts the B-Theory of time as "part and parcel of a bifurcationist view of man and nature that is at odds with theism."

2 It was so understood by the vast majority of early Church Fathers (*e.g.*, Tatian *Oratio ad graecos* 5.3; cf. 4.1ff, 12.1; Theophilus *Ad Autolycum* 1.4; 2.4, 10, 13; Irenaeus *Adversus haeresis* 3.10.3) and was finally promulgated as official Church teaching at the Fourth Lateran Council (1215).

3 John Polkinghorne, critical notice of *Cosmos as Creation*, ed. Ted Peters, *Expository Times* 101 (1990): 317. According to Polkinghorne, "To speak of God as Creator is not to attempt an answer to the question Who lit the blue touch paper of the Big Bang? To talk in that way belongs to deism and not to Christian theology" (John Polkinghorne, "Cosmology and Creation," address at Trinity Hall, Cambridge, undated photocopy). "There is general agreement that the Big Bang is nothing special from a theological point of view.... The idea of *creatio ex nihilo* asserts the total dependence of the universe upon the

states, "The doctrine of creation is not an assertion about what God did in the past to set things going; it is an assertion of what he is doing in the present to maintain the universe in being."[4] In fact, however, nearly the opposite is the case, biblically speaking. Creation in the Bible virtually always involves the notion of a temporal beginning (as is evident simply from the ubiquitous past-tense, rather than present-tense, verbs with respect to God's creating), and one will have to search for passages supporting the notion of the on-going ontological dependence of the universe upon God's sustaining will. Those passages are there to be found (Heb. 1.2); but we are everywhere confronted with the idea that at some point in the past God created the world. After surveying the data, George Hendry concludes that "Creation in the language of the Bible unquestionably connotes origination..., the bringing into existence of something that did not previously exist."[5] A robust doctrine of creation therefore involves both the affirmation that God brought the universe into being out of nothing at some moment in the finite past and the affirmation that He thereafter sustains it in being moment by moment.[6] The doctrine of *creatio ex nihilo* was codified by the Fourth Lateran Council in 1215 in its declaration that God is the "Creator of all things, visible and invisible,...who, by His almighty power, from the beginning of time has created both orders in the same way out of nothing."

Now the B-theorist can only ingenuously make such an affirmation. *Creatio ex nihilo* for the B-theorist means only that the world depends immediately upon God for its existence at every moment. The B-theorist's affirmation that God brought the universe into being out of nothing at some moment in the finite past can at best mean that there is (tenselessly) a moment which is separated from any other moment by a finite interval of time and before which no moment of comparable duration exists and that whatever exists at any moment, including the moments themselves, is tenselessly sustained in being immediately by God. All this adds to the doctrine of ontological dependence is that the tenselessly existing block universe has a front edge. It has a beginning only in the sense that a yardstick has a beginning. There is in the actual world no state of affairs of God existing alone without the spacetime

---

sustaining will of its Creator" (Polkinghorne, critical notice of *Cosmos as Creation*, p. 317). Whether the Big Bang represents the moment of creation is, however, irrelevant to the conceptual content of the doctrine of *creatio ex nihilo*. The biblical doctrine, like Deism, affirms a temporal beginning of the universe; moreover, Deists did not in fact deny God's conservation of the world in being, but rather His supernatural action in the world.

4 Polkinghorne, "Cosmology and Creation." So also Langdon Gilkey, *Maker of Heaven and Earth* (Ganden City, New York: Doubleday, 1959); Ian Barbour, *Issues in Science and Religion* (New York: Harper & Row, 1971), p. 384; Arthur Peacocke, *Creation and the World of Science* (Oxford: Clarendon Press, 1979), pp. 78-79. This watered-down doctrine of creation is the legacy of the father of modern theology, F. D. E. Schleiermacher, *The Christian Faith,* 2d ed., ed. H. R. Mackintosh and J. S. Stewart (Edinburgh: T. & T. Clark, 1928), sec. 36-41. While acknowledging that the biblical conception of creation involves a temporal beginning (sec. 36.2), Schleiermacher held that this component of the doctrine could be safely suppressed in favor of the absolute dependence of the creation on God (sec. 41). See remarks by Nelson Pike, *God and Timelessness*, Studies in Ethics and the Philosophy of Religion (New York: Schocken Books, 1970), pp. 107-110.

5 George S. Hendry, "Eclipse of Creation," *Theology Today* 28 (1972): 420. So also Paul Copan, "Is *Creatio ex nihilo* a Post-biblical Invention?" *Trinity Journal* 17 (1996): 77-93.

6 For more on this distinction, see William Lane Craig, "Creation and Conservation Once More," *Religious Studies* 34 (1998): 177-188.

universe. God never really brings the universe into being; as a whole it co-exists timelessly with Him.[7]

Leftow, whose theory of divine eternity entails a B-Theory of time, admits as much. He writes,

> So if God is timeless and a world or time exists, there is no phase of His life during which He is without a world or time or has not yet decided to create them, even if the world or time had a beginning.
>
> ...God need not *begin* to do anything, then, in order to create a world with a beginning. That action that from temporal perspectives is God's beginning time and the universe is in eternity just the timeless obtaining of a causal dependence or sustaining relation between God and a world whose time has a first moment.
>
> ...in eternity, God is changelessly the Lord: He timelessly coexists with His creatures.[8]

Leftow never addresses the theological objection that such an emasculated doctrine of *creatio ex nihilo* does not do justice to the biblical data, which give us clearly to understand that God and the universe do not timelessly co-exist, but that the actual world includes a state of affairs which is God's existing alone without the universe. Typically such a state is described in the ordinary language of the biblical authors as obtaining "before" the world began (Jn. 17.24; Eph. 1.4; I Pet. 1.20; cf. Mt. 13.35; 24.21; 25.34; Lk. 11.50; Heb. 9.6; Rev. 13.8; 17.8) or, even more boldly, "before all time" (II Tim. 1.9; Tit. 1.2 [*pro chronon aionion*]; Jude 25 [*pro pantos tou aionos*]). Jude's doxology is especially interesting: "to the only God...be glory, majesty, dominion, and authority, *before all time* and *now* and *forever*" (Jude 25). How these ordinary language expressions are to be formulated philosophically—the Bible is not, as Paul Helm reminds us, a philosophy book from which a doctrine of divine eternity may simply be read off the surface—I have addressed elsewhere; but their intent is clear and must be taken seriously.[9] The notion that God and the universe timelessly co-exist in an asymmetrical relation of ontological dependence is not only foreign to, but actually incompatible with the biblical writers' conception of *creatio ex nihilo*, of God's existing alone and bringing the world into being out of nothing.

Not only so, but the idea that God and creation tenselessly co-exist seems to negate God's triumph over evil. On the B-Theory of time, evil is never really

---

[7] Perhaps the B-theorist might hold that the beginning of the universe does not involve the origin of spacetime itself, but merely of all matter/energy and that God exists in an infinite, tenseless time prior to creating the material universe. Such a doctrine exempts time itself from being created *ex nihilo*, a compromise of the traditional doctrine, and faces other difficulties as well, as I explain in my *God, Time, and Eternity* (Dordrecht: Kluwer Academic Publishers, forthcoming).

[8] Brian Leftow, *Time and Eternity*, Cornell Studies in Philosophy of Religion (Ithaca, N. Y.: Cornell University Press, 1991), pp. 290-291, 310; cf. p. 322, where he affirms that God is eternally incarnate in Christ. Cf. also p. 239, where he affirms that in eternity events are "frozen" in an array of B-series positions. See also Yates's chapter on timeless creation in John C. Yates, *The Timelessness of God* (Lanham, Maryland: University Press of America, 1990), pp. 131-163.

[9] Consideration of *creatio ex nihilo* raises a nest of intriguing and difficult questions: Did God exist in time before the creation of the universe? Does *creatio ex nihilo* imply the creation of time itself? Can God's priority to time be understood in some way other than chronological? I address these questions in my *God, Time, and Eternity*. For now it is enough to realize that the biblical writers' expressions about God's existing and planning "before" creation clearly mean to affirm that in some sense God was alone and then brought the world into being out of nothing.

vanquished from the world: it exists just as sturdily as ever at its various locations in space-time, even if those locations are all earlier than some point in cosmic time. Creation is never really purged of evil on this view; at most it can be said that evil only infects those parts of creation which are earlier than certain other events. But the stain is indelible. It is unclear how we can say with Paul, "Death is swallowed up in victory!" (I Cor. 15.55) when Death is never really done away with on a B-Theory of time.

In order to affirm a robust doctrine of creation, about the only recourse available to the B-theorist would be to posit a hyper-time in which God brings into being the whole four-dimensional block universe at a moment of hyper-time. Not only would such a move be metaphysically extravagant, but then our problems would begin all over again. Is this hyper-time an A-theoretical time? Then the A-Theory is ultimately correct. Is it a B-theoretical time? Then the 4-D spacetime manifold exists tenselessly and never fails to be instantiated in being. Thus, the move to hyper-spacetime solves nothing.

In conclusion, we may say that essential to the biblical doctrine of *creatio ex nihilo* is the dual affirmation that God brought the universe into being out of nothing at some moment in the finite past and that He sustains it immediately in being moment by moment. The first part of this affirmation implies that there obtains in the actual world a state of affairs which may be described as God's existing timelessly alone without the universe. The B-theorist cannot affirm this truth, since on his theory of time the block universe co-exists timelessly with God, perhaps having tenselessly a first moment but never really coming into being out of non-being. The B-theorist thereby reduces the first component of the doctrine of creation to the second—tenseless ontological dependence—and thereby emasculates *creatio ex nihilo*. The only escape route would be the metaphysically extravagant postulation of a B-theoretical hyper-time in which the 4D universe originates, a view which requires a (hyper-)temporal deity. Since such a view is implausible, it follows that the B-Theory of time ultimately solves nothing.

CONCLUSION

In my book *The Tensed Theory of Time*, I argued that there are powerful reasons to affirm an A-Theory of time in the ineliminability of tense from language and in our experience of tense and temporal becoming and that objections to the A-Theory appealing to McTaggart's Paradox or the myth of temporal passage are not probative. Now in this study we have discovered that arguments for a B-Theory of time based on relativity theory or the mind-dependence of becoming turn out upon examination not to support such a tenseless theory, and we saw further that strong philosophical and theological objections stand against the B-Theory, springing from its "spatialization" of time, the incoherence of the thesis of the mind-dependence of becoming, the problem of temporary intrinsics, and the biblical doctrine of *creatio ex nihilo*. It follows, therefore, that it is the A-Theory of time which must be judged to be correct. Time is tensed.

BIBLIOGRAPHY

Alexander, H. G. "Introduction." In *The Leibniz-Clarke Correspondence*, pp. ix-lvi. Edited with an Introduction and Notes by H. G. Alexander. Manchester: Manchester University Press, 1956.

Angel, Roger B. *Relativity: The Theory and Its Philosophy.* Foundations and Philosophy of Science and Technology. Oxford: Pergamon Press, 1980.

Arthur, Richard T. W. "Newton's Fluxions and Equably Flowing Time." *Studies in History and Philosophy of Science* 26 (1995): 323-351.

Arzeliès, Henri. *Relativistic Kinematics.* Revised edition. Oxford: Pergamon Press, 1966.

Bachelard, Gaston. "The Philosophic Dialectic of the Concepts of Relativity." In *Albert Einstein: Philosopher-Scientist*, pp. 565-580. Edited by P. A. Schilpp. Library of Living Philosophers 7. LaSalle, Ill.: Open Court, 1949.

Baierlein, Ralph. *Newton to Einstein: The Trail of Light.* Cambridge: Cambridge University Press, 1992.

Baker, Lynn Rudder. "Temporal Becoming: The Argument from Physics." *Philosophical Forum* 6 (1975): 218-233.

Balashov, Yuri. "Enduring and Perduring Objects in Minkowski Space-Time." *Philosophical Studies* (forthcoming).

________. "Relativistic Objects." *Noûs* (forthcoming).

Banks, T. "*TCP*, Quantum Gravity, the Cosmological Constant, and All That...." *Nuclear Physics B* 249 (1985): 332-360.

Barbour, Julian B. *Absolute or Relative Motion?* Volume 1: *The Discovery of Dynamics*. Cambridge: Cambridge University Press, 1989.

________. "The Timelessness of Quantum Gravity: I." *Classical and Quantum Gravity* 11 (1994): 2853-2873.

________. "The Timelessness of Quantum Gravity: II." *Classical and Quantum Gravity* 11 (1994): 2875-2897.

Barrow, Isaac. *The Geometrical Lectures of Isaac Barrow.* Translated with notes by J. M. Child. Chicago: Open Court, 1916.

Barrow, John D. *The World within the World.* Oxford: Clarendon Press, 1988.

Barrow, John D. and Frank J. Tipler. *The Anthropic Cosmological Principle.* Oxford: Clarendon Press, 1986.

Bechler, Zev. "Introduction: Some Issues of Newtonian Historiography." In *Contemporary Newtonian Research*, pp. 1-20. Edited by Zev Bechler. Studies in the History of Modern Science 9. Dordrecht: D. Reidel, 1982.

________. *Newton's Physics and the Conceptual Structure of the Scientific Revolution.* Boston Studies in the Philosophy of Science 127. Dordrecht: Kluwer Academic Publishers, 1991.

Bell, J. S. "How to Teach Special Relativity." In *Speakable and Unspeakable in Quantum Mechanics*, pp. 66-80. Cambridge: Cambridge University Press, 1987.

Beller, Mara. "Bohm and the 'Inevitability' of Acausality." In *Bohmian Mechanics and Quantum Theory: An Appraisal*, pp. 211-230. Edited by James T. Cushing, Arthur Fine, and Sheldon Goldstein. Boston Studies in the Philosophy of Science 184. Dordrecht: Kluwer Academic Publishers, 1996.

Bilaniuk, Olexa-Myron, *et al.* "More about Tachyons." *Physics Today* (December 1969), pp. 47-52.

Black, Max. "The Direction of Time." *Analysis* 19 (1958-59): 54-63.

________. Review of *The Natural Philosophy of Time*, by G. J. Whitrow. *Scientific American* 206 (April 1962), 179-184.

Bohm, David. *The Special Theory of Relativity.* Lecture Notes and Supplements in Physics. New York: W. A. Benjamin, 1965.

Bondi, Herman. *Relativity and Common Sense.* New York: Dover Publications, 1964.

Born, Max. *Atomic Physics.* Fifth edition. Translated by J. Douglas. London: Blackie, 1951.

________. *Einstein's Theory of Relativity.* Translated by Henry L. Brose. New York: E. P. Dutton, no date.

Bradley, J. *Mach's Philosophy of Science.* London: Athlone Press, 1971.

Bricker, Phillip. "Absolute Time versus Absolute Motion: Comments on Lawrence Sklar." In *Philosophical Perspectives on Newtonian Science*, pp. 77-89. Edited by Phillip Bricker and R. I. G. Hughes. Cambridge, Mass.: MIT Press, 1990.

Bridgeman, P. W. "Einstein's Theories and the Operational Point of View." In *Albert Einstein: Philosopher-Scientist*, pp. 335-359. Edited by P. A. Schilpp. Library of Living Philosophers 7. LaSalle, Ill.: Open Court, 1949.

Broad, C. D. *An Examination of McTaggart's Philosophy.* 2 vols. Cambridge: Cambridge University Press, 1938. Reprint edition: New York: Octagon Books, 1976.

________. *Scientific Thought.* London: Kegan, 1925.

Brooke, John Hedley. "The God of Isaac Newton." In *Let Newton Be!*, pp. 169-183. Edited by John Fauvel, Raymond Flood, Michael Shortland, and Robin Wilson. Oxford: Oxford University Press, 1988.

________. "Science and Religion." In *Companion to the History of Modern Science*, pp. 763-782. Edited by R. C. Olby, G. N. Cantor, J. R. R. Christie, and M. J. S. Lodge. London: Routledge, 1990.

Brown, James Robert. "Einstein's Brand of Verificationism." *International Studies in the Philosophy of Science* 2 (1987): 33-35.

Buckley, Michael J. "God in the Project of Newtonian Mechanics." In *Newton and the New Direction in Science*, pp. 85-105. Edited by G. V. Coyne, M. Heller, and J. Zycinski. Vatican City: Specola Vaticana, 1988.

Builder, Geoffery. "The Constancy of the Velocity of Light." *Australian Journal of Physics* 11 (1958): 457-480. Reprinted in *Speculations in Science and Technology* 2 (1971): 421-437.

________. "Ether and Relativity." *Australian Journal of Physics* 11 (1958): 279-297. Reprinted in *Speculations in Science and Technology* 2 (1979): 230-242.

________. "The Resolution of the Clock Paradox." *Australian Journal of Physics* 10 (1957): 246-262.

Bunge, Mario. "Physical Time: The Objective and Relational Theory." *Philosophy of Science* 35 (1968): 355-388.

Burge, Tyler. "Philosophy of Language and Mind." *Philosophical Review* 101 (1992): 3-51.

Butterfield, J. "Spatial and Temporal Parts." *Philosophical Quarterly* 35 (1985): 32-44.

Capek, Milic. "Bergson's Theory of Matter and Modern Physics." In *Bergson and the Evolution of Physics*, pp. 297-330. Edited by P. A. Y. Gunter. Knoxville: University of Tennessee Press, 1969.

________. *The Concepts of Space and Time*. Boston Studies in the Philosophy of Science 22. Dordrecht: D. Reidel, 1976.

________. "Do the New Concepts of Space and Time Require a New Metaphysics?" In *The World View of Contemporary Physics*, pp. 90-104. Edited with an Introduction by Richard F. Kitchener. Albany: State University of New York Press, 1988.

________. "The Inclusion of Becoming in the Physical World." In *The Concepts of Space and Time*, pp. 501-524. Edited by Milic Capek. Boston Studies in the Philosophy of Science 22. Dordrecht: D. Reidel, 1976.

________. "Introduction." In *The Concepts of Space and Time*, pp. xvii-lvii. Edited by Milic Capek. Boston Studies in the Philosophy of Science 22. Dordrecht: D. Reidel, 1976.

________. *The New Aspects of Time*. Boston Studies in the Philosophy of Science 125. Dordrecht: Kluwer Academic Publishers, 1991.

________. *The Philosophical Impact of Contemporary Physics*. Princeton: D. Van Nostrand, 1961.

Capra, Fritjof. *The Tao of Physics*. Second edition. London: Fontana, 1983.

Carrier, Martin. "Physical Force or Geometrical Curvature?" In *Philosophical Problems of the Internal and External Worlds*, pp. 3-21. Edited by John Earman, Allen I. Janis, Gerald J. Massey, and Nicholas Rescher. Pittsburgh: University of Pittsburgh Press, 1993.

Carriero, John. "Newton on Space and Time: Comments on J. E. McGuire." In *Philosophical Perspectives on Newtonian Science*, pp. 109-133. Edited by Phillip Bricker and R. I. G. Hughes. Cambridge, Mass.: MIT Press, 1990.

Carter, William R. and H. Scott Hestevold. "On Passage and Persistence." *American Philosophical Quarterly* 31 (1994): 269-283.

Cartwright, Richard. "Scattered Objects." In *Analysis and Metaphysics*, pp. 153-171. Edited by Keith Lehrer. Philosophical Studies Series in Philosophy 4. Dordrecht: D. Reidel, 1975.

Cassirer, Ernst. *Substance and Function and Einstein's Theory of Relativity* [1923]. Translated by W. C. Swabey and M. C. Swabey. Reprint Edition: Dover, 1953.

Chisholm, Roderick. *Person and Object*. Muirhead Library of Philosophy. London: George Allen & Unwin, 1976.

Christensen, Ferrel M. "The Source of the River of Time." *Ratio* 18 (1976): 131-144.

________. *Space-like Time.* Toronto: University of Toronto Press, 1993.

Clarke, Samuel. *A Discourse Concerning the Being and Attributes of God, the Obligations of Natural Religion and the Truth and Certainty of the Christian Revelation* [1704-05]. In *The Works of Samuel Clarke*, vol. 2: *Sermons on Several Subjects*, pp. 513-578. London: John & Paul Knapton, 1738.

Clarke, Samuel, and Leibniz, G. W. *The Leibniz-Clarke Correspondence.* Edited with an Introduction and Notes by H. G. Alexander. Manchester: Manchester University Press, 1956.

Cleugh, Mary F. *Time and its Importance in Modern Thought.* With a Foreword by L. Susan Stebbing. London: Methuen, 1937.

Clifton, Rob and Mark Hogarth. "The Definability of Objective Becoming in Minkowski Spacefime." *Synthèse* 103 (1995): 355-387.

Cohen, I. Bernard. *Introduction to Newton's 'Principia'.* Cambridge: Cambridge University Press, 1971.

Cooper, Leon N. *An Introduction to the Meaning and Structure of Physics.* New York: Harper & Row, 1968.

Craig, William Lane. *Divine Foreknowledge and Human Freedom: The Coherence of Theism I: Omniscience.* Brill's Studies in Intellectual History 16. Leiden: E. J. Brill, 1991.

________. *God, Time, and Eternity.* Dordrecht: Kluwer Academic Publishers, forthcoming.

________. "In Defense of Presentism: A response to Smith's paper 'Reference to the Past and Future'." In *Time, Tense, and Reference.* Edited by Quentin Smith and Alexander Jokic. Forthcoming.

________. *The Tensed Theory of Time: a Critical Examination.* Synthèse Library. Dordrecht: Kluwer Academic Publishers, forthcoming.

________. *Time and the Metaphysics of Relativity.* Dordrecht: Kluwer Academic Publishers, forthcoming.

________. "'What Place, then, for a Creator?': Hawking on God and Creation." *British Journal for the Philosophy of Science* 41 (1990): 229-234.

Cunningham, E. *The Principle of Relativity.* Cambridge: Cambridge University Press, 1914.

Cushing, James T. *Philosophical Concepts in Physics.* Cambridge: Cambridge University Press, 1998.

D'Abro, A. *The Evolution of Scientific Thought* [1927]. Second Revised Edition. N.p.: Dover Publications, 1950.

D'Inverno, Ray. *Introducing Einstein's Relativity.* Oxford: Clarendon Press, 1992.

Davidson, Donald. "The Individuation of Events." In *Essays in Honor of Carl Hempel*, pp. 216-234. Edited by N. Rescher. Dordrecht: D. Reidel, 1969.

Davies, Paul C. W. "The Acceptance Speech of Professor Paul Davies." Westminster Abbey, London, 3 May 1995.

________. *God and the New Physics.* New York: Simon & Schuster, 1983.

________. *Space and Time in the Modern Universe.* Cambridge: Cambridge University Press, 1977.

________. "Spacetime Singularities in Cosmology and Black Hole Evaporations." In *The Study of Time III*, pp. 74-91. Edited by J. T. Fraser, N. Lawrence, and D. Park. Berlin: Springer, 1978.

De Beauregard, Olivier Costa. "The Principle of Relativity and the Spatialization of Time." *Revue des questions scientifiques* 7 (1949). Reprinted in *Bergson and the Evolution of Physics*, pp. 227-250. Edited by P. A. Y. Gunter. Knoxville: University of Tennessee Press, 1969.

________. *Time, the Physical Magnitude*. Boston Studies in the Philosophy of Science 99. Dordrecht: D. Reidel, 1987.

De Broglie, Louis. "The Concepts of Contemporary Physics and Bergson's Ideas on Time and Motion." In *Bergson and the Evolution of Physics*, pp. 45-62. Edited by P. A. Y. Gunter. Knoxville: University of Tennessee Press, 1969.

Denbigh, Kenneth G. *Three Concepts of Time*. Berlin: Springer Verlag, 1981.

Deshpande, D. Y. "Professor Ayer on the Past." *Mind* 65 (1956): 85-90.

Di Salle, Robert. "Spacetime Theory as Physical Geometry." *Erkenntnis* 42 (1995): 317-337.

Dieks, Dennis. "Newton's Conception of Time in Modern Physics and Philosophy." In *Newton's Scientific and Philosophical Legacy*, pp. 151-159. Edited by P. B. Scheurer and G. Debrock. Dordrecht: Kluwer Academic Publishers, 1988.

________. "The 'Reality' of the Lorentz Contraction." *Zeitschrift für allgemeine Wissenschaftstheorie* 15/2 (1984): 330-342.

Dingle, Herbert. *Science at the Crossroads*. London: Martin Brian & O'Keefe, 1972.

________. *The Special Theory of Relativity*. London: Methuen, 1940.

________. "Time in Philosophy and Physics." *Philosophy* 54 (1979): 99-104.

________. "Time in Relativity Theory: Measurement or Coordinate?" In *The Voices of Time*, pp. 455-472. Second edition. Edited with a new Introduction by J. T. Fraser. Amherst, Mass.: University of Massachusetts Press, 1981.

Dobbs, H. A. C. "The 'Present' in Physics." *British Journal for the Philosophy of Science* 19 (1969): 317-324.

________. "Reply to Prof. Grünbaum." *British Journal for the Philosophy of Science* 21 (1970): 275-278.

Dorato, Mauro. *Time and Reality: Spacetime Physics and the Objectivity of Temporal Becoming*. Collana di Studi Epistemologici II. Bologna: CLUEB, 1995.

Dorling, Jan. "Length Contraction and Clock Synchronization: The Empirical Equivalence of the Einsteinian and Lorentzian Theories." *British Journal for the Philosophy of Science* 19 (1968): 67-69.

Duffy, Michael C. "The Modified Vortex Sponge: a Classical Analogue for General Relativity." Paper presented at the International Conference of the British Society for the Philosophy of Science, "Physical Interpretations of Relativity Theory." Imperial College of Science and Technology. London, 16-19 September, 1988.

Dugas, René. *A History of Mechanics*. With a Foreword by Louis de Broglie. Translated by J. R. Maddox. New York: Central Book Company, 1955.

Earman, John. "Spacetime, or How to Solve Philosophical Problems and Dissolve Philosophical Muddles without Really Trying." *Journal of Philosophy* 67 (1970): 259-277.

________. "Who's Afraid of Absolute Space?" *Australasian Journal of Philosophy* 48 (1970): 287-319.

________. *World Enough and Space-Time*. Cambridge, Mass.: MIT Press, 1989.

Earman, John, and Friedman, Michael. "The Meaning and Status of Newton's Law of Inertia and the Nature of Gravitational Forces." *Philosophy of Science* 40 (1973): 329-359.

Eddington, Arthur S. *The Mathematical Theory of Relativity*. Second edition. Cambridge: Cambridge University Press, 1930.

________. *The Nature of the Physical World*. With an Introductory Note by Sir Edmund Whittaker. Everyman's Library. 1928. Reprint, London: J. M. Dent and Sons, 1964.

________. *Space, Time and Gravitation*. Cambridge Science Classics [1920]. Reprint Edition: Cambridge: Cambridge University Press, 1987.

Einstein, Albert. *Äether und Relativitätstheorie*. Berlin: Julius Springer Verlag, 1920.

________. "Autobiographical Notes." In *Albert Einstein: Philosopher-Scientist*, pp. 1-94. Third Edition. Edited by Paul Arthur Schilpp. Library of Living Philosophers 7. LaSalle, Ill.: Open Court, 1949.

________. "Comment on Meyerson's 'La déduction relativiste'." *Revue philosophique de la France et de l'étranger* 105 (1928): 161-166. Reprinted in *The Concepts of Space and Time*, pp. 363-367. Edited by Milic Capek. Boston Studies in the Philosophy of Science 22. Dordrecht: D. Reidel, 1976.

________. "Dialog über Einwände gegen die Relativitätstheorie." *Naturwissenschaften* (1918): 697-702.

________. "Die formale Grundlage der allgemeinen Relativitätstheorie." *Physikalische Zeitschrift* 14 (1913): 1249-1262.

________. "Die Relativitäts-Theorie." *Vierteljahrsschrift der naturforschenden Gesellschaft in Zürich* 56 (1911): 1-14.

________. "Ernst Mach." *Physikalische Zeitschrift* 17 (1916): 101. Reprinted in Ernst Mach, *Die Mechanik in ihrer Entwicklung historisch-kritisch dargestellt*, pp. 683-689. Edited by Renate Wahsner and Horst-Heino Borzeszkowski. Berlin, DDR: Akademie-Verlag, 1988.

________. "The Foundations of General Relativity Theory." In *General Theory of Relativity*, pp. 141-172. Edited by C. W. Kilmister. Selected Readings in Physics. Oxford: Pergamon Press, 1973.

________. "Fundamental Ideas and Problems of the Theory of Relativity." [1923]. In *Nobel Lectures, Physics: 1901-1921*, pp. 479-490. New York: Elsevier, 1967.

________. *Lettres à Maurice Solovine*. Paris: Gauthiers-Villars, 1956.

________. *The Meaning of Relativity* [1922]. Sixth edition. Reprint Edition: London: Chapman and Hall, 1967.

________. "On the Electrodynamics of Moving Bodies." Translated by Arthur I. Miller. In Appendix to Arthur Miller, *Albert Einstein's Special Theory of Relativity*, pp. 392-415. Reading, Mass.: Addison-Wesley, 1981.

________. "On the Methods of Theoretical Physics." In *The World as I See It*, pp. 30-40. New York: Covici-Friede, 1934.

________. "The Problem of Space, Ether, and the Field in Physics." [1934]. In *Ideas and Opinions*, pp. 276-285. Translated by Sonja Bargmann. New York: Crown Publishers, 1954.

________. *Relativity: The Special and the General Theory*. Translated by Robert W. Lauren. London: Methuen, 1920.

________. *Relativity: The Special and the General Theory*. Fifteenth edition. New York: Crown Trade Paperback, 1961.

________. "Remarks on Bertrand Russell's Theory of Knowledge." In *The Philosophy of Bertrand Russell*, pp. 277-291. Third edition. Edited by P. A. Schilpp. Library of Living Philosophers. New York: Tudor Publishing, 1951.

________. "Reply to Criticism." In *Albert Einstein: Philosopher-Scientist*, pp. 665-688. Edited by P. A. Schilpp. Library of Living Philosophers 7. LaSalle, Ill.: Open Court, 1949.

________. "Über das Relativitätsprinzip und die aus demselben gezogenen Folgerungen." *Jahrbuch der Radioaktivität und Elektronik* 4 (1907): 411-462.

________. "Über die Entwicklung unserer Anschauungen über das Wesen und die Konstitution der Strahlung." *Physikalische Zeitschrift* 10 (1909): 817-825.

________. "Zum Ehrenfestschen Paradoxon." *Physikalische Zeitschrift* 12 (1911): 509-510.

Einstein, Albert and Leopold Infeld. *The Evolution of Physics*. New York: Simon & Shuster, 1938.

Einstein, Albert and Besso, Michele. *Correspondance 1903-1955*. Translated with Notes and an Introduction by Pierre Speziali. Paris: Hermann, 1979.

Epstein, Paul S. "The Time Concept in Restricted Relativity." *American Journal of Physics* 10 (1942): 1-6.

Ferré, Frederick. "Grünbaum on Temporal Becoming: A Critique." *International Philosophical Quarterly* 12 (1972): 426-445.

________. "Grünbaum vs. Dobbs: the Need for Physical Transiency." *British Journal for the Philosophy of Science* 21 (1970): 278-280.

Fine, Arthur. *The Shaky Game: Einstein, Realism, and the Quantum Theory*. Chicago: University of Chicago Press, 1986.

Fitzgerald, Paul. "Nowness and the Understanding of Time." In *PSA 1972*, pp. 259-281. Boston Studies in the Philosophy of Science 20. Edited by Kenneth F. Schaffner and Robert S. Cohen. Dordrecht: D. Reidel, 1974.

________. "The Truth about Tomorrow's Sea Fight." *Journal of Philosophy* 66 (1969): 307-329.

Föppl, August. *Einführung in die Maxwell'sche Theorie der Elektrizität*. Leipzig: Teubner, 1894.

Force, James E. "Newton's God of Dominion: The Unity of Newton's Theological, Scientific, and Political Thought." In *Essays on the Context, Nature, and Influence of Isaac Newton's Theology*, pp. 75-102. Edited by James E. Force and Richard H. Popkin. International Archives of the History of Ideas 128. Dordrecht: Kluwer Academic Publishers, 1990.

Frank, Philipp. "Einstein, Mach, and Logical Positivism." In *Albert Einstein: Philosopher-Scientist*, pp. 271-286. Edited by P. A. Schilpp. Library of Living Philosophers 7. LaSalle, Ill.: Open Court, 1949.

________. "The Importance of Ernst Mach's Philosophy of Science for our Times." [1917]. Reprinted in *Ernst Mach: Physicist and Philosopher*, pp. 219-234. Edited by Robert S. Cohen and Raymond J. Seeger. Boston Studies in the Philosophy of Science 6. Dordrecht: D. Reidel, 1970.

________. *Interpretations and Misinterpretations of Modern Physics*. Actualités Scientifiques et Industrielles 587: Exposés de Philosophie Scientifique 2. Paris: Hermann & Cie., 1938.

________. *Philosophy of Science*. Englewood Cliffs, N. J.: Prentice-Hall, 1957.

Freddoso, Alfred J. "Accidental Necessity and Logical Determinism." *Journal of Philosophy* 80 (1983): 257-278.

________. "Accidental Necessity and Power over the Past." *Pacific Philosophical Quarterly* 63 (1982): 54-68.

Freundlich, Yehudah. "'Becoming' and the Asymmetries of Time." *Philosophy of Science* 40 (1973): 496-517.

Friedman, Michael. *Foundations of Spacetime Theories*. Princeton: Princeton University Press, 1983.

________. "Simultaneity in Newtonian Mechanics and Special Relativity." In *Foundations of Spacetime Theories*, pp. 403-432. Edited by John S. Earman, Clark N. Glymour, and John J. Stachel. Minnestota Studies in the Philosophy of Science 8. Minneapolis: University of Minnesota Press, 1977.

Gabbey, Alan. "Newton and Natural Philosophy." In *Companion to the History of Modern Science*, pp. 243-263. Edited by R. C. Olby, G. N. Cantor, J. R. R. Christie, and M. J. S. Lodge. London: Routledge, 1990.

Gale, George. "Cosmos and Conflict." Paper presented at the conference, "The Origin of the Universe." Colorado State University, Ft. Collins, Colorado, 22-25 September, 1988.

________. "Some Metaphysical Perplexities in Contemporary Physics." Paper presented at the 36th Annual Meeting of the Metaphysical Society of America. Vanderbilt University, 14-16 March, 1985.

Gale, Richard M. "Disanalogies between Space and Time." *Process Studies* 25 (1996): 1-18.

________. "'Here' and 'Now'." *Monist* 53 (1969): 396-409.

________. "Human Time: Introduction." In *The Philosophy of Time: a Collection of Essays*, pp. 293-303. Edited by R. M. Gale. New York: Humanities Press, 1968.

________. *The Language of Time*. International Library of Philosophy and Scientific Method. London: Routledge, Kegan & Paul, 1968.

Garson, James W. "Here and Now." *Monist* 53 (1969): 469-477.

Geach, Peter. "Some Problems about Time." In *Logic Matters*, pp. 302-318. By Peter Geach. Berkeley: University of California Press, 1972.

Geroch, Robert. *General Relativity from A to B.* Chicago: University of Chicago Press, 1978.

Glymour, Clark. "Topology, Cosmology, and Convention." In *Space, Time, and Geometry*, pp. 193-216. Edited by Patrick Suppes. Synthèse Library. Dordrecht: D. Reidel, 1973.

Gödel, Kurt. "A Remark About the Relationship Between Relativity Theory and Idealistic Philosophy." In *Albert Einstein: Philosopher-Scientist*, pp. 557-562. Edited by P. A. Schilpp. Library of Living Philosophers 7. LaSalle, Ill.: Open Court, 1949.

Goldberg, Stanley. "Putting New Wine in Old Bottles: The Assimilation of Relativity in America." In *The Comparative Reception of Relativity*, pp. 1-26. Edited by Thomas F. Glick. Boston Studies in the Philosophy of Science 103. Dordrecht: D. Reidel, 1987.

________. *Understanding Relativity: Origin and Impact of a Scientific Revolution.* Boston: Birkhäuser, 1984.

Goldstein, Sheldon. "Review Essay: Bohmian Mechanics and the Quantum Revolution." *Synthèse* 107 (1996): 145-165.

Goodman, Nelson. "Time and Language, and the Passage of Time." In *The Structure of Appearance*, chapter 11. Cambridge, Mass.: Harvard University Press, 1951. Reprinted in *Problems of Space and Time*, pp. 356-369. Edited by J. J. C. Smart. Problems of Philosophy. New York: Macmillan Co., 1964.

Graham, George. "Persons and Times." *Southern Journal of Philosophy* 15 (1977): 309-315.

Grieder, Alfons. "Relativity, Causality and the 'Substratum'." *British Journal for the Philosophy of Science* 28 (1977): 35-48.

Griffin, David Ray. "Introduction: Time and the Fallacy of Misplaced Concreteness." In *Physics and the Ultimate Significance of Time*, pp. 1-48. Edited by David R. Griffin. Albany, N. Y.: State University of New York Press, 1986.

________. *Physics and the Ultimate Significance of Time.* Albany, N. Y.: State University of New York Press, 1986.

Grünbaum, Adolf. "Are Physical Events Themselves Transiently Past, Present and Future? A Reply to H. A. C. Dobbs." *British Journal for the Philosophy of Science* 20 (1969): 145-162.

________. "Carnap's Views on the Foundations of Geometry." In *The Philosophy of Rudolf Carnap*, pp. 599-684. Edited by P. A. Schilpp. Library of Living Philosophers. LaSalle, Ill.: Open Court, 1963.

________. "The Exclusion of Becoming from the Physical World." Reprinted in *The Concepts of Space and Time*, pp. 471-500. Edited by Milic Capek. Boston Studies in the Philosophy of Science 22. Dordrecht: D. Reidel, 1976.

________. "Is There A 'Flow' of Time or Temporal 'Becoming'?" In *Philosophical Problems of Space and Time*, pp. 314-329. Second edition. Boston Studies for the Philosophy of Science 12. Dordrecht: D. Reidel, 1973.

________. "The Meaning of Time." In *Basic Issues in the Philosophy of Time*, pp. 196-227. Edited by E. Freeman and W. Sellars. LaSalle, Ill.: Open Court, 1971.

________. "Operationism and Relativity." In *The Validation of Scientific Theories*, pp. 84-94. Edited with an Introduction by Philipp G. Frank. Boston: Beacon Press, 1954.

________. *Philosophical Problems of Space and Time*. Second edition. Boston Studies in the Philosophy of Science 12. Dordrecht: D. Reidel, 1973.

________. "The Pseudo-Problem of Creation in Physical Cosmology." *Philosophy of Science* 56 (1989): 373-394.

________. "Science and Ideology." *Scientific Monthly* 79 (July 1954), pp. 13-19.

________. "The Status of Temporal Becoming." In *Modern Science and Zeno's Paradoxes*, pp. 7-36. Middletown, Conn.: Wesleyan University Press, 1967.

Gutting, Gary. "Einstein's Discovery of Special Relativity." *Philosophy of Science* 59 (1972): 51-68.

Hartle, James and Stephen W. Hawking. "Wave Function of the Universe." *Physical Review D* 28 (1983): 2960-2975.

Haslanger, Sally. "Endurance and Temporary Intrinsics." *Analysis* 49 (1989): 119-125.

Hawking, Stephen W. *A Brief History of Time: From the Big Bang to Black Holes.* With an Introduction by Carl Sagan. New York: Bantam Books, 1988.

Healey, Richard. "Introduction." In *Reduction, Time and Reality*, pp. vii-xi. Edited by Richard Healey. Cambridge: Cambridge University Press, 1981.

Heller, Mark. *The Ontology of Physical Objects: Four-Dimensional Hunks of Matter*. Cambridge: Cambridge University Press, 1990.

________. "Temporal Parts of Four Dimensional Objects." *Philosophical Studies* 46 (1984): 323-334.

________. "Things Change." *Philosophy and Phenomenological Research* 52 (1992): 695-704.

________. "Varieties of Four Dimensionalism." *Australasian Journal of Philosophy* 71 (1993): 47-59.

Helm, Paul. *Eternal God.* Oxford: Clarendon Press, 1988.

Herneck, Friedrich. "Nochmals über Einstein und Mach." *Physikalische Blätter* 17 (1961): 275-276.

Hinchliff, Mark. "McTaggart, Change, and Real Tense." In *Essays on Time and Related Topics*, pp. 43-49. Edited by L. Nathan Oaklander. Philosophy of Time Society Proceedings 1995-1999.

________. "The Puzzle of Change." Paper presented at the Pacific Division meeting of the American Philosophical Association, 2 April, 1994.

Hirosige, Tetu. "The Ether Problem, the Mechanistic Worldview, and the Origins of the Theory of Relativity." *Historical Studies in the Physical Sciences* 7 (1976): 3-82.

________. "Origins of Lorentz' Theory of Electrons and the Concept of the Electromagnetic Field." *Historical Studies in the Physical Sciences* 1 (1969): 151-209.

Hoefer, Carl and Nancy Cartwright. "Substantivalism and the Hole Argument." In *Philosophical Problems of the Internal and External Worlds*, pp. 23-43. Edited by John Earman, Alan I. Janis, Gerald J. Massey, and Nicholas Rescher. Pittsburgh-Konstanz Series in the Philosophy and History of Science. Pittsburgh: University of Pittsburgh Press, 1993.

Hoffmann, Banesh, with Dukas, Helen. *Albert Einstein: Creator and Rebel.* London: Hart-Davis, MacGibbon, 1972.

Holton, Gerald J. "Einstein's Scientific Program: The Formative Years." In *Some Strangeness in the Proportion*, pp. 49-68. Edited by Harry Woolf. Reading, Mass.: Addison-Wesley, 1980.

________. "Mach, Einstein, and the Search for Reality." In *Ernst Mach: Physicist and Philosopher*, pp. 165-199. Edited by Robert S. Cohen and Raymond J. Seeger. Boston Studies in the Philosophy of Science 6. Dordrecht: D. Reidel, 1970.

________. "On the Origin of the Special Theory of Relativity." In *Thematic Origins of Scientific Thought: Kepler to Einstein*, pp. 165-183. Edited by Gerald Holton. Cambridge, Mass.: Harvard University Press, 1973.

________. "Poincaré and Relativity." In *Thematic Origins of Scientific Thought: Kepler to Einstein*, pp. 185-195. Edited by Gerald Holton. Cambridge, Mass.: Harvard University Press, 1973.

________. "Where Is Reality? The Answers of Einstein." In *Science and Synthesis*, pp. 45-69. Edited by UNESCO. Berlin: Springer Verlag, 1971.

Horwich, Paul. *Asymmetries in Time.* Cambridge, Mass.: MIT Press, 1987.

________. "On the Existence of Time, Space, and Spacetime." *Noûs* 12 (1978): 397-419.

Hoy, Ronald C. "Becoming and Persons." *Philosophical Studies* 34 (1978): 269-280.

Hume, David. *An Enquiry concerning Human Understanding.* In *Enquiries concerning Human Understanding and concerning the Principles of Morals.* Edited by L. A. Selby-Bigge. Third edition. Revised with Notes by P. H. Nidditch. Oxford: Clarendon Press, 1975.

________. *A Treatise of Human Nature.* Edited by L. A. Selby-Bigge. Revised with Notes by P. H. Nidditch. Oxford: Clarendon Press, 1978.

Illy, Jozsef. "Einstein Teaches Lorentz, Lorentz Teaches Einstein. Their Collaboration in General Relativity, 1913-1920." *Archive for History of Exact Sciences* 39 (1989): 247-289.

Isham, Christopher. "Creation of the Universe as a Quantum Process." In *Physics, Philosophy, and Theology: a Common Quest for Understanding*, pp. 375-408. Edited by R. Russell, W. Stoeger, and G. V. Coyne. Vatican City State: Vatican Observatory, 1988.

________. "Quantum Theories of the Creation of the Universe." Unpublished paper. A preliminary version appears in *Interpreting the Universe as Creation*, pp. 37-64. Edited by V. Brümmer. The Netherlands: Pharos, 1991.

Ives, Herbert E. "Derivation of the Lorentz Transformations." *Philosophical Magazine* 36 (1945): 392-401. Reprinted in *Speculations in Science and Technology* 2 (1979): 247-257.

Jammer, Max. *Concepts of Space*. Cambridge, Mass.: Harvard University Press, 1954.

Jánossy, L. *Theory of Relativity Based on Physical Reality*. Budapest: Akadémiae Kiadó, 1971.

Jeans, James. *Physics and Philosophy*. Cambridge: University Press, 1942.

Johnston, Mark. "Is There a Problem about Persistence?" *Aristotelian Society Supplementary Volume* 61 (1987): 107-135.

Jubien, Michael. *Ontology, Modality, and the Fallacy of Reference*. Cambridge Studies in Philosophy. Cambridge: Cambridge University Press, 1993.

Kaku, Michio. *Introduction to Superstrings and M-Theory*. Second edition. Graduate Texts in Contemporary Physics. New York: Springer, 1999.

Kant, Immanuel. *Immanuel Kant's "Critique of Pure Reason"*. Translated by Norman Kemp Smith. London: Macmillan, 1970.

Katsumori, Makoto. "The Theories of Relativity and Einstein's Philosophical Turn." *Studies in History and Philosophy of Science* 23 (1992): 557-592.

Kaufmann, W. "Über die Konstitution des Elektrons." *Annalen der Physik* 19 (1906): 487-553.

Kennedy, John. Paper presented at the Central Division Meeting of the American Philosophical Association. Pittsburgh, Pennsylvania, 23-26 April, 1997.

Kilmister, C. W. "Why is Relativity Interesting?" *British Journal for the Philosophy of Science* 42 (1991): 413-423.

Kitchener, Richard F. "Introduction: The World View of Contemporary Physics: Does It Need a New Metaphysics?" In *The World View of Contemporary Physics*, pp. 3-24. Edited by Richard F. Kitchener. Albany: State University of New York Press, 1988.

Knox, A. J. "Hendrik Antoon Lorentz, the Ether, and the General Theory of Relativity." *Archive for History of Exact Sciences* 38 (1988): 67-78.

Kostro, Ludwig. "Einstein's New Conception of the Ether." Paper presented at the International Conference of the British Society for the Philosophy of Science, "Physical Interpretations of Relativity Theory." Imperial College of Science and Technology. London, 16-19 September, 1988.

Koyré, Alexandre and I. Bernard Cohen. "The Case of the Missing *Tanquam*." *Isis* 52 (1961): 555-566.

________. "Newton and the Leibniz-Clarke Correspondence." *Archives internationales d'histoires des sciences* 15 (1962): 63-126.

Kroes, Peter. "The Physical Status of Time Dilation within the Special Theory of Relativity." Paper presented at the International Conference of the British Society for the Philosophy of Science, "Physical Interpretations of Relativity Theory." Imperial College of Science and Technology. London, 16-19 September, 1988.

________. "Physics and the Flow of Time." In *Nature, Time, and History*, pp. 43-52. Edited by P. A. Kroes. Nijmegen Studies in the Philosophy of Nature and its Sciences 4/2. Nijmegen, The Netherlands: Catholic University of Nijmegen, 1985.

________. *Time: Its Structure and Role in Physical Theories*. Synthèse Library 179. Dordrecht: D. Reidel, 1985.

Laird, John. *Theism and Cosmology*. London: George Allen & Unwin, 1940.

Langevin, Paul. "L'aspect général de la théorie de la relativité." *Bulletin scientifique des étudiants de Paris* 2 (April-May, 1922), pp. 6ff.

Laue, Max von. *Das Relativitätsprinzip*. Sammlung Naturwissenschaftlicher und mathematischer Monographien 38. Braunschweig: Friedrich Vieweg & Sohn, 1911.

________. "Zwei Einwände gegen die Relativitätstheorie und ihre Widerlegung." *Physikalische Zeitschrift* 13 (1912): 118-120.

Leclerc, Ivor. "The Relation between Natural Science and Metaphysics." In *The World View of Contemporary Physics*, pp. 25-37. Edited with an Introduction by Richard F. Kitchener. Albany: State University of New York Press, 1988.

Lenzen, Victor F. "Einstein's Theory of Knowledge." In *Albert Einstein: Philosopher-Scientist*, pp. 357-384. Edited by P. A. Schilpp. Library of Living Philosophers 7. LaSalle, Ill.: Open Court, 1949.

Levison, Arnold B. "Events and Time's Flow." *Mind* 96 (1987): 341-352.

Lewis, David. *On the Plurality of Worlds*. Oxford: Basil Blackwell, 1986.

________. "Rearrangement of Particles: Reply to Lowe." *Analysis* 48 (1988): 65-72.

________. "Survival and Identity." In *Philosophical Papers*, vol. 1:55-71. Oxford: Oxford University Press, 1983.

Lewis, Delmas. "Persons, Morality, and Tenselessness." *Philosophy and Phenomenological Research* 47 (1986): 305-309.

Lockwood, Michael. *Mind, Brain, and the Quantum*. Oxford: Basil Blackwell, 1989.

Lodge, O. J. "Note on Mr. Sutherland's Objection to the Conclusiveness of the Michelson-Morley Aether Experiment." *Philosophical Magazine* 46 (1898): 343-344.

Loizou, Andros. *The Reality of Time*. Brookfield, Ver.: Gower, 1986.

Lombard, Lawrence Brian. "The Doctrine of Temporal Parts and the 'No-Change' Objection." *Philosophy and Phenomenological Research* 54 (1994): 365-372.

Lorentz, H. A. "Alte und neue Fragen der Physik." *Physikalische Zeitschrift* 11 (1910): 1234ff. Reprinted in H. A. Lorentz, *Collected Papers*, 7: 205-257. Edited by P. Zeeman and A. D. Fokker. The Hague: Martinus Nijhoff, 1934.

________. "Considérations élémentaires sur le principe de relativité." *Revue générale des Sciences* 25 (1914): 179ff. Reprinted in H. A. Lorentz, *Collected Papers*, 7: 147-165. Edited by P. Zeeman and A. D. Fokker. The Hague: Martinus Nijhoff, 1934.

________. "Deux mémoires de Henri Poincaré sur la physique mathématique." *Acta mathematica* 38 (1914): 293ff. Reprinted in H. A. Lorentz, *Collected Papers*, 7: 258-273. Edited by P. Zeeman and A. D. Fokker. The Hague: Martinus Nijhoff, 1934.

________. *The Einstein Theory of Relativity*. New York: Brentano's, 1920.

________. "Het relativiteitsbeginsel. Voordrachten in Teyler's Stichtung." *Archives Musée Teyler* 2 (1914): 1-60.

________. *Problems of Modern Physics*. Edited by H. Bateman. Boston: Ginn, 1927.

________. *The Theory of Electrons* [1909]. Reprint Edition: New York: Dover, 1952.

Lorentz, H. A.; Einstein, A.; and Minkowski. H. *Das Relativitätsprinzip*. Fortschritte der mathematischen Wissenschaften 2. Mit Anmerkungen von A. Sommerfeld und Vorwort von O. Blumenthal. Leipzig: B. G. Teubner, 1913.

Lorenz, Dieter. "Über die Realität der FitzGerald-Lorentz Kontraction." *Zeitschrift für allgemeine Wissenschaftstheorie* 13/2 (1982): 294-319.

Lotze, Hermann. *Metaphysic*. 2 vols. Second edition. Edited by Bernard Bosanquet. Oxford: Clarendon Press, 1887.

Lowe, E. J. "The Indexical Fallacy in McTaggart's Proof of the Unreality of Time." *Mind* 96 (1987): 62-70.

________. "The Problems of Intrinsic Change: Rejoinder to Lewis." *Analysis* 48 (1988): 72-77.

________. "Substance Identity and Time." *Aristotelian Society Supplementary Volume* 62 (1988): 61-78.

Lucas, John R. *The Future: an Essay on God, Temporality and Truth*. Oxford: Basil Blackwell, 1989.

________. *Space, Time, and Causality*. Oxford: Clarendon Press, 1985.

________. *A Treatise on Time and Space*. London: Methuen, 1973.

Lucas, John R. and Peter E. Hodgson. *Spacetime and Electromagnetism*. Oxford: Clarendon Press, 1990.

Mayo, Bernard. "Space and Time Re-assimilated." *Mind* 85 (1976): 576-580.

McCormmach, Russell. "H. A. Lorentz and the Electromagnetic View of Nature." *Isis* 61 (1970): 459-497.

McGilvray, James A. "A Defense of Physical Becoming." *Erkenntnis* 14 (1979): 275-299.

McGuire, J. E. "Existence, Actuality, and Necessity: Newton on Space and Time." *Annals of Science* 35 (1978): 463-508.

________. "Force, Active Principles, and Newton's Invisible Realm." *Ambix* 15/3 (1968): 154-208.

________. "Newton on Place, Time, and God: An Unpublished Source." *British Journal for the History of Science* 11 (1978): 114-129.

________. "Predicates of Pure Existence: Newton on God's Space and Time." In *Philosophical Perspectives on Newtonian Science*, pp. 91-108. Edited by Phillip Bricker and R. I. G. Hughes. Cambridge, Mass.: MIT Press, 1990.

________. "Space, Infinity and Indivisibility: Newton on the Creation of Matter." In *Contemporary Newtonian Research*, pp. 145-190. Edited by Zev Bechler. Studies in the History of Modern Science 9. Dordrecht: D. Reidel, 1982.

McTaggart, John McTaggart Ellis. *The Nature of Existence*. 2 vols. Edited by C. D. Broad. Cambridge: Cambridge University Press, 1927. Reprint Edition: 1968.

________. "The Unreality of Time." *Mind* 17 (1908): 457-474.

Mach, Ernst. *Die Geschichte und die Wurzel des Satzes von der Erhaltung der Arbeit*. Prague: J. G. Calvé, 1872. In Ernst Mach, *Abhandlungen*. Edited by J. Thiele. Amsterdam: E. J. Bonset, 1969.

________. *History and Root of the Principle of the Conservation of Energy*. Translated by Philip E. B. Jourdain. Chicago: Open Court, 1911.

________. *Erkenntnis und Irrtum*. Second edition. Leipzig: D. Reidel, 1906.

________. *Die Mechanik in ihrer Entwicklung historisch-kritisch dargestellt*. Edited by Renate Wahsner and Horst-Heino Borzeszkowski. Berlin, DDR: Akademie-Verlag, 1988.

________. *The Science of Mechanics: A Critical and Historical Account of Its Development*. Translated by Thomas J. McCormack. LaSalle, Ill.: Open Court, 1960.

________. *Prinzipien der Wärmelehre*. Leipzig: J. A. Barth, 1896.

________. *Principles of the Theory of Heat*. Translated by T. J. McCormack. Revised and compl. by P. E. B. Jourdain and A. E. Heath. With an Introduction by M. J. Klein. Edited by Brian McGuiness. Vienna Circle Collection 17. Dordrecht: D. Reidel, 1986.

Maciel, A. K. A., and Tiomno, J. "Analysis of Absolute Spacetime Lorentz Theories." *Foundations of Physics* 19 (1989): 505-519.

Malament, David. "Causal Theories of Time and the Conventionality of Simultaneity." *Noûs* 11 (1977): 293-300.

Marder, L. *Time and the Space-Traveller*. London: George Allen & Unwin, 1971.

Margenau, Henry. "Metaphysical Elements in Physics." *Reviews of Modern Physics* 13 (1941): 176-189.

Margenau, Henry, and Mould, Richard A.. "Relativity: an Epistemological Appraisal." *Philosophy of Science* 24 (1957): 297-307.

Maudlin, Tim. *Quantum Non-Locality and Relativity*. Aristotelian Society Series 13. Oxford: Blackwell, 1994.

Mehlberg, Henry. "Philosophical Aspects of Physical Time." *Monist* 53 (1969): 340-384.

________. *Time, Causality, and the Quantum Theory*. 2 vols. Edited by Robert S. Cohen. Boston Studies in the Philosophy of Science 19. Dordrecht: D. Reidel, 1980.

Mellor, D. H. "McTaggart, Fixity and Coming True." In *Reduction, Time and Reality*, pp. 79-97. Edited by Richard Healey. Cambridge: Cambridge University Press, 1981.

________. *Real Time*. Cambridge: Cambridge University Press, 1981.

________. *Real Time II*. London: Routledge, 1999.

________. "Reply to Hinchliff." In *Essays on Time and Related Topics*, pp. 70-74. Edited by L. Nathan Oaklander. Philosophy of Time Society Proceedings 1995-1999.

________. "Special Relativity and Present Truth." *Analysis* 34 (1973-74): 74-77.

Merricks, Trenton. "Endurance and Indiscernibility." *Journal of Philosophy* 91 (1994): 165-184.

________. "Fission and Personal Identity over Time." *Philosophical Studies* 88/2 (1997): 163-186.

________. "On the Incompatibility of Enduring and Perduring Entities." *Mind* 104 (1995): 523-531.

________. "Persistence, Parts, and Presentism." *Noûs* 33 (1999): 421-438.

Merrill, A. A. "The *t* of Physics." *Journal of Philosophy* 19 (1922): 238-241.

Meyerson, Émile. *La déduction relativiste.* Payot: Paris, 1925. Reprinted in *The Concepts of Space and Time*, pp. 353-362. Edited by Milic Capek. Boston Studies in the Philosophy of Science 22. Dordrecht: D. Reidel, 1976.

Miller, Arthur I. *Albert Einstein's Special Theory of Relativity.* Reading, Mass.: Addison-Wesley,1981.

________. "On Some Other Approaches to Electrodynamics in 1905." In *Some Strangeness in the Proportion*, pp. 66-91. Edited by Harry Woolf. Reading, Mass.: Addison-Wesley, 1980.

Milne, E. A. "Presidential Address to the Royal Astronomical Society." *Monthly Notices of the Royal Astronomical Society* 104 (1944): 120-134.

Minkowski, H. "Space and Time." In *The Principle of Relativity*, pp. 75-91. By A. Einstein, *et al.* With Notes by A. Sommerfeld. Translated by W. Perrett and G. B. Jeffery. 1923. Reprint Edition: New York: Dover Publications, 1952.

Misner, C.; Thorne, K. S.; and Wheeler, J. A. *Gravitation.* San Francisco: W. H. Freeman, 1973.

Møller, C. *The Theory of Relativity.* Second Edition. Oxford: Clarendon Press, 1972.

More, Henry. "The Immortality of the Soul." In *Philosophical Writings of Henry More*, pp. 57-180. Edited with an Introduction and Notes by Flora Isabel MacKinnon. Wellesley Semi-Centennial Series. New York: Oxford University Press, 1925.

Motz, Lloyd and Jefferson Hane Weaver. *The Concepts of Science from Newton to Einstein.* New York: Plenum Press, 1988.

Nagel, Ernst. "Relativity and Twentieth-Century Intellectual Life." In *Some Strangeness in the Proportion*, pp. 38-45. Edited by Harry Woolf. Reading, Mass.: Addison-Wesley Publishing Co., 1980.

Nerlich, Graham. *The Shape of Space.* Cambridge: Cambridge University Press, 1976.

________. "Time as Spacetime." In *Questions of Time and Tense*, pp. 119-134. Edited by R. LePoidevin. Oxford: Oxford University Press, 1998.

________. *What Spacetime Explains.* Cambridge: Cambridge University Press, 1994.

Newton, Isaac. *Isaac Newton's Papers and Letters on Natural Philosophy.* Second edition. Edited with a General Introduction by I. Bernard Cohen. Cambridge, Mass.: Harvard University Press, 1978.

________. "On the Gravity and Equilibrium of Fluids." In *Unpublished Scientific Papers of Isaac Newton*, pp. 89-156. Edited by A. Rupert Hall and Marie Boas Hall. Cambridge: Cambridge University Press, 1962.

________. *Opticks, or a Treatise of the Reflections, Refractions, Inflections and Colours of Light.* Based on fourth edition [1730]. With a Foreword by Albert Einstein, an Introduction by Sir Edmund Whittaker, and a Preface by I. Bernard Cohen. New York: Dover, 1952.

________. *Philosophiae Naturalis Principia Mathematica.* Third edition [1726]. Edited by Alexandre Koyré and I. Bernard Cohen. 2 vols. Cambridge, Mass.: Harvard University Press, 1972.

________. "Place, Time, and God." In J. E. McGuire, "Newton on Place, Time, and God: An Unpublished Source." *British Journal for the History of Science* 11 (1978): 114-129.

________. *Sir Isaac Newton's 'Mathematical Principles of Natural Philosophy' and his 'System of the World.'* Translated by Andrew Motte. Revised with an Appendix by Florian Cajori. 2 vols. Los Angeles: University of California Press, 1966.

Newton-Smith, W.-H. "Space, Time, and Spacetime: A Philosopher's View." In *The Nature of Time*, pp. 22-35. Edited by Raymond Flood and Michael Lockwood. Oxford: Basil Blackwell, 1986.

Norton, John D. "Philosophy of Space and Time." In *Introduction to the Philosophy of Science*, pp. 179-232. Edited by Merilee Salmon. New Jersey: Prentice-Hall, 1992.

Oaklander, L. Nathan. "Bigelow, Possible Worlds, and the Passage of Time." *Analysis* 54 (1994): 244-248.

________. "Comments on William Craig's 'McTaggart's Paradox and the Problem of Temporary Intrinsics'." Central Division Meeting of the American Philosophical Association, Pittsburgh, Pennsylvania, 23-26 April, 1997.

________. "Delmas Lewis and Responsibility: a Critique." *Philosophy Research Archives* 13 (1987-88): 181-187.

________. "McTaggart's Paradox and Smith's Tensed Theory of Time." *Synthèse* 107 (1996): 205-221.

________. "McTaggart's Paradox and the Infinite Regress of Temporal Attributions: a Reply to Smith." *Southern Journal of Philosophy* 25 (1987): 425-431.

________. "Temporal Passage and Temporal Parts." *Noûs* 26 (1992): 79-84.

________. *Temporal Relations and Temporal Becoming.* Lanham, Md: University Press of America, 1984.

Oderberg, David S. *The Metaphysics of Identity over Time.* New York: St. Martin's Press, 1993.

Ostwald, Wilhelm. *Chemische Energie: Lehrbuch der allgemeinen Chemie.* Second edition. Leipzig: Verlag von Wilhelm Engelman, 1893.

Padgett, Alan G. *God, Eternity and the Nature of Time.* New York: St. Martin's Press, 1992.

Pagels, Heinz. *The Cosmic Code.* London: Michael Joseph, 1982.

Pais, Abraham. *'Subtle is the Lord...': The Science and Life of Albert Einstein.* Oxford: Oxford University Press, 1982.

Pap, Arthur. *Introduction to the Philosophy of Science.* London: Eyre & Spottiswoode, 1963.

Park, David. *The Image of Eternity.* Amherst, Mass.: University of Massachusetts Press, 1980.

Petkov, Vesselin. "Simultaneity, Conventionality, and Existence." *British Journal for the Philosophy of Science* 40 (1989): 69-76.

Petzoldt, Joseph. "Das Verhältnis der Machschen Gedankenwelt zur Relativitätstheorie." In *Die Mechanik in ihrer Entwicklung*, pp. 490-517. By Ernst Mach. Eighth edition. Leipzig: F. A. Brockhaus, 1921.

________. "Positivistische Philosophie." *Zeitschrift für positivistische Philosophie* 1 (1913): 1-16.

Plantinga, Alvin. *The Nature of Necessity.* Clarendon Library of Logic and Philosophy. Oxford: Clarendon Press, 1974.

________. "Reply to Robert Adams." In *Alvin Plantinga*, pp. 371-382. Edited by James Tomberlin and Peter Van Inwagen. Profiles 5. Dordrecht, Holland: D. Reidel, 1985.

Podlaha, M. F. "Length Contraction and Time Dilatation in the Special Theory of Relativity—Real or Apparent Phenomena?" *Indian Journal of Theoretical Physics* 25 (1975): 69-75.

Poincaré, Henri. "The Measure of Time." In *The Value of Science* [1905]. Translated by G. B. Halstead. In *The Foundations of Science*, pp. 223-234. Science Press: 1913. Reprint Edition: Washington, D. C.: University Press of America, 1982.

________. *Science and Hypothesis* [1902]. In *The Foundations of Science*, pp. 1-197. Science Press, 1913. Reprint Edition: Washington, D. C.: University Press of America, 1982.

Popper, Karl R. "A Note on the Difference between the Lorentz-Fitzgerald Contraction and the Einstein Contraction." *British Journal for the Philosophy of Science* 16 (1966): 332-333.

Powers, Jonathan. *Philosophy and the New Physics.* Ideas. London: Methuen, 1982.

Price, Huw. *Time's Arrow and Archimedes' Point.* New York: Oxford University Press, 1996.

Prior, Arthur N. "Changes in Events and Changes in Things." In *Papers on Time and Tense*, pp. 1-14. Oxford: Clarendon Press, 1968.

________. "On Spurious Egocentricity." In *Papers on Time and Tense*, pp. 15-25. Oxford: Clarendon Press, 1968.

________. "Some Free Thinking about Time." In *Logic and Reality*, pp. 47-51. Edited by B. J. Copeland. Oxford: Clarendon Press, 1996.

________. "Tense Logic and the Logic of Earlier and Later." In *Papers on Time and Tense*, pp. 116-134. Oxford: Clarendon Press, 1968.

Prokhovnik, S. J. "An Introduction to the Neo-Lorentzian Relativity of Builder." *Speculations in Science and Technology* 2 (1979): 225-242.

________. *Light in Einstein's Universe.* Dordrecht, Holland: D. Reidel, 1985.

________. "The Logic of the Clock Paradox." Paper presented at the International Conference of the British Society for the Philosophy of Science, "Physical Interpretations of Relativity Theory." London, 16-19 September, 1988.

Putnam, Henry. "Time and Physical Geometry." *Journal of Philosophy* 64 (1967): 240-247.

Quine, Willard Van Orman. *Word and Object.* Cambridge, Mass.: MIT Press, 1960.

Ramige, Eldon Albert. *Contemporary Concepts of Time and the Idea of God.* Boston: Stratford, 1935.

Rea, Michael. "Temporal Parts Unmotivated." *Noûs* (forthcoming).

Redhead, Michael. "The Conventionality of Simultaneity." In *Philosophical Problems of the Internal and External Worlds*, 103-128. Edited by John Earman, Alan I. Janis, Gerald J. Massey, and Nicholas Rescher. Pittsburgh-Konstanz Series in the Philosophy and History of Science. Pittsburgh: University of Pittsburgh Press, 1993.

________. "Nonlocality and Peaceful Coexistence." In *Space, Time and Causality*, pp. 151-189. Edited by Richard Swinburne. Synthèse Library 157. Dordrecht: D. Reidel, 1983.

Reichenbach, Hans. *Elements of Symbolic Logic.* London: Collier-Macmillan, 1947.

________. "Les fondements logiques de la mécanique des quanta." *Annales de l'Institut Henri Poincaré* 13 (1952): 109-157.

________. "The Philosophical Significance of the Theory of Relativity." In *Albert Einstein: Philosopher-Scientist*, pp. 289-311. Edited by P. A. Schilpp. Library of Living Philosophers 7. LaSalle, Ill.: Open Court, 1949.

________. "The Present State of the Discussion on Relativity." *Logos* 10 (1921): 316-378. Reprinted in *Modern Philosophy of Science*, pp. 1-45. Translated and edited by Maria Reichenbach. London: Routledge, Kegan & Paul, 1959.

Rindler, Wolfgang. "Einstein's Priority in Recognizing Time Dilation Physically." *American Journal of Physics* 38 (1970): 1111-1115.

Robb, A. A. *A Theory of Time and Space.* Cambridge: Heffer and Sons, 1913.

Rohrlich, Fritz. *From Paradox to Reality.* Cambridge: Cambridge University Press, 1987.

Rompe, R. and H.-J. Treder. "Is Physics at the Threshold of a New Stage of Evolution?" In *Quantum, Space and Time—The Quest Continues*, pp. 595-609. Edited by Asim O. Barut, Alwyn van der Merwe, and Jean-Pierre Vigier. Cambridge Monographs on Physics. Cambridge: Cambridge University Press, 1984.

Rosser, W. G. V. *An Introduction to the Theory of Relativity.* London: Butterworths, 1964.

Rothman, Tony, and Ellis, George. "Has Astronomy become Metaphysical?" *Astronomy* (February 1987), 6-22.

Rovelli, Carlo. "Halfway through the Woods: Contemporary Research on Space and Time." In *The Cosmos of Science*, pp. 180-223. Edited J. Norton and J. Earman. Pittsburgh: University of Pittsburgh Press, 1998.

________. "What Does Present Days [*sic*] Physics Tell Us about Time and Space?" Lecture presented at the Annual Series Lectures of the Center for Philosophy of Science of the University of Pittsburgh, 17 September 1993.

Ruderfer, Martin. "Introduction to Ives' 'Derivation of the Lorentz Transformations'." *Speculations in Science and Technology* 2 (1979): 243-246.

Ryckman, T. A. "'P(oint)-C(oincidence) Thinking': The Ironical Attachment of Logical Empiricism to General Relativity (and some lingering Consequences)." *Studies in History and Philosophy of Science* 23 (1992): 471-497.

Salmon, Wesley C. *Space, Time, and Motion.* Encino, Calif.: Dickenson Publishing, 1975.

Savitt, Steven F. "The Replacement of Time." *Australasian Journal of Philosophy* 72 [1994]: 463-474.

Schaffner, Kenneth. "Einstein versus Lorentz: Research Programmes and the Logic of Comparative Theory Evaluation." *British Journal for the Philosophy of Science* 25 (1974): 45-78.

Schlesinger, George N. *Aspects of Time.* Indianapolis: Hackett, 1980.

________. "Comparing Space and Time Once More." *Mind* 87 (1978): 264-266.

________. "The Similarities between Space and Time." *Mind* 84 (1975): 161-176.

Schuster, Melvin M. "On the Denial of Past and Future Existence." *Review of Metaphysics* 21 (1967-68): 447-467.

Seelig, C. *Albert Einstein: A Documentary Biography.* London: Staples Press, 1956.

Sellars, Wilfrid. "Time and the World Order." In *Scientific Explanation, Space, and Time*, pp. 527-616. Edited by Herbert Feigl and Grover Maxwell. Minnesota Studies in the Philosophy of Science 3. Minneapolis: University of Minnesota Press, 1962.

Shaw, R. "The Length Contraction Paradox." *American Journal of Physics* 30/1 (1962): 72

Silberstein, L. *The Theory of Relativity.* London: Macmillan, 1914.

Sinks, John. "On Some Accounts about the Future." *Journal of Critical Analysis* 2 (1971): 8-16.

Sjödin, T. "On the One-Way Velocity of Light and its Possible Measurability." Paper presented at the International Conference of the British Society for Philosophy of Science, "Physical Interpretations of Relativity Theory." Imperial College of Science and Technology. London, 16-19 September, 1988.

Sklar, Lawrence. "Comments on Mark Hinchliff: 'The Puzzle of Change'." Comment at the Pacific Division meeting of the American Philosophical Association, 2 April, 1994.

________. *Philosophy and Spacetime Physics.* Berkeley: University of California Press, 1985.

________. "Real Quantities and their Sensible Measures." In *Philosophical Perspectives on Newtonian Science*, pp. 57-75. Edited by Phillip Bricker and R. I. G. Hughes. Cambridge, Mass.: MIT Press, 1990.

________. *Space, Time, and Spacetime.* Berkeley: University of California Press, 1976.

________. "Time in Experience and in Theoretical Description of the World." In *Time's Arrows Today*, pp. 217-229. Edited by Steven F. Savitt. Cambridge: Cambridge University Press, 1995.

________. "Time, Reality and Relativity." In *Reduction, Time and Reality*, pp. 129-142. Edited by Richard Healey. Cambridge: Cambridge University Press, 1981.

Smart, J. J. C. *Philosophy and Scientific Realism.* London: Routledge, Kegan & Paul, 1963.

________. "The Reality of the Future." *Philosophia* 10 (1981): 141-150.

________. "Spacetime and Individuals." In *Logic and Art*, pp. 3-20. Edited by Richard Rudner and Israel Scheffler. Indianapolis: Bobbs-Merrill, 1972.

________. "Spatialising Time." *Mind* 64 (1955): 239-241.

Smith, Quentin. "The Infinite Regress of Temporal Attributions." *Southern Journal of Philosophy* 24 (1986): 383-396.

________. "The Logical Structure of the Debate about McTaggart's Paradox." *Philosophy Research Archives* 14 (1988-89): 371-379.

________. "The Mind-Independence of Temporal Becoming." *Philosophical Studies* 47 (1985): 109-119.

________. "The Multiple Uses of Indexicals." *Synthèse* 78 (1989): 167-191.

________. "Personal Identity and Time." *Philosophia* 22 (1993): 155-167.

Squires, Euan. *The Mystery of the Quantum World.* Bristol: Adam Hilger, 1986.

Stein, Howard. Critical notice of *The Language of Time*, by Richard M. Gale. *Journal of Philosophy* 66 (1969): 350-355.

________. "Newtonian Spacetime." *Texas Quarterly* 10 (1967): 174-200.

________. "On Einstein-Minkowski Spacetime." *Journal of Philosophy* 65 (1968): 5-23.

Strong, E. W. "Newton and God." *Journal of the History of Ideas* 13 (1952): 147-167.

Suppe, Frederick. "The Search for Philosophic Understanding of Scientific Theories." In *The Structure of Scientific Theories*, pp. 3-118. Second edition. Edited by F. Suppe. Urbana, Ill.: University of Illinois Press, 1977.

Swinburne, Richard. *Space and Time.* Second edition. London: Macmillan, 1981.

________. "Verificationism and Theories of Spacetime." In *Space, Time, and Causality*, pp. 63-76. Edited by Richard Swinburne. Synthèse Library 157. Dordrecht: D. Reidel, 1983.

Taylor, Edwin F. and John Archibald Wheeler. *Spacetime Physics.* San Francisco: W. H. Freeman, 1966.

Taylor, J. G. *Special Relativity.* Oxford Physical Series. Oxford: Clarendon Press, 1975.

Taylor, Richard. *Metaphysics*, chap. 7. Second Edition. Foundations of Philosophy. Englewood Cliffs, N. J.: Prentice-Hall, 1974.

________. "Moving about in Time." *Philosophical Quarterly* 9 (1959): 289-301.

________. "Spatial and Temporal Analogies and the Concept of Identity." *Journal of Philosophy* 52 (1955): 599-611.

Thomson, Judith Jarvis. "Parthood and Identity over Time." *Journal of Philosophy* 80 (1983): 201-220.

________. "Time, Space, and Objects." *Mind* 74 (1965): 1-27.

Tiwari, S. C. "Fresh Look on Relativistic Time and Lifetime of an Unstable Particle." Paper presented at the International Conference of the British Society for the Philosophy of Science, "Physical Interpretations of Relativity Theory." Imperial College of Science and Technology. London, 16-19 September, 1988.

Tonnelat, Marie-Antoinette. *Histoire du principe de relativité.* Nouvelle Bibliothèque Scientifique. Paris: Flammarion, 1971.

Tooley, Michael. *Time, Tense, and Causation.* Oxford: Clarendon Press, 1997.

Torrance, Thomas F. *Transformation and Convergence: Explorations in the Interrelations of Scientific and Theological Enterprise.* Grand Rapids: William B. Eerdmans Publishing Company, 1984.

Van Inwagen, Peter. "The Doctrine of Arbitrary Undetached Parts." *Pacific Philosophical Quarterly* 62 (1981): 123-137.

________. "Four-Dimensional Objects." *Noûs* 24 (1990): 245-255.

________. *Material Beings*. Ithica, New York: Cornell University Press, 1990.

________. *Metaphysics*. Dimensions of Philosophy Series. Oxford: Oxford University Press, 1993.

Varicak, V. "Zum Ehrenfestschen Paradoxon." *Physikalische Zeitschrift* 12 (1911): 169-170.

Weinberg, Steven. *Gravitation and Cosmology: Principles and Explications of the General Theory of Relativity*. New York: John Wiley & Sons, 1972.

Weingard, Robert. "Relativity and the Reality of Past and Future Events." *British Journal for the Philosophy of Science* 23 (1972): 119-121.

Weisskopf, Victor. "Elementary Particle Physics in the Very Early Universe." In *Astrophysical Cosmology*, pp. 503-528. Edited by H. A. Brück, G. V. Coyne, and M. S. Longair. Pontificiae Academiae Scientiarvm Scripta Varia 48. Vatican City: Pontificia Academia Scientiarvm, 1982.

Wenzl, A. "Einstein's Theory of Relativity, Viewed from the Standpoint of Critical Realism, and its Significance for Philosophy." In *Albert Einstein: Philosopher-Scientist*, pp. 583-606. Edited by P. A. Schilpp. Library of Living Philosophers 7. LaSalle, Ill.: Open Court, 1949.

Wertheimer, Max. *Productive Thinking*. Enlar. Ed. Edited by Michael Wertheimer. London: Tavistock, 1961.

Westfall, Richard S. *Never at Rest: A Biography of Isaac Newton*. Cambridge: Cambridge University Press, 1980.

Wheeler, J. A. "From Relativity to Mutability." In *The Physicist's Conception of Nature*, pp. 202-247. Edited by J. Mehra. Dordrecht: D. Reidel, 1973.

________. *Frontiers of Time*. Amsterdam: North-Holland, 1979.

Whitrow, G. J. *The Natural Philosophy of Time*. Second edition. Oxford: Clarendon Press, 1980.

Whitt, L. A. "Absolute Space: Did Newton Take Leave of His (Classical) Empirical Senses?" *Canadian Journal of Philosophy* 12 (1982): 709-724.

Wiggins, David. *Identity and Spatio-Temporal Continuity*. Oxford: Basil Blackwell, 1971.

Will, Clifford M. *Was Einstein Right?* New York: Basic Books, 1986.

Winnie, John A. "The Causal Theory of Spacetime." In *Foundations of Spacetime Theories*, pp. 134-205. Edited by John S. Earman, Clark N. Glymour, and John J. Stachel. Minnestota Studies in the Philosophy of Science 8. (Minneapolis: University of Minnesota Press, 1977.

________. "The Twin-Rod Thought Experiment." *American Journal of Physics* 40 (1972): 1091-1094.

Worrall, John. "How to Remain (Reasonably) Optimistic: Scientific Realism and the 'Luminiferous Ether'." In *PSA 1994*, pp. 334-342. Edited by David Hull, Micky Forbes, and Richard M. Burian. East Lansing, Mich.: PSA, 1994.

Yourgrau, Palle. *The Disappearance of Time*. Cambridge: Cambridge University Press, 1991.

Zahar, Elie. "Why Did Einstein's Programme Supersede Lorentz's? (I)." *British Journal for the Philosophy of Science* 24 [1973]: 95-123.

________. "Why Did Einstein's Programme Supersede Lorentz's? (II)." *British Journal for the Philosophy of Science* 24 [1973]: 223-262.

Zeeman, E. C. "Causality Implies the Lorentz Group." *Journal of Mathematical Physics* 5 (1964): 490-493.

Zemach, Eddy M. "Time and Self." *Analysis* 39 (1979): 143-147.

Zwart, P. J. "The Flow of Time." *Synthèse* 24 (1972): 133-158.

SUBJECT INDEX

PROPER NAME INDEX

163. D. Mayr and G. Süssmann (eds.), *Space, Time, and Mechanics*. Basic Structures of a Physical Theory. 1983 ISBN 90-277-1525-4
164. D. Gabbay and F. Guenthner (eds.), *Handbook of Philosophical Logic*. Vol. I: Elements of Classical Logic. 1983 ISBN 90-277-1542-4
165. D. Gabbay and F. Guenthner (eds.), *Handbook of Philosophical Logic*. Vol. II: Extensions of Classical Logic. 1984 ISBN 90-277-1604-8
166. D. Gabbay and F. Guenthner (eds.), *Handbook of Philosophical Logic*. Vol. III: Alternative to Classical Logic. 1986 ISBN 90-277-1605-6
167. D. Gabbay and F. Guenthner (eds.), *Handbook of Philosophical Logic*. Vol. IV: Topics in the Philosophy of Language. 1989 ISBN 90-277-1606-4
168. A. J. I. Jones, *Communication and Meaning*. An Essay in Applied Modal Logic. 1983 ISBN 90-277-1543-2
169. M. Fitting, *Proof Methods for Modal and Intuitionistic Logics*. 1983 ISBN 90-277-1573-4
170. J. Margolis, *Culture and Cultural Entities*. Toward a New Unity of Science. 1984 ISBN 90-277-1574-2
171. R. Tuomela, *A Theory of Social Action*. 1984 ISBN 90-277-1703-6
172. J. J. E. Gracia, E. Rabossi, E. Villanueva and M. Dascal (eds.), *Philosophical Analysis in Latin America*. 1984 ISBN 90-277-1749-4
173. P. Ziff, *Epistemic Analysis*. A Coherence Theory of Knowledge. 1984 ISBN 90-277-1751-7
174. P. Ziff, *Antiaesthetics*. An Appreciation of the Cow with the Subtile Nose. 1984 ISBN 90-277-1773-7
175. W. Balzer, D. A. Pearce, and H.-J. Schmidt (eds.), *Reduction in Science*. Structure, Examples, Philosophical Problems. 1984 ISBN 90-277-1811-3
176. A. Peczenik, L. Lindahl and B. van Roermund (eds.), *Theory of Legal Science*. Proceedings of the Conference on Legal Theory and Philosophy of Science (Lund, Sweden, December 1983). 1984 ISBN 90-277-1834-2
177. I. Niiniluoto, *Is Science Progressive?* 1984 ISBN 90-277-1835-0
178. B. K. Matilal and J. L. Shaw (eds.), *Analytical Philosophy in Comparative Perspective*. Exploratory Essays in Current Theories and Classical Indian Theories of Meaning and Reference. 1985 ISBN 90-277-1870-9
179. P. Kroes, *Time: Its Structure and Role in Physical Theories*. 1985 ISBN 90-277-1894-6
180. J. H. Fetzer, *Sociobiology and Epistemology*. 1985 ISBN 90-277-2005-3; Pb 90-277-2006-1
181. L. Haaparanta and J. Hintikka (eds.), *Frege Synthesized*. Essays on the Philosophical and Foundational Work of Gottlob Frege. 1986 ISBN 90-277-2126-2
182. M. Detlefsen, *Hilbert's Program*. An Essay on Mathematical Instrumentalism. 1986 ISBN 90-277-2151-3
183. J. L. Golden and J. J. Pilotta (eds.), *Practical Reasoning in Human Affairs*. Studies in Honor of Chaim Perelman. 1986 ISBN 90-277-2255-2
184. H. Zandvoort, *Models of Scientific Development and the Case of Nuclear Magnetic Resonance*. 1986 ISBN 90-277-2351-6
185. I. Niiniluoto, *Truthlikeness*. 1987 ISBN 90-277-2354-0
186. W. Balzer, C. U. Moulines and J. D. Sneed, *An Architectonic for Science*. The Structuralist Program. 1987 ISBN 90-277-2403-2
187. D. Pearce, *Roads to Commensurability*. 1987 ISBN 90-277-2414-8
188. L. M. Vaina (ed.), *Matters of Intelligence*. Conceptual Structures in Cognitive Neuroscience. 1987 ISBN 90-277-2460-1

189. H. Siegel, *Relativism Refuted.* A Critique of Contemporary Epistemological Relativism. 1987 ISBN 90-277-2469-5
190. W. Callebaut and R. Pinxten, *Evolutionary Epistemology.* A Multiparadigm Program, with a Complete Evolutionary Epistemology Bibliograph. 1987 ISBN 90-277-2582-9
191. J. Kmita, *Problems in Historical Epistemology.* 1988 ISBN 90-277-2199-8
192. J. H. Fetzer (ed.), *Probability and Causality.* Essays in Honor of Wesley C. Salmon, with an Annotated Bibliography. 1988 ISBN 90-277-2607-8; Pb 1-5560-8052-2
193. A. Donovan, L. Laudan and R. Laudan (eds.), *Scrutinizing Science.* Empirical Studies of Scientific Change. 1988 ISBN 90-277-2608-6
194. H.R. Otto and J.A. Tuedio (eds.), *Perspectives on Mind.* 1988 ISBN 90-277-2640-X
195. D. Batens and J.P. van Bendegem (eds.), *Theory and Experiment.* Recent Insights and New Perspectives on Their Relation. 1988 ISBN 90-277-2645-0
196. J. Österberg, *Self and Others.* A Study of Ethical Egoism. 1988 ISBN 90-277-2648-5
197. D.H. Helman (ed.), *Analogical Reasoning.* Perspectives of Artificial Intelligence, Cognitive Science, and Philosophy. 1988 ISBN 90-277-2711-2
198. J. Woleński, *Logic and Philosophy in the Lvov-Warsaw School.* 1989 ISBN 90-277-2749-X
199. R. Wójcicki, *Theory of Logical Calculi.* Basic Theory of Consequence Operations. 1988 ISBN 90-277-2785-6
200. J. Hintikka and M.B. Hintikka, *The Logic of Epistemology and the Epistemology of Logic.* Selected Essays. 1989 ISBN 0-7923-0040-8; Pb 0-7923-0041-6
201. E. Agazzi (ed.), *Probability in the Sciences.* 1988 ISBN 90-277-2808-9
202. M. Meyer (ed.), *From Metaphysics to Rhetoric.* 1989 ISBN 90-277-2814-3
203. R.L. Tieszen, *Mathematical Intuition.* Phenomenology and Mathematical Knowledge. 1989 ISBN 0-7923-0131-5
204. A. Melnick, *Space, Time, and Thought in Kant.* 1989 ISBN 0-7923-0135-8
205. D.W. Smith, *The Circle of Acquaintance.* Perception, Consciousness, and Empathy. 1989 ISBN 0-7923-0252-4
206. M.H. Salmon (ed.), *The Philosophy of Logical Mechanism.* Essays in Honor of Arthur W. Burks. With his Responses, and with a Bibliography of Burk's Work. 1990 ISBN 0-7923-0325-3
207. M. Kusch, *Language as Calculus vs. Language as Universal Medium.* A Study in Husserl, Heidegger, and Gadamer. 1989 ISBN 0-7923-0333-4
208. T.C. Meyering, *Historical Roots of Cognitive Science.* The Rise of a Cognitive Theory of Perception from Antiquity to the Nineteenth Century. 1989 ISBN 0-7923-0349-0
209. P. Kosso, *Observability and Observation in Physical Science.* 1989 ISBN 0-7923-0389-X
210. J. Kmita, *Essays on the Theory of Scientific Cognition.* 1990 ISBN 0-7923-0441-1
211. W. Sieg (ed.), *Acting and Reflecting.* The Interdisciplinary Turn in Philosophy. 1990 ISBN 0-7923-0512-4
212. J. Karpiński, *Causality in Sociological Research.* 1990 ISBN 0-7923-0546-9
213. H.A. Lewis (ed.), *Peter Geach: Philosophical Encounters.* 1991 ISBN 0-7923-0823-9
214. M. Ter Hark, *Beyond the Inner and the Outer.* Wittgenstein's Philosophy of Psychology. 1990 ISBN 0-7923-0850-6
215. M. Gosselin, *Nominalism and Contemporary Nominalism.* Ontological and Epistemological Implications of the Work of W.V.O. Quine and of N. Goodman. 1990 ISBN 0-7923-0904-9
216. J.H. Fetzer, D. Shatz and G. Schlesinger (eds.), *Definitions and Definability.* Philosophical Perspectives. 1991 ISBN 0-7923-1046-2
217. E. Agazzi and A. Cordero (eds.), *Philosophy and the Origin and Evolution of the Universe.* 1991 ISBN 0-7923-1322-4

218. M. Kusch, *Foucault's Strata and Fields*. An Investigation into Archaeological and Genealogical Science Studies. 1991 ISBN 0-7923-1462-X
219. C.J. Posy, *Kant's Philosophy of Mathematics*. Modern Essays. 1992 ISBN 0-7923-1495-6
220. G. Van de Vijver, *New Perspectives on Cybernetics*. Self-Organization, Autonomy and Connectionism. 1992 ISBN 0-7923-1519-7
221. J.C. Nyíri, *Tradition and Individuality*. Essays. 1992 ISBN 0-7923-1566-9
222. R. Howell, *Kant's Transcendental Deduction*. An Analysis of Main Themes in His Critical Philosophy. 1992 ISBN 0-7923-1571-5
223. A. García de la Sienra, *The Logical Foundations of the Marxian Theory of Value*. 1992 ISBN 0-7923-1778-5
224. D.S. Shwayder, *Statement and Referent*. An Inquiry into the Foundations of Our Conceptual Order. 1992 ISBN 0-7923-1803-X
225. M. Rosen, *Problems of the Hegelian Dialectic*. Dialectic Reconstructed as a Logic of Human Reality. 1993 ISBN 0-7923-2047-6
226. P. Suppes, *Models and Methods in the Philosophy of Science: Selected Essays*. 1993 ISBN 0-7923-2211-8
227. R. M. Dancy (ed.), *Kant and Critique: New Essays in Honor of W. H. Werkmeister*. 1993 ISBN 0-7923-2244-4
228. J. Woleński (ed.), *Philosophical Logic in Poland*. 1993 ISBN 0-7923-2293-2
229. M. De Rijke (ed.), *Diamonds and Defaults*. Studies in Pure and Applied Intensional Logic. 1993 ISBN 0-7923-2342-4
230. B.K. Matilal and A. Chakrabarti (eds.), *Knowing from Words*. Western and Indian Philosophical Analysis of Understanding and Testimony. 1994 ISBN 0-7923-2345-9
231. S.A. Kleiner, *The Logic of Discovery*. A Theory of the Rationality of Scientific Research. 1993 ISBN 0-7923-2371-8
232. R. Festa, *Optimum Inductive Methods*. A Study in Inductive Probability, Bayesian Statistics, and Verisimilitude. 1993 ISBN 0-7923-2460-9
233. P. Humphreys (ed.), *Patrick Suppes: Scientific Philosopher*. Vol. 1: Probability and Probabilistic Causality. 1994 ISBN 0-7923-2552-4
234. P. Humphreys (ed.), *Patrick Suppes: Scientific Philosopher*. Vol. 2: Philosophy of Physics, Theory Structure, and Measurement Theory. 1994 ISBN 0-7923-2553-2
235. P. Humphreys (ed.), *Patrick Suppes: Scientific Philosopher*. Vol. 3: Language, Logic, and Psychology. 1994 ISBN 0-7923-2862-0
Set ISBN (Vols 233–235) 0-7923-2554-0
236. D. Prawitz and D. Westerståhl (eds.), *Logic and Philosophy of Science in Uppsala*. Papers from the 9th International Congress of Logic, Methodology, and Philosophy of Science. 1994 ISBN 0-7923-2702-0
237. L. Haaparanta (ed.), *Mind, Meaning and Mathematics*. Essays on the Philosophical Views of Husserl and Frege. 1994 ISBN 0-7923-2703-9
238. J. Hintikka (ed.), *Aspects of Metaphor*. 1994 ISBN 0-7923-2786-1
239. B. McGuinness and G. Oliveri (eds.), *The Philosophy of Michael Dummett*. With Replies from Michael Dummett. 1994 ISBN 0-7923-2804-3
240. D. Jamieson (ed.), *Language, Mind, and Art*. Essays in Appreciation and Analysis, In Honor of Paul Ziff. 1994 ISBN 0-7923-2810-8
241. G. Preyer, F. Siebelt and A. Ulfig (eds.), *Language, Mind and Epistemology*. On Donald Davidson's Philosophy. 1994 ISBN 0-7923-2811-6
242. P. Ehrlich (ed.), *Real Numbers, Generalizations of the Reals, and Theories of Continua*. 1994 ISBN 0-7923-2689-X

243. G. Debrock and M. Hulswit (eds.), *Living Doubt.* Essays concerning the epistemology of Charles Sanders Peirce. 1994 ISBN 0-7923-2898-1
244. J. Srzednicki, *To Know or Not to Know.* Beyond Realism and Anti-Realism. 1994 ISBN 0-7923-2909-0
245. R. Egidi (ed.), *Wittgenstein: Mind and Language.* 1995 ISBN 0-7923-3171-0
246. A. Hyslop, *Other Minds.* 1995 ISBN 0-7923-3245-8
247. L. Pólos and M. Masuch (eds.), *Applied Logic: How, What and Why.* Logical Approaches to Natural Language. 1995 ISBN 0-7923-3432-9
248. M. Krynicki, M. Mostowski and L.M. Szczerba (eds.), *Quantifiers: Logics, Models and Computation.* Volume One: Surveys. 1995 ISBN 0-7923-3448-5
249. M. Krynicki, M. Mostowski and L.M. Szczerba (eds.), *Quantifiers: Logics, Models and Computation.* Volume Two: Contributions. 1995 ISBN 0-7923-3449-3
Set ISBN (Vols 248 + 249) 0-7923-3450-7
250. R.A. Watson, *Representational Ideas from Plato to Patricia Churchland.* 1995 ISBN 0-7923-3453-1
251. J. Hintikka (ed.), *From Dedekind to Gödel.* Essays on the Development of the Foundations of Mathematics. 1995 ISBN 0-7923-3484-1
252. A. Wiśniewski, *The Posing of Questions.* Logical Foundations of Erotetic Inferences. 1995 ISBN 0-7923-3637-2
253. J. Peregrin, *Doing Worlds with Words.* Formal Semantics without Formal Metaphysics. 1995 ISBN 0-7923-3742-5
254. I.A. Kieseppä, *Truthlikeness for Multidimensional, Quantitative Cognitive Problems.* 1996 ISBN 0-7923-4005-1
255. P. Hugly and C. Sayward: *Intensionality and Truth.* An Essay on the Philosophy of A.N. Prior. 1996 ISBN 0-7923-4119-8
256. L. Hankinson Nelson and J. Nelson (eds.): *Feminism, Science, and the Philosophy of Science.* 1997 ISBN 0-7923-4162-7
257. P.I. Bystrov and V.N. Sadovsky (eds.): *Philosophical Logic and Logical Philosophy.* Essays in Honour of Vladimir A. Smirnov. 1996 ISBN 0-7923-4270-4
258. Å.E. Andersson and N-E. Sahlin (eds.): *The Complexity of Creativity.* 1996 ISBN 0-7923-4346-8
259. M.L. Dalla Chiara, K. Doets, D. Mundici and J. van Benthem (eds.): *Logic and Scientific Methods.* Volume One of the Tenth International Congress of Logic, Methodology and Philosophy of Science, Florence, August 1995. 1997 ISBN 0-7923-4383-2
260. M.L. Dalla Chiara, K. Doets, D. Mundici and J. van Benthem (eds.): *Structures and Norms in Science.* Volume Two of the Tenth International Congress of Logic, Methodology and Philosophy of Science, Florence, August 1995. 1997 ISBN 0-7923-4384-0
Set ISBN (Vols 259 + 260) 0-7923-4385-9
261. A. Chakrabarti: *Denying Existence.* The Logic, Epistemology and Pragmatics of Negative Existentials and Fictional Discourse. 1997 ISBN 0-7923-4388-3
262. A. Biletzki: *Talking Wolves.* Thomas Hobbes on the Language of Politics and the Politics of Language. 1997 ISBN 0-7923-4425-1
263. D. Nute (ed.): *Defeasible Deontic Logic.* 1997 ISBN 0-7923-4630-0
264. U. Meixner: *Axiomatic Formal Ontology.* 1997 ISBN 0-7923-4747-X
265. I. Brinck: *The Indexical 'I'.* The First Person in Thought and Language. 1997 ISBN 0-7923-4741-2
266. G. Hölmström-Hintikka and R. Tuomela (eds.): *Contemporary Action Theory.* Volume 1: Individual Action. 1997 ISBN 0-7923-4753-6; Set: 0-7923-4754-4

267. G. Hölmström-Hintikka and R. Tuomela (eds.): *Contemporary Action Theory.* Volume 2: Social Action. 1997 ISBN 0-7923-4752-8; Set: 0-7923-4754-4
268. B.-C. Park: *Phenomenological Aspects of Wittgenstein's Philosophy.* 1998 ISBN 0-7923-4813-3
269. J. Paśniczek: *The Logic of Intentional Objects.* A Meinongian Version of Classical Logic. 1998 Hb ISBN 0-7923-4880-X; Pb ISBN 0-7923-5578-4
270. P.W. Humphreys and J.H. Fetzer (eds.): *The New Theory of Reference.* Kripke, Marcus, and Its Origins. 1998 ISBN 0-7923-4898-2
271. K. Szaniawski, A. Chmielewski and J. Woleński (eds.): *On Science, Inference, Information and Decision Making.* Selected Essays in the Philosophy of Science. 1998 ISBN 0-7923-4922-9
272. G.H. von Wright: *In the Shadow of Descartes.* Essays in the Philosophy of Mind. 1998 ISBN 0-7923-4992-X
273. K. Kijania-Placek and J. Woleński (eds.): *The Lvov–Warsaw School and Contemporary Philosophy.* 1998 ISBN 0-7923-5105-3
274. D. Dedrick: *Naming the Rainbow.* Colour Language, Colour Science, and Culture. 1998 ISBN 0-7923-5239-4
275. L. Albertazzi (ed.): *Shapes of Forms.* From Gestalt Psychology and Phenomenology to Ontology and Mathematics. 1999 ISBN 0-7923-5246-7
276. P. Fletcher: *Truth, Proof and Infinity.* A Theory of Constructions and Constructive Reasoning. 1998 ISBN 0-7923-5262-9
277. M. Fitting and R.L. Mendelsohn (eds.): *First-Order Modal Logic.* 1998 Hb ISBN 0-7923-5334-X; Pb ISBN 0-7923-5335-8
278. J.N. Mohanty: *Logic, Truth and the Modalities from a Phenomenological Perspective.* 1999 ISBN 0-7923-5550-4
279. T. Placek: *Mathematical Intiutionism and Intersubjectivity.* A Critical Exposition of Arguments for Intuitionism. 1999 ISBN 0-7923-5630-6
280. A. Cantini, E. Casari and P. Minari (eds.): *Logic and Foundations of Mathematics.* 1999 ISBN 0-7923-5659-4 set ISBN 0-7923-5867-8
281. M.L. Dalla Chiara, R. Giuntini and F. Laudisa (eds.): *Language, Quantum, Music.* 1999 ISBN 0-7923-5727-2; set ISBN 0-7923-5867-8
282. R. Egidi (ed.): *In Search of a New Humanism.* The Philosophy of Georg Hendrik von Wright. 1999 ISBN 0-7923-5810-4
283. F. Vollmer: *Agent Causality.* 1999 ISBN 0-7923-5848-1
284. J. Peregrin (ed.): *Truth and Its Nature (if Any).* 1999 ISBN 0-7923-5865-1
285. M. De Caro (ed.): *Interpretations and Causes.* New Perspectives on Donald Davidson's Philosophy. 1999 ISBN 0-7923-5869-4
286. R. Murawski: *Recursive Functions and Metamathematics.* Problems of Completeness and Decidability, Gödel's Theorems. 1999 ISBN 0-7923-5904-6
287. T.A.F. Kuipers: *From Instrumentalism to Constructive Realism.* On Some Relations between Confirmation, Empirical Progress, and Truth Approximation. 2000 ISBN 0-7923-6086-9
288. G. Holmström-Hintikka (ed.): *Medieval Philosophy and Modern Times.* 2000 ISBN 0-7923-6102-4
289. E. Grosholz and H. Breger (eds.): *The Growth of Mathematical Knowledge.* 2000 ISBN 0-7923-6151-2

SYNTHESE LIBRARY

290. G. Sommaruga: *History and Philosophy of Constructive Type Theory*. 2000 ISBN 0-7923-6180-6
291. J. Gasser (ed.): *A Boole Anthology*. Recent and Classical Studies in the Logic of George Boole. 2000 ISBN 0-7923-6380-9
292. V.F. Hendricks, S.A. Pedersen and K.F. Jørgensen (eds.): *Proof Theory*. History and Philosophical Significance. 2000 ISBN 0-7923-6544-5
293. W.L. Craig: *The Tensed Theory of Time*. A Critical Examination. 2000 ISBN 0-7923-6634-4
294. W.L. Craig: *The Tenseless Theory of Time*. A Critical Examination. 2000 ISBN 0-7923-6635-2

Previous volumes are still available.

KLUWER ACADEMIC PUBLISHERS – DORDRECHT / BOSTON / LONDON

www.ingramcontent.com/pod-product-compliance
Ingram Content Group UK Ltd.
Pitfield, Milton Keynes, MK11 3LW, UK
UKHW020136160926
475607UK00003B/70
9780792366355